The
Quantum
Structure
of Space and
Time

INTERNATIONAL SOLVAY INSTITUTES
BRUSSELS

Proceedings of the 23rd Solvay Conference on Physics

Brussels, Belgium   1 - 3 December 2005

The Quantum Structure of Space and Time

EDITORS

DAVID GROSS
Kavli Institute, University of California, Santa Barbara, USA

MARC HENNEAUX
Université Libre de Bruxelles & International Solvay Institutes, Belgium

ALEXANDER SEVRIN
Vrije Universiteit Brussel & International Solvay Institutes, Belgium

World Scientific

NEW JERSEY · LONDON · SINGAPORE · BEIJING · SHANGHAI · HONG KONG · TAIPEI · CHENNAI

Published by

World Scientific Publishing Co. Pte. Ltd.

5 Toh Tuck Link, Singapore 596224

USA office: 27 Warren Street, Suite 401-402, Hackensack, NJ 07601

UK office: 57 Shelton Street, Covent Garden, London WC2H 9HE

British Library Cataloguing-in-Publication Data

A catalogue record for this book is available from the British Library.

THE QUANTUM STRUCTURE OF SPACE AND TIME
Proceedings of the 23rd Solvay Conference on Physics

ISBN 978-981-256-952-3
ISBN 978-981-256-953-0 (pbk)

Printed in Singapore

The International Solvay Institutes

Guest Members

Prof. Albert Goldbeter
Professor at the ULB and Scientific Secretary of the Committee for Chemistry

Mr Pascal De Wit
Adviser Solvay S.A.

Prof. Niceas Schamp
Secretary of the Royal Flemish Academy for Science and the Arts

Prof. Alexandre Sevrin
Professor at the VUB and Scientific Secretary of the Committee for Physics

Director

Prof. Marc Henneaux
Professor at the ULB

Solvay Scientific Committee for Physics

Professor Herbert Walther (chair)
Max-Planck Institut (Munich, Germany)

Professor Tito Arecchi
Università di Firenze and INOA (Firenze, Italy)

Professor Jocelyn Bell Burnell
University of Bath (Bath, UK)

Professor Claude Cohen-Tannoudji
Ecole Normale Supérieure (Paris, France)

Professor Ludwig Faddeev
V.A Steklov Mathematical Institute (Saint-Petersburg, Russia)

Professor David Gross
Kavli Institute (Santa Barbara, USA)

Professor Gerard 't Hooft
Spinoza Instituut (Utrecht, The Netherlands)

Professor Klaus von Klitzing
Max-Planck-Institut (Stuttgart, Germany)

Professor Pierre Ramond
University of Florida (Gainesville, USA)

Professor Alexandre Sevrin (Scientific Secretary)
Vrije Universiteit Brussel (Belgium)

23rd Solvay Conference on Physics

Hotel Métropole (Brussels), 1-3 December 2005

The Quantum Structure of Space and Time
Chair: Professor David Gross

The 23rd Solvay Conference on Physics took place in Brussels from December 1 through December 3, 2005 according to the tradition initiated by Lorentz at the 1st Solvay Conference on Physics in 1911 ("Premier Conseil de Physique Solvay"). It was followed on December 4 by a public event co-organized with the European Commission, during which R. Dijkgraaf and B. Greene delivered public lectures and a panel of scientists (T. Damour, R. Dijkgraaf, B. Greene, D. Gross, G.'t Hooft, L. Randall, G. Veneziano) answered questions from the audience.

The Solvay Conferences have always benefitted from the support and encouragement of the Royal Family. His Royal Highness Prince Philippe of Belgium attended the first session on December 1 and met some of the participants.

The organization of the 23rd Solvay Conference has been made possible thanks to the generous support of the Solvay Family, the Solvay Company, the Belgian National Lottery, the "Université Libre de Bruxelles", the "Vrije Universeit Brussel", the "Communauté française de Belgique", the David and Alice Van Buuren Foundation and the Hôtel Métropole.

Participants

Nima	**Arkani-Hamed**	Harvard
Abhay	**Ashtekar**	Penn State
Michael	**Atiyah**	Edinburgh
Constantin	**Bachas**	Paris
Tom	**Banks**	Rutgers
Lars	**Brink**	Göteborg
Robert	**Brout**	Brussels
Claudio	**Bunster**	Valdivia
Curtis	**Callan**	Princeton
Thibault	**Damour**	Bures-sur-Yvette
Jan	**de Boer**	Amsterdam
Bernard	**de Wit**	Utrecht
Robbert	**Dijkgraaf**	Amsterdam
Michael	**Douglas**	Rutgers
Georgi	**Dvali**	New York
François	**Englert**	Brussels
Ludwig	**Faddeev**	St-Petersburg
Pierre	**Fayet**	Paris
Willy	**Fischler**	Austin
Peter	**Galison**	Harvard
Murray	**Gell-Mann**	Santa Fe
Gary	**Gibbons**	Cambridge (UK)
Michael	**Green**	Cambridge (UK)
Brian	**Greene**	Columbia
David	**Gross**	Santa Barbara
Alan	**Guth**	Cambridge (USA)
Jeffrey	**Harvey**	Chicago
Gary	**Horowitz**	Santa Barbara
Bernard	**Julia**	Paris
Shamit	**Kachru**	Stanford
Renata	**Kallosh**	Stanford
Elias	**Kiritsis**	Palaiseau
Igor	**Klebanov**	Princeton
Andrei	**Linde**	Stanford
Dieter	**Lüst**	Munich
Juan	**Maldacena**	Princeton
Nikita	**Nekrasov**	Bures-sur-Yvette
Hermann	**Nicolai**	Potsdam
Hirosi	**Ooguri**	Pasadena
Joseph	**Polchinski**	Santa Barbara
Alexander	**Polyakov**	Princeton
Eliezer	**Rabinovici**	Jerusalem
Pierre	**Ramond**	Gainesville
Lisa	**Randall**	Harvard
Valery	**Rubakov**	Moscow
John	**Schwarz**	Pasadena
Nathan	**Seiberg**	Princeton

Ashoke	**Sen**	Allahabad
Stephen	**Shenker**	Stanford
Eva	**Silverstein**	Stanford
Paul	**Steinhardt**	Princeton
Andrew	**Strominger**	Harvard
Gerard	**´t Hooft**	Utrecht
Neil	**Turok**	Cambridge
Gabriele	**Veneziano**	Paris
Steven	**Weinberg**	Austin
Frank	**Wilczek**	Cambridge(USA)
Paul	**Windey**	Paris
Shing-Tung	**Yau**	Harvard

Auditors

Riccardo	**Argurio**	ULB
Glenn	**Barnich**	ULB
Ben	**Craps**	VUB
Frank	**Ferrari**	ULB
Jean-Marie	**Frère**	ULB
Raymond	**Gastmans**	KUL
Marc	**Henneaux**	ULB
Thomas	**Hertog**	CERN
Laurent	**Houart**	ULB
Franklin	**Lambert**	VUB
Christiane	**Schomblond**	ULB
Alexander	**Sevrin**	VUB
Philippe	**Spindel**	UMH
Peter	**Tinyakov**	ULB
Walter	**Troost**	KUL
Jan	**Troost**	ENS
Michel	**Tytgat**	ULB
Antoine	**Van Proeyen**	KUL

Opening Session

Opening Address by Marc Henneaux

Your Royal Highness,
Mrs. and Mr. Solvay,
Ladies and Gentlemen,
Dear Colleagues,
Dear Friends,

About one hundred years ago, at the invitation of Ernest Solvay, the leading physicists of the time gathered in the hotel Métropole for a 6-day mythical meeting. This meeting began a tradition of unique conferences that shaped modern science. A total of 22 conferences have taken place every three years except during war periods, covering most aspects at the frontiers of physics.

This year's meeting continues the tradition. It is a great honor for me and a moving moment to welcome all of you to the 23rd Solvay Conference in Physics, in the building where the first meeting took place.

A distinctive feature of the Solvay Conferences is that they benefit from the support and encouragement of the Royal Family. The pictures of Einstein with the Queen Elisabeth of Belgium are in the mind of all physicists. It is with a respectful gratitude that we acknowledge the continuation of this tradition today, at the time of the celebration of the 175th anniversary of Belgium.

At the start of the 23rd Solvay Conference, I would like to have a thought for Ernest Solvay, the Founder of the Institutes, and for the men who assisted him in this enterprise, Paul Héger and Hendrik Lorentz.

Born in 1838, Ernest Solvay exhibited a passion for physics and chemistry from a very young age. He developed a new process for industrial production of sodium carbonate (Na_2CO_3 - "soude" in French, not "sodas" (in the plural) as I could see on some uncontrolled web side). This was at the origin of his wealth, which he used to create many charitable foundations. In particular, after the success of the 1911 Solvay Conference, he founded the International Institute of Physics. The mission of the Institute was "to promote research, the purpose of which is to enlarge and deepen our understanding of natural phenomena, without excluding problems belonging to other areas of science provided that these are connected with physics".

Mr. Solvay, we are very fortunate that the same interest in fundamental science has been transmitted to the following generations.

Paul Héger and Hendrik Lorentz played also a central role in the foundation of the Institutes. Héger was professor at the Université Libre de Bruxelles and a close collaborator of Ernest Solvay. He wrote in 1912 the rules of the Institute of Physics with the Dutch theoretical physicist Hendrik Lorentz, 1902 Nobel Laureate. Lorentz was the first scientific chair of the Institute until his death in 1928. He "governed" (if I can say so) the Institute with an iron hand in a velvet glove. His exceptional vision of physics and his diplomatic skills were the keys to the lasting scientific success of the Solvay Institutes.

*****

It is instructive to read the original rules of the Institutes that prevailed until 1949. They tell us a lot about the remarkable personalities and foresights of Ernest Solvay and of Hendrik Lorentz. These rules are astonishingly modern and most of them are still valid today.

Ernest Solvay and Hendrik Lorentz clearly saw, for instance that science is international. There is no Dutch or French or German science. Science is universal and an activity that elevates mankind without border. This absence of compromise to nationalisms put the Solvay Institutes in a unique position after the First World War and enabled them to play a leading role in the reconciliation of French and German scientists.

One can also see some distrust of Lorentz to fashion or well-established schools. There is an explicit paragraph in the rules that states that invitations to the Solvay Conference should be made on the sole basis of scientific merit and originality, irrespective of whether the scientists work in a well-established institution or in an obscure place, and with no account taken of their official recognitions.

Another interesting rule - but this is more on the anecdotic side - is that Lorentz designated himself, in 1912, scientific chair of the Institutes until 1930! If one recalls that he died in 1928, this is really a life self-crowning that tells a lot about his strong personality.

*****

There are, however, two points on which we deviate from the original rules. One is the limit on the number of invited participants to the Solvay Conferences. Knowing the natural tendency of humanity to inflation, this number was fixed, by rule, to be 25. And that rule was strictly adhered to until the last Solvay Conference chaired by Lorentz, in 1927. We are today 60. Is this a signal of weakness on the part of the organizers? I do not think so. It is just the sign that science and the number of scientists have exploded in the 20th century. I was told that if one counts the total number of physicists that ever did research since the beginning of humanity, about 90 percents are living now. I therefore think that it is as difficult today to pick 60 physicists as it was in the early days to pick 25.

Another striking paragraph in the text written in 1912 concerns the lifetime of the Institutes, which was fixed to be 30 years. This rule reflects, I believe, the optimism that prevailed at the time of the foundation of the Institute, that all problems of physics would be solved after a finite amount of time.

We clearly deviate from this rule, and for good reasons. Science, like history, is not finished. Today's conference will not be a replay of the 1911 conference in which we will reproduce, on the occasion of the international year of physics, the discussions between Einstein, Lorentz and Marie Curie. No, we will be interested in new questions, in new challenges, in new physics. These new questions can be asked now thanks to the new knowledge gained by the discoveries made by our predecessors and could not have been anticipated in 1911.

I am convinced that Ernest Solvay and Hendrik Lorentz would not be disappointed in learning that the end of physics did not occur 30 years after the foundation of the Institutes. They would share our excitement in trying to answer the new questions that enlarge further our vision of the universe. They would be delighted to see that the same enthusiasm and passion for understanding natural phenomena animate today's researchers as they inhabited the participants of the 1911 Conference.

*****

We can now start our work. I would like to thank all of you for your positive answer to our invitation and for being with us today. I would also like to thank the rapporteurs, the session chairs and, of course, our conference chair David Gross, for all the preparation work that went into this meeting. Finally, as you can see, we have tried to arrange the conference room in a way that recalls the setting of the early days.

Thank you very much for your attention.

Opening Address by David Gross

Your Royal Highness,
Mrs. and Mr. Solvay,
Dear Friends and Colleagues,

For me it is a great pleasure and a great honor to chair this 23rd Solvay Conference in Physics. For all of us who have grown up in the 20th century, these conferences have played such an important role in our collective memory of physics that we hope that the revived and re-invigorated Solvay Institutes will continue this tradition of Solvay Conferences in the same spirit. Perhaps they will play a role as important in the 21st century.

As Marc has indicated, we have tried to preserve those traditions of the Solvay Conferences that are worth preserving, in particular, the concentration during these few days on the most deep and central questions that face us in fundamental physics. The first conference held in 1911 was devoted to theories of radiation and quanta. The famous meeting in 1927 was entitled "Electrons and Photons". We decided to title this one and to concentrate on "The Quantum Structure of Space and Time", which, in its broadest sense encompasses many of the deep questions we face today in dealing with the stormy marriage of quantum mechanics and relativity. With the new ideas in string theory and in cosmology, we are faced, as we all know, with questions as perplexing and as deep as those that were faced almost a century ago.

In the tradition of the Solvay Conferences, we have tried to organize this conference in a similar fashion, with rapporteur talks whose purpose it is to survey a given area and to lay the stage for subsequent discussion. There will be in addition, as you know, short presentations, prepared presentations, and hopefully much spontaneous discussion.

Hopefully, if everyone sticks as instructed to the time allotted, we will have much time for discussion and, in a way, that is the heart of the meeting. Solvay Conferences are famous for what went on in these spontaneous discussions, either in the hall or outside the hall. It is our hope that this will be as exciting a meeting as any in the past and that it will perhaps help us understanding the quantum structure of space and time.

At this point, on schedule, I turn over the chair of the first half of the first session to Marc.

Contents

Session 1

History

Chair: *Marc Henneaux*, Université Libre de Bruxelles, Belgium
Rapporteur: *Peter Galison*, Pellegrino University Professor of History of Science and of Physics, Harvard University, USA
Scientific secretaries: *Riccardo Argurio* (Université Libre de Bruxelles) and *Glenn Barnich* (Université Libre de Bruxelles)

1.1 Rapporteur talk: Solvay Redivivus, by Peter Galison

1.1.1 *Three Miracles*

It seems impossible: that knowledge-transforming confrontations occurred not just once but several times within a series of small, highly-planned scientific meetings. Yet this is what made the "Solvay Councils" - a few of them at least - turning points in the history of modern science. Of course such luminous moments are not all that the Solvay meetings accomplished. These assemblies served as sites for powerful reviews of the field, and catalysts for intellectual and social networks that helped advance research. Importantly, the Solvay meetings also played a crucial role in setting and maintaining international scientific exchange after some of the worst times of a deeply troubled twentieth century.

But now, in these first moments of a revived Conseil Solvay, we do not need the spirit or structure of this century-long tradition to propel forward either international scientific exchange or the comprehensive reviews of specialized domains. We have other, larger and more efficacious means to advance these goals. There are journals like Physics Reports or Reviews of Modern Physics that commission effective, easily accessible summaries of fields. We have international meetings that effectively disseminate the current state of action in the many subfields of physics. And of course, in successful pursuit of coordinated international scientific activity, we have regional and worldwide agencies, institutes, and laboratories. Assembling fifty or so participants, however remarkable they may be, would not, in any case, be the best way to advance the flow of talent and ideas from country to country and continent to continent.

No, when, as a historian of modern science, I look back on the Solvay phenomenon, what is so remarkable is that, once in a while, something powerful happened in the concatenation of differing points of view, something that might not be anticipated. By way of historical introduction to this 23rd Solvay Conference, I would like to call a few of those moments to mind [1]. Of course, pace Santillana, history is not going to repeat itself - whether we study it or not - but it might be worthwhile to see just how remarkable a few of those moments were: a few days in October-November 1911, and then several more in the Octobers of 1927 and 1930. First, a few words about the beginning.

The successful inauguration of the Solvay meetings required at least three (pre-string) miracles. A first miracle demanded a precise balance between two philanthropic forces. On one side, this required that the powerful Belgian industrialist Ernest Solvay was passionate enough about science (including his own pet theory) to put astonishing resources into the Conseil. On the other side, again because he had a favored theory, it is all the more impressive that he was willing to step aside intellectually, and to leave the physics to the leading physicists of his time with no interference. Without resources - or with interference - the convocation could never have succeeded.

Solvay's boundless hopes for science built on his central enterprise: his development - and industrial prosecution - of a new way to synthesize soda using limestone, salt and ammonia. Soda Ash (sodium carbonate) had a vast range of uses, from the manufacture of glass and medicines to soaps and photography. What Solvay discovered was a method to produce the sodium carbonate from salt (sodium chloride) and limestone (calcium carbonate) - this replaced an earlier method that required the same inputs along with sulfuric acid and coal, a more wasteful and expensive product. Until the massive Wyoming deposits were discovered years later, the Solvay method, applied to a far-flung network of production and distribution, garnered a significant share of the world market and won him his fortune.

Solvay was a liberal, in its several senses, someone who held a scientific hope for a new world in which individualist politics, technically-based nutritional reform, and a novel sociology would transform the whole and alleviate suffering. He created institutes (in physiology [1892], and sociology [1902], school of commerce [1904]) that he saw as instrumental to ground the movement of societal change.

Despite the range of these ambitious pursuits, none of these institutionalized programs got at the depth of what Solvay sought. He also wanted a physics. His own theory, of which he was quite proud, carried the title "Gravitique" (1887), and put gravity at the source of all processes; it was to be more basic than energy. It would embrace the historical contributions of Kepler, Newton; it would include molecular contact and ethers ... If one had to form a slogan to capture Solvay's ambition, it might be this: he wanted both a politics of physics (that would use the physical sciences as the basis for a reformation of society), and a physics of politics (in which conduct of politics would be stripped down to its law-like, technical aspects).

Through a mutual friend, Solvay met Walther Nernst (then director of the Second Chemical Institute at the University of Berlin) in the spring of 1910 and told Nernst of his ideas about physics and his hope that it would be possible to assemble some of the greats of the discipline to discuss how things stood [2]. Nernst liked the idea and let Solvay know that he (Nernst) had already tried to interest colleagues in a discussion of the new quantum ideas - unfortunately, Max Planck had demurred, suggesting that such an assembly was premature. No doubt encouraged by Solvay, Nernst reiterated to the industrialist that far from being premature, "there seems that there could hardly have been a time as the present when such a Conseil could more favorably influence the development of physics and chemistry ... " [3] Nernst's idea was to focus on a set of seven problems that included (among others): the derivation of the "Rayleigh formula" of radiation; the Planck radiation law, the theory of quantized energy, and the relation of specific heats and the theory of quanta [4].

The first Conseil de Physique gathered in the luxurious Hotel Metropole in Brussels on Monday 30 October 1911 for a meeting that lasted through Friday 3 November-call it Solvay-1. Ernest Solvay welcomed the assembled luminaries with his theory (every delegate had already received a reprint of his views). Even if Nernst had not chosen the topic, as he had, Solvay himself might have : he expected to produce on his own an "exact and therefore definitive" account of the finite fundamental elements of the active universe. Be that as it may, the combination of Nernst and Solvay put the program in good stead-funded and scientifically connected.

The second miracle began when Solvay concluded his preamble with the words: "I am now happy to cede my place to our eminent president, M[onsieur] Lorentz." From that instant forward, Hendrik-Antoon Lorentz took charge of the meeting. Lorentz was gracious about Solvay's support and intervention, and gently but firmly guided the conversation even when the consensus violated Lorentz's most deeply-held convictions about the direction of the field. Indeed, to understand Lorentz's response to the new quantum physics, it is foolish to represent him as a reactionary, but instead we need to read his response to the new through the lens of his earlier achievements.

When Lorentz entered the electrodynamic world, it was saturated with complex theories of the ether, in which this most subtle of substances could be dragged, moved, compressed, sheared, and spun. There was a long-standing tradition that sought, since the time of Maxwell, to derive the existence of charged particles from stable flows in the ether-like smoke rings in the air. Ether models proliferated - mathematical models, physical models, analogy models. From this baroque and confusing mix, Lorentz extracted a theory of extraordinary simplicity: a rigid ether in which particulate electrons moved by a simple, (now eponymous) force law coupled to Maxwell's equations. The Dutch theorist produced miracles from this combination: he could explain myriad effects from reflection and refraction to the magnetic splitting of spectral lines in the Zeemann Effect; for the last of these he received the Nobel Prize in 1902. As an encore in 1904, he continued his reasoning about electrons, electrodynamics, and the ether, producing first an approximate and then the exact form of his transformations.

I'll come back to this notion later, but to anticipate - what Lorentz did was to present a principled, focused vision of what physics might be based on a kind of "radical conservatism" - a pushing of the electron plus rigid-ether program that he himself had followed with such stunning results. But both earlier (and at Solvay) his own commitments never stopped from encouraging views that were orthogonal to it. By doing so, the participants never had far to look to catch a glimpse of where physics was taking on a new complexion, and where it had been.

In fact, I do not think that anyone else but Lorentz could have guided these first Solvay Councils beginning in October 1911. Einstein certainly could not have - at the time of the first Council, at age 32, he was far from presenting an ecumenical, sage-like demeanor to the world. Remember, at that point he'd held a "real" academic job for just two years. He was driven, impatient, biting in his sarcasm, and wouldn't or couldn't hide his disdain for bad or wrong-headed approaches. The great mathematician-physicist Poincaré certainly couldn't have guided the 1911 discussions. True, by then, at various times, he had executed high-level administrative functions with great aplomb, true too he was learned beyond measure even in this illustrious crowd. But Poincaré had a blind spot toward young Einstein (not facilitated by Einstein's refusal to cite any of Poincaré's relativity work), and toward Einstein's new, heuristic quantum ideas. Max Planck, of course, had launched reasoning about the quantum discontinuity when he proposed, back in 1900, his conditions on the energy oscillators could have in the walls of a black-body cavity; but he had already shown himself uneasy with various aspects of the quantum phenomena - and indeed found premature Nernst's very idea for the Solvay event. Perhaps Nernst himself could have taken the lead but, as great a scientist as he was, in 1910-11 Nernst commanded neither the scientific authority of Lorentz (Nernst's best quantum work was just coming into view, his Nobel Prize a decade away), nor the personal admiration so many scientists had for the Dutch theorist [5].

The third miracle, of course, was the presence and prior contributions of Albert Einstein. Remember, this was not the Einstein of world-historical fame - that Einstein did not yet exist, and wouldn't until Einstein had finished his general relativistic work and the public had gazed over the large-type headlines of November 1915, announcing the results of Arthur Eddington's eclipse expedition. Nor was the "Solvay Einstein" the "molecular Einstein" - the Einstein who had cracked the Brownian motion problem, extended the Boltzmannian science of statistical mechanics, and provided a remarkable analysis of molecular dimensions. For these accomplishments he was, in the physics community, quickly and widely hailed. Finally, surprising as it might be in retrospect, the Einstein of Solvay-1 was also not the Einstein of special relativity. Relativity, and its second cousins the electron theories, and the ether theories were not on the intellectual order of the day that Nernst had circulated. Instead, the relevant Einstein, the Einstein whose work set much of the agenda of Solvay-1 was the light-quantum Einstein.

Back in 1905, Max Planck and Wilhelm Wien were as well known as Albert Einstein was obscure. After a period of quite painful professional marginality (one friend wrote his father that Einstein was half starving), in 1902, Einstein very happily shed his unemployment [for] a job in the Bern patent office. It was from there that Einstein wrote his friend Conrad Habicht in May 1905: "So, what are you up to, you frozen whale, you smoked, dried, canned piece of soul, or whatever else I would like to hurl at your head ... !" Why haven't you sent your dissertation, Einstein demanded. " ... I promise you four papers in return, the first of which I might send you soon ... The paper deals with radiation and the energy properties of light and is very revolutionary." (For the second and third, Einstein told Habicht he would report on atomic sizes using diffusion and dilute solutions, and the third would analyze Brownian motion.) "The fourth paper is only a rough draft at this point, and is an electrodynamics of moving bodies which employs a modification of the theory of space and time ... ." [6]

Einstein's remark about the "very revolutionary" light quantum was the only time he ever referred to any part of his own work in such strong terms. For startling as his contribution to relativity (the electrodynamics of moving bodies) was, there was something deeply disturbing about the light quantum - far more disturbing even than Max Planck's ambivalent stance toward the quantization of oscillator energy levels. Reaction to Einstein's work on molecular sizes and Brownian motion came quickly, responses to the light quantum were slower and much more reserved. But by the end of the decade, the idea had begun to catch fire. Nernst wrote to the English physicist Arthur Schuster on 17 March 1910: "I believe that, as regards the development of physics, we can be very happy to have such an original young thinker, a 'Boltzmann redivivus'; the same certainty and speed of thought; great boldness in theory, which however cannot harm, since the most intimate contact with experiment is preserved. Einstein's 'quantum hypothesis' is probably among the most remarkable thought [constructions] ever; if it is correct, then it indicates completely new paths [for the ether and molecular theories;] if it false, well, then it will remain for all times 'a beautiful memory'." [7] It was but a short time after penning this encomium that, in July 1910, Nernst wrote to Solvay: "It would appear that we currently find ourselves in the middle of a new revolution in the principles on which the kinetic theory of matter is based ... As has been shown, most notably by Planck and Einstein ... contradictions are eliminated if ... the postulate of quanta of energy ... [is] imposed on the motion of electrons and atoms. [It] unquestionably mean[s] a radical reform of current fundamental theories." [8]

In 1909-1910, the quantum discontinuity finally hit home among experts - and it was then that Einstein took on a new stature within the physics world. Nernst played a key role in that recognition, but he was not alone. Here then was the third miracle: at just the time that Solvay was willing to fund, Nernst to organize, and Lorentz to lead, Einstein had vastly deepened the quantum discontinuity, and in so doing had launched a research program that during the organization of the first Solvay Council, was on the cusp of recognition by many of the world's most illustrious theoretical physicists [9].

Shuttling between these high-ranking theorists, Nernst's set agenda of the first Solvay meeting. It was Nernst who early and powerfully recognized the importance of the "Boltzmann redivivus" who had just emerged from the patent office; it was Nernst who served as the lead contact with Solvay, and it was Nernst who induced Lorentz to preside over the whole. But once the conference began, the exchanges among physicists threw into relief the novelty of what was afoot. I want to focus on a few of those pivotal interactions, those involving (in the main) Einstein in conversation with Lorentz, Poincaré, and Bohr.

1.1.2   *SOLVAY -1: Einstein-Lorentz, Einstein-Poincaré*

In relativity theory we rightly see a transformation of space and time, a shift from absolutes of space and time to the quasi-operationalized concepts of ruler - measured distances and light-coordinated clocks. But seen from another angle, Einstein's greatest contribution in the paper was his introduction of a way of thinking, an invitation to reason toward symmetries in the explanatory structure of the theory that he demanded match the symmetries of the phenomena. If the phenomena were symmetric with respect to changes in the inertial frame of reference (magnet moves toward coil versus coil moves toward magnet) then the theory should show that same invariance. Similarly, we rightly attend to the quantum discontinuity of Einstein and Planck as a founding document in the history of quantum mechanics. But again, looked at from the point of view of the history of physical reasoning, we can see Einstein's paper differently: not just as a contribution to the nature of light but to the broader idea that in physics sometimes what is needed is not a full-blown theory but instead a heuristic, a provisional step, one that might not even appear consistent with other dearly-held tenets.

Einstein himself put it this way at the Council: "We all agree that the so-called quantum theory of today, although a useful device, is not a theory in the usual sense of the word, in any case not a theory that can be developed coherently at present. On the other hand ... classical mechanics...can no longer be considered a sufficient schema for the theoretical representation of all physical phenomena." [10]

Precisely this "incoherence" that did nothing to stop Einstein struck the great mécanicien, Henri Poincaré, as disastrous. It fell to Poincaré to summarize one session in Brussels. And having heard Einstein and his colleagues pronounce on the quanta - having heard them try to navigate a corpuscular as well as wave-theoretical notion of light - his view was dim indeed: "What the new research seems to put in question is not only the fundamental principles of mechanics; it is something that seems to us up to now inseparable from the very notion of a natural law. Can we still express these laws in the form of differential equations? Furthermore, what struck me in the discussions that we just heard was seeing one same theory based in one place on the principles of the old mechanics, and in another on the new hypotheses that are their negation; one should not forget that there's no proposition that one can't easily prove through the use of two contradictory premises." [11] Poincaré's points are two. First, differential equations give the moment-to-moment unfolding of phenomena that Poincaré took to define the very object of physics. This was what Newtonian gravitational physics had bequeathed us. But Poincaré surely also had in mind the world that issued from Maxwell's electrodynamics and all its subsequent modifications. For out of that mix had come the "new mechanics" embracing the electrodynamics of moving bodies that Lorentz and Poincaré himself had fought so hard to create - along with (though Poincaré never much liked his contributions) young Einstein. But this newfangled quantum hypothesis was something else. Just insofar as it was not representable in terms of a differential equation it threatened to depart from "the very notion of a natural law."

Second, and in some ways even more distressing, Poincaré pointed in his peroration to the flat-out contradiction that seemed to be the rule in the discussions he'd just heard about the quantum hypothesis. On the one hand his colleagues were happy to invoke "the principles of the old mechanics" - namely electrodynamics and the wave theory of light alongside Lorentz's law for the motion of particulate charges. On the other hand, Einstein and those who were following him were invoking "the new hypotheses that are their negation" - the quantum of light. As logic dictates, from contradictory premises follows anything at all. This invocation of light-as-wave and light-as-particle threatened not only to undermine itself but the very idea of science.

What to make of Poincaré's response to the deliberations? All too often he is depicted as a crusty reactionary, but the characterization is far from helpful. He was perfectly willing to accept, even to celebrate quite radical changes in physical theory; he helped invent and embraced the "new mechanics" of a modified electrodynamics that included Lorentz's hypothesis of contraction - and Poincaré's own version of the "local time." In thinking about the three-body problem, Poincaré helped usher in what became non-linear dynamics. But he would not countenance a mechanics that defied representation in differential-equation form, nor embrace simultaneously what he considered to be the proposition A and the proposition not-A.

It is telling, for example, that, in the discussion at Solvay, Poincaré was perfectly willing to consider modifying the very foundation of the electrodynamics of moving bodies: "Before accepting these discontinuities which force the abandonment of our usual expression of natural laws through differential equations, it would be better to try to make mass depend not only on speed as in electromagnetic theory, but also on acceleration." Poincaré took his suggestion to heart and set to work. Not long after the meeting adjourned, he reported back to his colleagues, as printed in the minutes, that "on my return to Paris, I tried calculations in this direction; they led me to a negative result. The hypothesis of quanta appears to be the only one that leads to the experimental law of radiation, if one accepts the formula usually adopted for the relation between the energy of resonators and the ether, and if one supposes that the exchange of energy can occur between resonators by the mechanical shock of atoms or electrons." It is a remarkable concession, indicative not just of his stance toward the particular physics question, but also of the engagement that occurred in Brussels [12]. Perhaps Poincaré's reversal on such a central matter could be a model for our and future Solvay Councils: fight hard, calculate hard - and concede defeat when the work demands it.

It was not just Poincaré who understood and recoiled at the upset heralded by the quantum of light. Lorentz too registered the conundrum faced by physics: "At this moment, we are far from a full [spiritual] satisfaction that the kinetic theory of gases, extended to fluids and dilute solutions and to systems of electrons, gave ten or twenty years ago. Instead, we have the sense of being at a dead end, the old theories having shown themselves more and more impotent to pierce the shadows that surround us on all sides. In this state of things, the beautiful hypothesis of energy elements, put forward for the first time by M. Planck and applied to numerous phenomena by Einstein and Nernst and others, has been a precious glimmer of light." However hard it was to grasp fully the implications of the physics of quanta, he agreed that they were not contradiction with older ideas like actions or forces. Yet "I understand perfectly that we have no right to believe that in the physical theories of the future all will conform to the rules of classical mechanics." [13] Not classical mechanics, but some kind of mechanics. For Lorentz insisted that some "mode of action" be uncovered that would explain the discrete acquisition of energy - only such an understanding would lead to "the New Mechanics which will take the place of the old one." [14]

Lorentz and Poincaré were flexible enough to consider another "new" mechanics. That mechanics might have the form of a mechanics reflecting the dynamics of the electrodynamics of moving bodies. Or it might be an as-yet uncovered system of mechanics appropriate to the quantum steps of energy allowed in the molecular oscillators. What was clear, however, was that they wanted above all a mechanics, some mechanics that would be expressed through a definite, visualizable microphysics (Lorentz) or differential equations (Poincaré). Not unreasonably, both wanted a theory. In fact, back in 1903, when Poincaré addressed the graduates of Ecole Polytechnique he had rhetorically asked his audience what they, this extraordinary group of scientists, military men, industrialists, and national leaders had in common. The answer: mechanics. Mechanics, modifiable, improvable - not at all frozen the age of Newton - was the centerpiece of reason about the world [15].

Knowing the importance that Lorentz and Poincaré attached to mechanics (new or old), it is instructive to register what may be the only "conversation" ever recorded between Poincaré and Einstein. It went like this:

Poincaré to Einstein: "What mechanics are you using?"
Einstein: "No Mechanics."
de Broglie: "[This] appeared to surprise his interlocutor."

"Surprise" wasn't the half of it. Poincaré had modified his mechanics for his take on the electrodynamics of moving bodies, he had altered mechanics to make room for his work on the three-body problem. And he was willing to contemplate a transformation here, in the excruciatingly difficult domain of the quantum. But Einstein's answer, "no mechanics" was, for Poincaré, an impossible one. For it was precisely the heuristic non-theory that marked one of the key aspects of Einstein's intervention. Here was an aspect of light - so Einstein was telling his contemporaries - that was not yet built into anything worthy of the name theory. And yet this discrete aspect must, in the long run, become part of our account of the physical world. How it should be incorporated remained an open question, but after reasoning about scattering, photoelectric, and specific heat effects that the quantal aspect of the world would stay remained for Einstein, and an expanding circle of others, a deep conviction.

After Friday 3 November 1911, the participants scattered. Einstein wrote to one of his friends, "H. A. Lorentz is a marvel of intelligence and tact. He is a living work of art! In my opinion he was the most intelligent among the theoreticians present." But of Poincaré, Einstein had a much dimmer view: " ... Poincaré was simply negative in general, and all his acumen notwithstanding, he showed little grasp of the situation." But for all their disagreements, amicable with Lorentz, more strained with Poincaré, Einstein left an indelible impression. A formal, rather distant Lorentz wrote to Einstein in February 1912 about the invitation he was extending to Einstein to succeed him as professor of theoretical physics at Leiden: "Personally, I cannot tell you how tempting the perspective to work in constant contact with you would be. If it were granted to me to welcome you here as a successor and at the same time as a colleague, it would fulfill a wish I have cherished in silence for a long time, but unfortunately could not express earlier. As one becomes older and the power of creativity slowly fails, one admires even more the good spirits and enthusiastic creative impulse of a younger man." [16] Nothing in the clash of ideas entered into the domain of the personal - on the contrary, it led Lorentz to appreciate Einstein's distinctive approach all the more.

What did Poincaré make of Einstein? Despite their non-meeting of minds, the older physicist wrote to Pierre Weiss in November 1911, recommending Einstein to the Swiss Federal Institute of Technology. Einstein, Poincaré wrote, was "one of the most original minds I have known." He had "already taken a very honorable rank among the leading scholars of his time. ... He does not remain attached to classical principles, and, in the presence of a proble of physics, is prompt to envision all the possibilities." Not all of Einstein's ideas would bear fruit, Poincaré, added, but if even one did, that would suffice. "The future will show more and more the value of Mr. Einstein", Poincaré ended, "and the university that finds a way to secure this young master is assured of drawing from it great honor." [17] As they weighed the discussion unfolding before them at Solvay-1, Lorentz and Poincaré were shocked at what was happening to physics, to the physics that they had done so much to develop. But they were listening - as few have done any time in the face of something so disturbing to everything for which they stood.

1.1.3 *Ignoramus, Ignorabimus at Solvay-5 and Solvay-6*

The calamity of World War I crashed through the physics community. French and Belgian scientists in particular dug a trench between them and their German homologues, neither forgetting nor forgiving. During the Great War, Max Planck, Ernst Haeckel, and Wilhelm Roentgen (joining ninety other luminaries) had issued a fierce defense of the destruction at Louvain, the invasion of Belgium - and linked German high culture to the iron will of the military. Einstein and two colleagues responded with a blast of their own with their plea for European civilization. At some risk, Einstein struggled to maintain relations between belligerents. As late as 1927 there were difficulties inside the Council itself as Solvay-5 entered its final planning stages - one handwritten note in the Solvay archives noted one participant would not be there "puisqu'il y a des allemands"; another memo, just before Solvay-5 opened, came to Lorentz on 14 October 1927: "I know that patriotism is intransigent, as much in those who attack as those who defend; these are territories of an infinite sensitivity." [18]

If the postwar political scene was overheated, the scientific one was as well. After two years of extraordinarily intense work, by mid-1927, its creators were celebrating a triumphant quantum theory. Werner Heisenberg had extended the Niels Bohr's work into his "matrix mechanics," eschewing the visual elements that he found too redolent of a dead classical physics. Erwin Schroedinger had hoped to counter the Heisenbergian anti-visual with his wave mechanics - and Max Born had but recently offered the probabilistic interpretation of the theory's wave function.

Lorentz's commentary as Solvay-5 advanced offers us an extraordinary ring-side seat, not because he was a participant in assembling the new quantum mechanics - but precisely because he shows us what one of the great theoretical physicists of all times thought of a theory that had departed so radically from the microphysics of fields and electrons that he worked so hard to put in place. Indeed, in many ways the new quantum theory departed from Lorentzian precepts even more dramatically than Einstein's heuristic light particle had back at Solvay-1 in 1911. And now, in earnest, Einstein leveled his own criticisms at the new theory in one of the greatest dialogues that has ever taken place in the history of science - the battle between Einstein and Bohr. Though in brief compass one cannot possibly do justice to this long story, it is worth recalling some of what happened as Lorentz and now Einstein faced a very different kind of physics.

To set things in perspective, it is useful to begin with Lorentz's reaction to

Bohr and his young quantum mechanicians' work. But one has to see it from Lorentz's angle of vision - and for that one has to recall Lorentz's extraordinary several decades of success with electron theorizing as the charged particle moved, oscillated, radiated. Only then can we grasp just how sensible it must have seemed for him to urge caution before abandoning the tools that so very recently had yielded such extraordinary structures. Microphysics was new, visualizable, calculable: it taught us how to think all the way down to the micro-structures that explained the splitting of spectral lines in the presence of a magnetic field. It gave a clear picture of what was happening in the reflection and refraction - it was at long last a way to put paid to the promise Maxwell's theory made to join optics to electrodynamics. This was a theory and a way of doing science worth fighting for. Here's Lorentz as he reflected at Solvay-5 on the new work by his young quantum colleagues: "We want ... to make an image in our imagination [esprit]. Until now we have always wanted to form images through ordinary notions of time and space. These notions may be innate; in any case, they were developed by our personal experience, by our everyday observations. For me, these notions are clear and I admit that I cannot form an idea of physics without these notions. The image that I want to form of phenomenon must be absolutely distinct and defined, and ... we cannot form such an image except in space and in time. " [19] To read these words of Lorentz is to see the innovation of Bohr, Schrödinger, and Heisenberg from the shadow it cast on turn-of-the-century electron physics. "What Mr. Bohr does is this: after an observation he limits anew the wave packet in a way that will represent for him that which the observation taught us on the position and movement of the electron. Then begins a new period during which the packet diffuses again, up to the moment when a new observation permits us to effect the reduction once more. But I would like an image of all this during an unlimited time." [20] Yet it was exactly this image that the new quantum mechanics would not - could not - provide. Lorentz looked at the physical description and said: ignoramus - we do not know (but we could). Bohr and Heisenberg replied, essentially, ignorabimus: we cannot know.

Einstein too sought a way out of a description of nature that to him seemed too impoverished to catch nature in the fullness that it, in principle, should allow: "In my opinion, the difficulty can only be resolved in this way: one does not only describe the process using Schroedinger's wave, but at the same time one localizes the particle during propagation. I think that de Broglie is right to look in this direction. If one works only with Schroedinger's waves, in my opinion, the second interpretation of [psi squared] implies a contradiction with the postulate of relativity." [21]

Lorentz reckoned that if one wanted to have an idea of an electron at one moment and then at another, one had to think of its trajectory, "a line in space." "And if this electron encounters an atom and penetrates it, and after several adventures it leaves the atom, I make a theory in which this electron maintains its individuality; that is I imagine a line following which that electron passes through the atom."

Now it could be very difficult - but it should be possible. Could electrons suffer transformations? Fine. Could an electron melt into a cloud? Fine too. But then as far as Lorentz was concerned it was our duty to figure out how that transformation takes place. But one could not, a priori, forbid ourselves to conduct research into such questions. Put another way, Lorentz could perfectly well allow that we could not answer the question now - that was acceptable. But banish it forever? That seemed to him absurd. "If we abandon old ideas one can always maintain the old names. I want to conserve this ideal from older times of describing everything that happens in the world by distinct images." Lorentz welcomed new theories (be they wave-like or particulate) - so long as it was possible to keep clear and distinct images of the underlying process. He was never one to say, as some physicists had, that the older knowledge was "in principle" complete. That we don't know (ignoramus) was fine. That we can't know (ignorabimus) was too much.

Lorentz was even willing to have that probability calculable by the square of the wave function. "But the examples given by Mr. Heisenberg teach me that I would have attained all that experience allow me to attain." This limitation was what was at stake for the Council's leader. It was the idea that this notion of probability should be put at the beginning of our physics that bothered Lorentz. At the conclusion of our calculations, a probabilistic result would be no more consequential for the meaning of physics than other results that issued from a calculation. But make probability part of the axiomatic, the a priori, and Lorentz bridled: "I can always guard my determinist faith for fundamental phenomena ... .Couldn't one keep determinism in making it the object of a belief? Must we necessarily establish indeterminism in principle?"

Though they may have split in various particulars, Lorentz and Einstein both were bothered by in-principle ignorance. And here, toward the end of Solvay-5, Einstein advanced a picturable thought experiment (first figure). Imagine particles entering the device from point 0 and then spreading toward a circular screen. Einstein then posits that there are two imaginable roles that the theory might play. Possibility 1: the theory with its psi-squared only claims to describe an ensemble of particles, not each particle one-by-one; Possibility 2: "the theory claims to be a complete theory of single processes. Every particle which moves towards the screen has a position and a speed, insofar as they can be determined by a packet of de Broglie-Schroedinger waves with small wavelength and angular opening." Bohr rejected Einstein's choice, noting that the particles could not be considered in isolation - it is only permitted for us to consider the system as a whole, diaphragm 0 and particles. The position, (and therefore the momentum and momentum transfer to the particles), of the diaphragm matters in what we can say about the particle and its subsequent path [22].

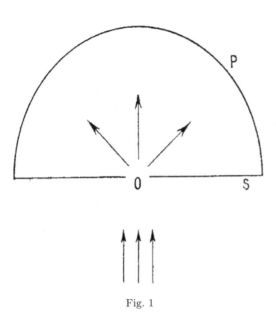

Fig. 1

Poincaré died not long after the first Solvay Council; Lorentz just a few months after Solvay-5. Marie Curie wrote for the volume in honor of what "the soul of our meetings" had meant to the assembled physicists: "The illustrious master teacher [maître] and physicist, H.-A. Lorentz, was taken from us [4] February 1928 by a sudden sickness - when we had just admired, one more time, his magnificent intellectual gifts that age had not diminished." He had, in Curie's view, brought to the meetings diplomacy, students, followers, and collaborators - he was in all senses master of the field. Lorentz was "the real creator of the theoretical edifice that explained optical and electromagnetic phenomena by the exchange of energy between electrons and radiation in accordance with Maxwell's theory. Lorentz retained a devotion to this classical theory. It was therefore all the more remarkable that his flexibility of mind was such that he followed the disconcerting evolution of quantum theory and the new mechanics." [23]

Langevin took over for Solvay-6 following Lorentz's death; it was Langevin who guided discussion in October 1930. By then, the themes we have been discussing split Einstein ever further from Bohr, and their struggle continued long after they left the Metropole - through Einstein-Podolsky-Rosen and beyond. Throughout, Einstein famously pressed ahead in his quest to show that the problem with contemporary quantum mechanics was that it was incomplete - that our state was one of ignorance in fact, not in principle: ignoramus not ignorabimus. One Gedankenexperiment followed another, for which just one, perhaps the most remarkable, must stand for many.

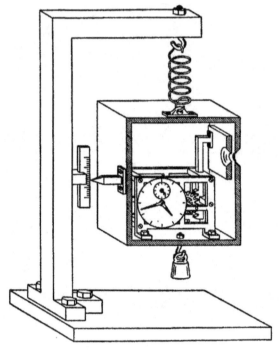

Fig. 2

Einstein: imagine a clock-like device like that shown in the second figure. Einstein: Bohr and his quantum mechanics forbid knowing both time and energy. What if we measure the time a photon is released (using a clocklike release mechanism of figure 2) and we weigh the source of light before and after photon released? Then we would have the time of the photon's launch and its energy using the weight change of the mechanism and E = mc2. Wouldn't that outsmart the uncertainty principle showing we could measure quantities more accurately than the theory allows? In other words, ignoramus: we can be clever, and get both quantities where the theory tells us we must choose.

This, one contemporary observer recorded, left Bohr miserable. It did indeed look as if Einstein had shown the theory to be incomplete because it could not fully represent physically measurable quantities. "During the whole evening [Bohr] was extremely unhappy, going from one to the other and trying to persuade them that it couldn't' be true, that it would be the end of physics if Einstein were right; but he couldn't produce any refutation." [24] Finally Bohr found a solution: To weigh the box is to fix its position vertically. But the uncertainty principle then requires the box have an uncertainty in vertical momentum. Reweighing the box requires a time T for the box to settle, and a corresponding uncertainty in height. But this uncertainty in height corresponds, by the gravitational red-shift, to an uncertainty in the clock speed (as you, Einstein, proved!). So because we fixed photon energy we cannot know exactly when the photon is launched. Ignorabimus!, Bohr, in essence, replied. One cannot, in fact, have both the time of photon release and its energy. The theory is not incomplete.

Einstein's and Bohr's debate did not end there, of course. A proper account would wend its way through the rest of their Solvay debate, to the Einstein-Podolsky-Rosen thought experiment and eventually up to Bell's theorem. But even this one snapshot captures the great stakes involved for each of the antagonists as they faced off at the Solvay Council: the shape and even the existence of physics. [25]

1.1.4   *Solvay Redivivus*

The casualties of World War I were so terrible, the furor over militant nationalism so deep, that it took all of Hendrik Lorentz's - and Albert Einstein's - force and good will to repair the damage to international science. The meetings of the early 1920s, but especially of Solvay-5 (1927) and Solvay-6 (1930) were a salve to these wounds. No doubt the collective achievement of quantum mechanics, and the role that the Solvay Councils played in its interpretation, were and were seen to be lasting, international accomplishments. The composite nature of the work was visible down to the traces any physicist made as he or she calculated anything: Heisenberg's matrices, Schroedinger's wave equation, Dirac's notation and relativistic extension, Bohr's twin doctrines of complementarity and correspondence.

But where World War I only provisionally damaged the Solvay Councils, the rise, rule and ruin of Nazism did much worse. Putting together a serious Continental European conference in physics after Hitler's election in January 1933 became almost impossible. And by the time the war ended, twelve and a half bloody years later, the return address of physics had changed. The great institutes of Born and Bohr lay shattered. Many of the younger European physicists were gone - deported, killed, or driven into exile. In the United States the physics community was entirely restructured by the vast war projects of radar and the atomic bomb; theoretical physics underwent a tremendous expansion. All this meant that at just the time European physics lay most devastated, American physics stood poised on a vast armamentarium of theoretical and experimental technique, accompanied by a budget beyond anyone's wildest pre-war dreams. Working conferences like that of Shelter Island (June 1947) or Pocono Manor (March-April 1948) stood as exemplars for the new generation of theorists now beginning to take their skills to civilian issues: Richard Feynman, Julian Schwinger among them [26].

Consequently, by the time the Solvay Council met in late 1948, the world of physics had turned. Reading the pages of the proceedings, one senses a tone more of elegy than of excitement. J. Robert Oppenheimer is there, but no one breathed a word about a transformed role for physics and physicists that necessarily accompanied a world with nuclear weapons. Meanwhile, the major figures from the pre-war era issued one cautionary note after another. Dirac worried about the infinities, hoping that they could be done away with in a formulation of the physics that would by-pass perturbative methods altogether. For his part, Bohr was turning to quantum philosophy beyond physics - epistemic worries far from the pressing concerns of younger, more pragmatic theorists who wanted to dig in, build accelerators and calculate things. Bohr's worry was to set limits to visual, and he slammed those who persisted in the search for the visual, even in a disastrous encounter at the Pocono meeting where he ripped into Feynman for trying with diagrams to visualize the unvisualizable. Heisenberg too added his warning: fundamental lengths could limit the validity of all present theory.

When Feynman did speak at Solvay-12 (1961), by which time his diagrams were to physicists what hammers and saws were to carpenters, he took the opportunity to push on his own theory of QED as hard as possible. Not for him a wished-for revolution based on fundamental length, hoped-for mathematical by-passes or philosophical introspection: "In writing this report on the present state of [QED], I have been converted from a long-held strong prejudice that it must fail [other than by being incomplete] at around 1 Gev virtual energy. The origin of this feeling lies in the belief that the mass of the electron ... and its charge must be ultimately computable and that Q.E.D must play some part in this future analysis. I still hold this belief and do not subscribe to the philosophy of renormalization. But I now realize that there is much to be said for considering [Q.E.D.] exact ... to suggest definite theoretical research. This is Wheeler's principle of 'radical conservatism'." [27] Writing for the Solvay record seems to push people to think hard about even long-held views; Feynman was no exception. But in retrospect, even Feynman's radical conservatism wasn't radically conservative enough - there was a huge amount of structure still to be plumbed in renormalization (starting with the renormalization group).

Today we begin deliberations at Solvay-23 on the quantum structure of space and time. Solvay-1 had before it the problem of the light quantum, Solvay-5 and Solvay-6 confronted the brand-new quantum mechanics. We too have our agenda before us - emergent spacetime, singularities, and new structures perched between physics and mathematics. No doubt basic and high-stakes questions will arise: what do we want from our explanations in matters of cosmology? What kind of singularities do we face and what will they mean for the current campaign in theoretical physics? What is the proper place for the anthropic principle? Is the hunt for a physics that will pick out the masses, charges, and coupling constants the right goal for physics- and is our present inability to do so a matter of our not knowing (ignoramus)? Or is it rather that it is not given to us to know these things - that our ability to ask the question presupposes that we are in this universe and no other? Is it the case that we can never know those values as deduced from first principles (ignorabimus)? These and other questions of similar difficulty about the right place of string theory are just the kind of hard task that Nernst, Lorentz, Einstein, and Bohr had in mind when they gathered in this grand site almost a hundred years ago.

Western philosophy is often said, in only part exaggeration, to be a long footnote to Plato. The physics of this last century, documented - and now re-awakened - in the Conseils de Solvay, might be similarly seen as a long elaboration of the great dialogue Einstein initiated between relativity and the quantum. It remains our ground. Colleagues: Welcome back to Solvay.

Bibliography

[1] For two fine examples of studies of the effect of the Solvay Councils in other domains, see e.g. Roger Stuewer, "The Seventh Solvay Conference: Nuclear Physics at the Crossroads," in A. J. Kox and Daniel M. Siegel, No Truth Except in the Details (Dordrecht: Kluwer, 1995), pp. 333-362; Mary Jo Nye, "Chemical Explanation and Physical Dynamics: Two Research Schools at the First Solvay Chemistry Conferences, 1922-1928," Annals of Science 46 (1989), pp. 461-480. Christa Jungnickel and Russell McCormmach set the Solvay Council in the larger frame of the history of theoretical physics: Mastery of Nature. Theoretical Physics from Ohm to Einstein (Chicago: University of Chicago Press, 1986), esp. pp. 309-21. For an excellent discussion of Nernst's role in the first Solvay Council, see Diana Kormos Barkan, "The Witches' Sabbath: The First International Solvay Congress in Physics," Science in Context 6.

[2] (1993): 59-82, hereafter, "Witches Sabbath"; on Nernst more broadly see D. Barkan, Walther Nernst and the Transition to Modern Physical Science, (Cambridge: Cambridge University Press, Cambridge, 1999).

[3] Louis d'Or and Anne-Marie Wirtz-Cordier, Ernest Solvay (Académie Royale de Belgique, Mémoires de la classe des sciences XLIV Fascicule 2 (1981), p. 50-51; citation on p. 51 (Nernst to Solvay, 26 July 1910). For more on Ernest Solvay the following are very helpful and contain many further references: Jacques Bolle, Solvay: L'homme, la découverte, l'entreprise industrielle (Bruxelles: SODI); Andrée Despy-Meyer and Didier Devriese, Ernest Solvay et son temps (Bruxelles: Archives de l'université de Bruxelles, 1997); and Pierre Marage and Grégoire Wallenborn, The Solvay Councils and the Birth of Modern Physics (Basel: Birkhäuser Verlag, 1999).

[4] D'Or and Wirtz-Cordier, Solvay, pp. 51-52.

[5] In addition to the Barkan work cited above, there is a very interesting piece by Elisabeth Crawford, "The Solvay Councils and the Nobel Institution," in Marage and Wallenborn, Birth, pp. 48-54 in which she shows how Nernst positioned the Solvay Council as a competitor to Svante Arrhenius's Nobel Institute.

[6] Einstein to Habicht, 18 or 25 May 1905, in The Collected Papers of Albert Einstein, vol. 5, transl. Anna Beck (Princeton: Princeton University Press, 1995), pp. 19-20, on p. 20.

[7] Nernst to Schuster, 17 March 1910, cited in Barkan, "Witches [Sabbath]," p. 62.

[8] Cited in Didier Derviese and Grégoire Wallenborn, "Ernest Solvay: The System, The Law and the Council," in Birth, p. 14.

[9] On the timing of recognition of the quantum discontinuity see T.S. Kuhn, Black-Body Theory and the Quantum Discontinuity, 1894-1912. (Chicago: University of Chicago Press, 1978, 1987).

[10] Einstein, Solvay-1, in P. Langevin and M. de Broglie, "La Théorie du Rayonnement et les Quanta," Paris: Gauthier-Villars, 1912), 30 October-3 November, 1911, p. 436. Hereafter Solvay-1.

[11] Poincaré: Solvay-1, p. 451.

[12] Poincaré, Solvay-1, p. 453.

[13] Lorentz, Solvay-1, pp.6-7.

[14] Lorentz, Solvay-1, p. 7.

[15] On Poincaré, Polytechnique, and mechanics, see Galison, Einstein's Clocks (New York: W.W. Norton, 2003), pp. 48-50.

[16] Cited in A.J. Kox, Einstein and Lorentz, "More than Just Good Colleagues," Science in Context 6 (1993): 43-56, on p. 44.

[17] Poincaré to Weiss, cited in Galison, Einstein's Clocks, p. 300.

[18] "des Allemands" from handwritten note on letter to Lefebure, 6 October 1927, document 2545 in Solvay Council Archives, Brussels; "patriotisme" from Lefébure to Lorentz, 14 October 1927, document 2534 in Solvay Council Archives, Brussels.

[19] Lorentz, Solvay-5 (1927), 248.

[20] Lorentz, Solvay-5 (1927), 287.

[21] Einstein, in Solvay-5, pp. 255-56, cited in Marage and Wallenborn, Birth, p. 168.

[22] Einstein, from Solvay-5, cited by Marage and Wallenborn in Marage and Wallenborn, Birth, pp. 167-68.

[23] Institut International de Physique Solvay, Electrons et Photons, Paris: Gauthier-Villars, 1928), n.p. Hereafter Solvay-5.

[24] L. Rosenfeld, "Some Concluding Remarks and Reminiscences," Fundamental Problems in Elementary Particle Physics (14th Solvay Council of Physics, held in Brussels in 1967; New York: John Wiley Interscience, 1968), p. 232, cited in Marage and Wallenborn, "The Birth of Modern Physics," in Marage and Wallenborn, Birth, p. 171.

[25] There is a nearly infinite large literature on the Einstein-Bohr debate, but the locus classicus is in the contribution by Bohr and Einstein's reply in P. Schilpp, Albert Einstein Philosopher and Scientist (Northwester University and Southern Illinois University: Open Court, 1951).

[26] On postwar expansion in American physics see Galison, Image and Logic: A Material Culture of Microphysics (Chicago: University of Chicago Press, 1997), esp. ch. 4. For an excellent account of the development of quantum electrodynamics, see S. S. Schweber, QED and the Men Who Made It (Princeton: Princeton University Press, 1994); and a splendid account of the spread of Feynman diagrams, David Kaiser, Drawing Theories Apart: The Dispersion of Feynman Diagrams in Postwar Physics (Chicago: University of Chicago Press, 2005).

[27] Richard Feynman, "The Present Status of Quantum Electrodynamics" in The Quantum Theory of Fields (New York: John Wiley Interscience, [1961]), p. 89.

1.2 Discussion

T. Damour I was surprised by what you said concerning the only conversation between Poincaré and Einstein. My understanding from the text of Maurice de Broglie was that the conversation did not concern quantum mechanics as you seem to convey, but "la mécanique nouvelle." Although later "meécanique nouvelle" meant quantum mechanics, I think this conversation makes more sense if it means relativistic mechanics, which Poincaré always called "mécanique nouvelle." This is the difference between Poincaré always having to assume some microscopic mechanics to discuss relativistic physics and Einstein making general postulates and not having to make dynamical postulates about microscopic physics. Do you really intend to challenge this view?

P. Galison I am quite certain it is not about the new mechanics. All of Poincaré's comments, both in the session and afterwards, concern the problem that the quantum goes outside the bounds of the description under differential equations. In the context of that discussion, it very directly concerns the quantum of light. It is true there was a misreading of that conversation in Banesh Hoffmann's book where it was attributed to a discussion about the new mechanics and relativity. In the context of the discussion however, not only where it's located, but in view of Poincaré's other comments about differential equations and the nature of mechanics, it is clear that what he is referring to is an absence of description under differential equations. This is what he considers to be a necessary, if not sufficient, condition for having a mechanics at all. That seems to be the basis on which that particular exchange is framed. On relativity I looked long and hard for direct exchanges between Einstein and Poincaré, but they just do not talk about that to each other. The session that Poincaré was running, and in which Einstein participated, had nothing to do with relativity, it was only about the quantum of light.

T. Damour But that quote by Maurice de Broglie is separate from the Solvay context. It is in a text which is not in the Solvay proceedings.

P. Galison It is about a confrontation that occurred at the Solvay conference in 1911. That is the only time Poincaré and Einstein met.

T. Damour The whole point is whether "mécanique nouvelle" is relativistic mechanics or quantum mechanics. I think it makes more sense if it is relativistic mechanics, it really gives meaning to this conversation.

P. Galison I think that if you look at the context you will see it is about the quantum, and it has to do with this question of differential equations which was so crucial. I reviewed the citations for that. I would love it to be about relativity, it would be much more interesting to me for other reasons, but it is not.

S. Weinberg In the story about Einstein and Bohr's famous argument, what bothers me is that Bohr is supposed to have won by invoking general relativity. But what if general relativity were wrong? Then, does that mean quantum mechanics would be inconsistent? I suspect, although I have not studied the debate in detail, that the issue actually arises only within a framework in which there is a gravitational redshift and it does not really depend on the validity of general relativity.

P. Galison That seems right. That is to say, what is required for the clock speed to depend on the gravitational potential is much less than the full structure of general relativity.

G. Gibbons Basically, it is energy conservation and nothing more. In fact in that little thought experiment, you are not using gravity, it could be any force that is holding up the clock. I think it is completely decoupled from general relativity, I agree entirely with what Weinberg has said.

P. Galison Rhetorically what Bohr profits from is that he is referring back to Einstein's own work.

G. Gibbons It is certainly true that Einstein discovered the gravitational redshift, but it is decoupled from general relativity.

Session 2

Quantum Mechanics

Chair: *David Gross*, KITP, Santa Barbara, USA
Rapporteur: *James B. Hartle*, UCSB, USA
Scientific secretaries: *Riccardo Argurio* (Université Libre de Bruxelles) and *Glenn Barnich* (Université Libre de Bruxelles)

2.1 Rapporteur talk: Generalizing Quantum Mechanics, by James B. Hartle

Note: The rapporteur talk was prepared by James Hartle but delivered by David Gross and Murray Gell-Mann as Jim was unable to attend the conference. The text below has been prepared by James Hartle.

2.1.1 *Abstract*

Familiar textbook quantum mechanics assumes a fixed background spacetime to define states on spacelike surfaces and their unitary evolution between them. Quantum theory has changed as our conceptions of space and time have evolved. But quantum mechanics needs to be generalized further for quantum gravity where spacetime geometry is fluctuating and without definite value. This paper reviews a fully four-dimensional, sum-over-histories, generalized quantum mechanics of cosmological spacetime geometry. This generalization is constructed within the framework of generalized quantum theory. This is a minimal set of principles for quantum theory abstracted from the modern quantum mechanics of closed systems, most generally the universe. In this generalization, states of fields on spacelike surfaces and their unitary evolution are emergent properties appropriate when spacetime geometry behaves approximately classically. The principles of generalized quantum theory allow for the further generalization that would be necessary were spacetime not fundamental. Emergent spacetime phenomena are discussed in general and illustrated with the example of the classical spacetime geometries with large spacelike surfaces that emerge from the 'no-boundary' wave function of the universe. These must be Lorentzian with one, and only one, time direction. The essay concludes by raising the question of whether quantum mechanics itself is emergent.

2.1.2 *Introduction*

Does quantum mechanics apply to spacetime? This is the question the organizers asked me to address. It is an old issue. The renowned Belgian physicist Léon Rosenfeld wrote one of the first papers on quantum gravity [1], but late in his career came to the conclusion that the quantization of the gravitational field would be meaningless[1] [3, 4]. Today, there are probably more colleagues of the opinion that quantum theory needs to be replaced than there are who think that it doesn't apply to spacetime. But in the end this is an experimental question as Rosenfeld stressed .

This lecture will answer the question as follows: *Quantum mechanics can be applied to spacetime provided that the usual textbook formulation of quantum theory is suitably generalized.* A generalization is necessary because, in one way or another, the usual formulations rely on a fixed spacetime geometry to define states on spacelike surfaces and the time in which they evolve unitarily one surface to another. But in a quantum theory of gravity, spacetime geometry is generally fluctuating and without definite value. The usual formulations are emergent from a more general perspective when geometry is approximately classical and can supply the requisite fixed notions of space and time.

A framework for investigating generalizations of usual quantum mechanics can be abstracted from the modern quantum mechanics of closed systems [5–7] which enables quantum mechanics to be applied to cosmology. The resulting framework — generalized quantum theory [8–10] — defines a broad class of generalizations of usual quantum mechanics.

A generalized quantum theory of a physical system (most generally the universe) is built on three elements which can be very crudely characterized as follows:

- The possible fine-grained descriptions of the system.
- The coarse-grained descriptions constructed from the fine-grained ones.
- A measure of the quantum interference between different coarse-grained descriptions incorporating the principle of superposition.

We will define these elements more precisely in Section 6, explain how they are used to predict probabilities, and provide examples. But, in the meantime, the two-slit experiment shown in Figure 1 provides an immediate, concrete illustration.

A set of possible fine-grained descriptions of an electron moving through the two-slit apparatus are its Feynman paths in time (histories) from the source to the

---

[1]Rosenfeld considered the example of classical geometry curved by the expected value of the stress-energy of quantum fields. Some of the difficulties with this proposal, including experimental inconsistencies, are discussed by Page and Geilker [2].

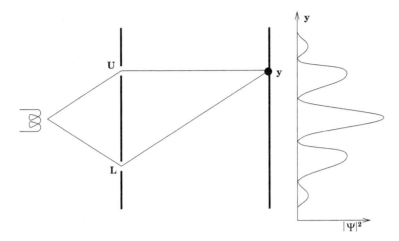

Fig. 2.1 The two-slit experiment. An electron gun at left emits an electron traveling towards a screen with two slits, U and L, its progress in space recapitulating its evolution in time. The electron is detected at a further screen in a small interval Δ about the position y. It is not possible to assign probabilities to the alternative histories of the electron in which it went through the upper slit U on the way to y, or through the lower slit L on the way to y because of the quantum interference between these two histories.

detecting screen. One coarse-grained description is by which slit the electron went through on its way to detection in an interval Δ about a position y on the screen at a later time. Amplitudes $\psi_U(y)$ and $\psi_L(y)$ for the two coarse-grained histories where the electron goes through the upper or lower slit and arrives at a point y on the screen can be computed as a sum over paths in the usual way (Section 4). The natural measure of interference between these two histories is the overlap of these two amplitudes integrated over the interval Δ in which the electron is detected. In this way usual quantum mechanics is a special case of generalized quantum theory.

Probabilities cannot be assigned to the two coarse-grained histories illustrated in Figure 1 because they interfere. The probability to arrive at y should be the sum of the probabilities to go by way of the upper or lower slit. But in quantum theory, probabilities are squares of amplitudes and

$$|\psi_U(y) + \psi_L(y)|^2 \neq |\psi_U(y)|^2 + |\psi_L(y)|^2 . \tag{1}$$

Probabilities can only be predicted for sets of alternative coarse-grained histories for which the quantum interference is negligible between every pair of coarse-grained histories in the set (decoherence).

Usual quantum mechanics is not the only way of implementing the three elements of generalized quantum theory. Section 7 sketches a sum-over-histories generalized quantum theory of spacetime. The fine-grained histories are the set of four-dimensional cosmological spacetimes with matter fields on them. A coarse graining is a partition of this set into (diffeomorphism invariant) classes. A natural measure of interference is described. This is a fully four-dimensional quantum theory without an equivalent 3+1 formulation in terms of states on spacelike surfaces and their unitary evolution between them. Rather, the usual 3+1 formulation is emergent for those situations, and for those coarse grainings, where spacetime geometry behaves approximately classically. The intent of this development is not to propose a new quantum theory of gravity. This essentially low energy theory suffers from the usual ultraviolet difficulties. Rather, it is to employ this theory as a model to discuss how quantum mechanics can be generalized to deal with quantum geometry.

A common expectation is that spacetime is itself emergent from something more fundamental. In that case a generalization of usual quantum mechanics will surely be needed and generalized quantum theory can provide a framework for discovering it (Section 8). Emergence in quantum theory is discussed generally in Section 9. Section 10 describes the emergence of Lorentz signatured classical spacetimes from the no-boundary quantum state of the universe.

Section 11 concludes with some thoughts about whether quantum mechanics itself could be emergent from something deeper. But before starting on the path of extending quantum theory so far we first offer some remarks on where it is today in Section 2.

2.1.3 *Quantum Mechanics Today*

Three features of quantum theory are striking from the present perspective: its success, its rejection by some of our deepest thinkers, and the absence of compelling alternatives.

Quantum mechanics must be counted as one of the most successful of all physical theories. Within the framework it provides, a truly vast range of phenomena can be understood and that understanding is confirmed by precision experiment. We perhaps have little evidence for peculiarly quantum phenomena on large and even familiar scales, but there is *no* evidence that all the phenomena we do see, from the smallest scales to the largest of the universe, cannot be described in quantum mechanical terms and explained by quantum mechanical laws. Indeed, the frontier to which quantum interference is confirmed experimentally is advancing to ever larger, more 'macroscopic' systems[2]. The textbook electron two-slit experiment shown schematically in Fig. 1 has been realized in the laboratory [12]. Interference has been confirmed for the biomolecule tetraphenylporphyrin ($C_{44}H_{30}N_4$) and the flurofullerine ($C_{60}F_{48}$) in analogous experiments [13] (Figure 2). Experiments with superconducting squids have demonstrated the coherent superposition of macroscopic currents [14–16]. In particular, the experiment of Friedman, *et al.* [16] exhibited the coherent superposition of two circulating currents whose magnetic moments were of order $10^{10}\mu_B$ (where $\mu_B = e\hbar/2m_e c$ is the Bohr magneton). Experiments under development will extend the boundary further [17]. Experiments of increasing

---

[2]For an insightful and lucid review see [11].

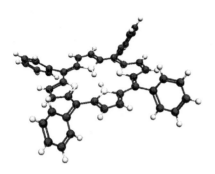
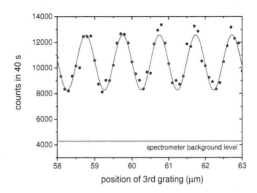

Fig. 2.2 Interference of Biomolecules. The molecule tetraphenylporphyrin ($C_{44}H_{30}N_4$) is shown at left. Its quantum interference fringes in a Talbot-Lau interferometer are shown at right from experiment carried out in Anton Zeilinger's group (Hackermüller, et al (2002)).

ingenuity and sophistication have extended the regime in which quantum mechanics has been tested. No limit to its validity has yet emerged.

Even while acknowledging its undoubted empirical success, many of our greatest minds have rejected quantum mechanics as a framework for fundamental theory. Among the pioneers, the names of Einstein, Schrödinger, DeBroglie, and Bohm stand out in this regard. Among our distinguished contemporaries, Adler, Leggett, Penrose, and 't Hooft could probably be counted in this category. Much of this thought has in common the intuition that quantum mechanics is an effective approximation of a more fundamental theory built on a notion of reality closer to that classical physics.

Remarkably, despite eighty years of unease with its basic premises, and despite having been tested only in a limited, largely microscopic, domain, no fully satisfactory alternative to quantum theory has emerged. By fully satisfactory we mean not only consistent with existing experiment, but also incorporating other seemingly secure parts of modern physics such as special relativity, field theory, and the standard model of elementary particle interactions. As Steve Weinberg summarized the situation, "It is striking that it has not so far been possible to find a logically consistent theory that is close to quantum mechanics other than quantum mechanics itself" [18]. Alternatives to quantum theory meeting the above criteria would be of great interest if only to guide experiment.

There are several directions under investigation today which aim at a theory from which quantum mechanics would be emergent. Neither space nor the author's competence permit an extensive discussion of these ideas. But we can mention some of the more important ones.[3]

Bohmian mechanics [20] in its most representative form is a deterministic but

---

[3]The references to these ideas are obviously not exhaustive, nor are they necessarily current. Rather, they are to typical sources. For an encyclopedic survey of different interpretations and alternatives to quantum mechanics, see [19].

highly non-classical theory of particle dynamics whose statistical predictions largely coincide with quantum theory [21]. Fundamental noise [22] or spontaneous dynamical collapse of the wave function [23, 24] are the underlying ideas of another class of model theories whose predictions are distinguishable from those of quantum theory, in principle. Steve Adler has proposed a statistical mechanics of deterministic matrix models from which quantum mechanics is emergent [25]. Gerard 't Hooft has a different set of ideas for a determinism beneath quantum mechanism that are explained in his article in this volume [26]. Roger Penrose has championed a role for gravity in state vector reduction [27, 28]. This has not yet developed into a detailed alternative theory, but has suggested experimental situations in which the decay of quantum superpositions could be observed [28, 17].

In the face of an increasing domain of confirmed predictions of quantum theory and the absence as yet of compelling alternatives, it seems natural to extend quantum theory as far as it will go — to the largest scales of the universe and the smallest of quantum gravity. That is the course we shall follow in this paper. But as mentioned in the introduction, usual quantum theory must be generalized to apply to cosmology and quantum spacetime. We amplify on the reasons in the next section.

2.1.4 *Spacetime and Quantum Theory*

Usual, textbook quantum theory incorporates definite assumptions about the nature of space and time. These assumptions are readily evident in the two laws of evolution for the quantum state Ψ. The Schrödinger equation describes its unitary evolution between measurements.

$$i\hbar\,\frac{\partial\Psi}{\partial t} = H\Psi \ . \tag{2}$$

At the time of an ideal measurement, the state is projected on the outcome and renormalized

$$\Psi \to \frac{P\Psi}{\|P\Psi\|} \ . \tag{3}$$

The Schrödinger equation (2) assumes a fixed notion of time. In the non-relativistic theory, t is the absolute time of Newtonian mechanics. In the flat spacetime of special relativity, it is the time of any Lorentz frame. Thus, there are many times but results obtained in different Lorentz frames, are unitarily equivalent.

The projection in the second law of evolution (3) is in Hilbert space. But in field theory or particle mechanics, the Hilbert space is constructed from configurations of fields or position in physical *space*. In that sense it is the state on a spacelike surface that is projected (3).

Because quantum theory incorporates notions of space and time, it has changed as our ideas of space and time have evolved. The accompanying table briefly summarizes this co-evolution. It is possible to view this evolution as a process of increasing

Table 2.1 **A Short History of Spacetime and Quantum Theory**

Newtonian Physics	Fixed 3-d space and a single universal time t.	**Non-relativistic Quantum Theory:** The Schrödinger equation $$i\hbar(\partial\Psi/\partial t) = H\Psi$$ holds between measurements in the Newtonian time t.
Special Relativity	Fixed flat, 4-d spacetime with many different timelike directions.	**Relativistic Quantum Field Theory:** Choose a Lorentz frame with time t. Then (between measurements) $$i\hbar(\partial\Psi/\partial t) = H\Psi.$$ The results are unitarily equivalent to those from any other choice of Lorentz frame.
General Relativity	Fixed, but curved spacetime geometry.	**Quantum Field Theory in Curved Spacetime:** Choose a foliating family of spacelike surfaces labeled by t. Then (between measurements) $$i\hbar(\partial\Psi/\partial t) = H\Psi.$$ But the results are *not* generally unitarily equivalent to other choices.
Quantum Gravity	Geometry is *not* fixed, but rather a quantum variable	**The Problem of Time:** What replaces the Schrödinger equation when there is no fixed notion of time(s)?
M-theory, Loop quantum gravity, Posets, etc.	Spacetime is not even a fundamental variable	**?**

generalization of the concepts in the usual theory. Certainly the two laws of evolution (2) and (3) have to be generalized somehow if spacetime geometry is not fixed. One such generalization is offered in this paper, but there have been many other ideas [29]. And if spacetime geometry is emergent from some yet more fundamental description, we can certainly expect that a further generalization — free of any reference to spacetime — will be needed to describe that emergence. The rest of this article is concerned with these generalizations.

2.1.5 *The Quantum Mechanics of Closed Systems*

This section reviews, very briefly, the elements of the modern quantum mechanics of closed systems[4] aimed at a quantum mechanics for cosmology. To keep the present

---

[4]See, *e.g.* [5–7] for by now classic expositions at length or [30] for a shorter summary.

discussion manageable we focus on a simple model universe of particles moving in a very large box (say $\gtrsim 20{,}000$ Mpc in linear dimension). Everything is contained within the box, in particular galaxies, stars, planets, observers and observed (if any), measured subsystems, and the apparatus that measures them.

We assume a fixed background spacetime supplying well-defined notions of time. The usual apparatus of Hilbert space, states, operators, Feynman paths, etc. can then be employed in a quantum description of the contents of the box. The essential theoretical inputs to the process of prediction are the Hamiltonian H and the initial quantum state $|\Psi\rangle$ (the 'wave function of the universe'). These are assumed to be fixed and given.

The most general objective of a quantum theory for the box is the prediction of the probabilities of exhaustive sets of coarse-grained alternative time histories of the particles in the closed system. For instance, we might be interested in the probabilities of an alternative set of histories describing the progress of the Earth around the Sun. Histories of interest here are typically very coarse-grained for at least three reasons: They deal with the position of the Earth's center-of-mass and not with the positions of all the particles in the universe. The center-of-mass position is not specified to arbitrary accuracy, but to the error we might observe it. The center-of-mass position is not specified at all times, but typically at a series of times.

But, as described in the Introduction, not every set of alternative histories that may be described can be assigned consistent probabilities because of quantum interference. Any quantum theory must therefore not only specify the sets of alternative coarse-grained histories, but also give a rule identifying which sets of histories can be consistently assigned probabilities as well as what those probabilities are. In the quantum mechanics of closed systems, that rule is simple: probabilities can be assigned to just those sets of histories for which the quantum interference between its members is negligible as a consequence of the Hamiltonian H and the initial state $|\Psi\rangle$. We now make this specific for our model universe of particles in a box.

Three elements specify this quantum theory. To facilitate later discussion, we give these in a spacetime sum-over-histories formulation.

(1) *Fine-grained histories*: The most refined description of the particles from the initial time $t = 0$ to a suitably large final time $t = T$ gives their position at all times in between, *i.e.* their Feynman paths. We denote these simply by $x(t)$.

(2) *Coarse-graining*: The general notion of coarse-graining is a partition of the fine-grained paths into an exhaustive set of mutually exclusive classes $\{c_\alpha\}, \alpha = 1, 2, \cdots$. For instance, we might partition the fine-grained histories of the center-of-mass of the Earth by which of an exhaustive and exclusive set of position intervals $\{\Delta_\alpha\}, \alpha = 1, 2, \cdots$ the center-of-mass passes through at a series of times $t_1, \cdots t_n$. Each coarse-grained history consists of the bundle of fine-grained paths that pass through a specified sequence of intervals at the series of times.

Each coarse-grained history specifies an orbit where the center-of-mass position is localized to a certain accuracy at a sequence of times.

(3) *Measure of Interference*: Branch state vectors $|\Psi_\alpha\rangle$ can be defined for each coarse-grained history in a partition of the fine-grained histories into classes $\{c_\alpha\}$ as follows

$$\langle x|\Psi_\alpha\rangle = \int_{c_\alpha} \delta x \ \exp(iS[x(t)]/\hbar) \langle x'|\Psi\rangle \,. \tag{4}$$

Here, $S[x(t)]$ is the action for the Hamiltonian H. The integral is over all paths starting at x' at $t = 0$, ending at x at $t = T$, and contained in the class c_α. This includes an integral over x'. (For those preferring the Heisenberg picture, this is equivalently

$$|\Psi_\alpha\rangle = e^{-iHT/\hbar} P^n_{\alpha_n}(t_n) \cdots P^1_{\alpha_1}(t_1) |\Psi\rangle \tag{5}$$

when the class consists of restrictions to position intervals at a series of times and the P's are the projection operators representing them.)

The measure of quantum interference between two coarse-grained histories is the overlap of their branch state vectors

$$D(\alpha', \alpha) \equiv \langle \Psi_{\alpha'}|\Psi_\alpha\rangle \,. \tag{6}$$

This is called the *decoherence functional*.

When the interference between each pair of histories in a coarse-grained set is negligible

$$\langle \Psi_\alpha|\Psi_\beta\rangle \approx 0 \text{ all } \alpha \neq \beta \,, \tag{7}$$

the set of histories is said to *decohere*[5]. The probability of an individual history in a decoherent set is

$$p(\alpha) = \| \, |\Psi_\alpha\rangle \|^2 \,. \tag{8}$$

The decoherence condition (6) is a sufficient condition for the probabilities (7) to be consistent with the rules of probability theory. Specifically, the p's obey the sum rules

$$p(\bar{\alpha}) \approx \sum_{\alpha \in \bar{\alpha}} p(\alpha) \tag{9}$$

where $\{\bar{c}_{\bar{\alpha}}\}$ is any coarse-graining of the set $\{c_\alpha\}$, *i.e.* a further partition into coarser classes. It was the failure of such a sum rule that prevented consistent probabilities from being assigned to the two histories previously discussed in the two-slit experiment (Figure 1). That set of histories does not decohere.

Decoherence of familiar quasiclassical variables is widespread in the universe. Imagine, for instance, a dust grain in a superposition of two positions, a multimeter apart, deep in intergalactic space. The 10^{11} cosmic background photons that scatter

---

[5]This is the *medium* decoherence condition. For a discussion of other conditions, see, *e.g.* [31–33].

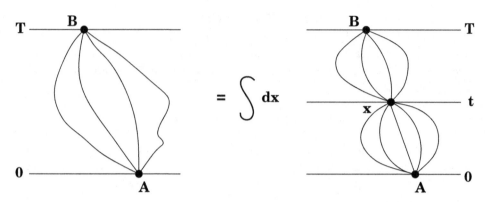

Fig. 2.3   The origin of states on a spacelike surface. These spacetime diagrams are a schematic representation of Eq. (10). The amplitude for a particle to pass from point A at time $t = 0$ to a point B at $t = T$ is a sum over all paths connecting them weighted by $\exp(iS[x(t)])$. That sum can be factored across an intermediate constant time surface as shown at right into product of a sum from A to x on the surface and a sum from x to B followed by a sum over all x. The sums in the product define states on the surface of constant time at t. The integral over x defines the inner product between such states, and the path integral construction guarantees their unitary evolution in t. Such factorization is possible only if the paths are single valued functions of time.

off the dust grain every second dissipate the phase coherence between the branches corresponding to the two locations on the time scale of about a nanosecond [34].

Measurements and observers play no fundamental role in this generalization of usual quantum theory. The probabilities of measured outcomes can, of course, be computed and are given to an excellent approximation by the usual story.[6] But, in a set of histories where they decohere, probabilities can be assigned to the position of the Moon when it is not being observed and to the values of density fluctuations in the early universe when there were neither measurements taking place nor observers to carry them out.

2.1.6   Quantum Theory in 3+1 Form

The quantum theory of the model universe in a box in the previous section is in fully 4-dimensional spacetime form. The fine-grained histories are paths in spacetime, the coarse-grainings were partitions of these, and the measure of interference was constructed by spacetime path integrals. No mention was made of states on spacelike surfaces or their unitary evolution.

However, as originally shown by Feynman [35, 36], this spacetime formulation is equivalent to the familiar 3+1 formulation in terms of states on spacelike surfaces and their unitary evolution through a foliating family of such surfaces. This section briefly sketches that equivalence emphasizing properties of spacetime and the fine-grained histories that are necessary for it to hold.

The key observation is illustrated in Figure 3. Sums-over-histories that are

---

[6]See, *e.g.* [8], Section II.10.

single-valued in time can be factored across constant time surfaces. A formula expressing this idea is

$$\int_{[A,B]} \delta x \, e^{iS[x(t)]/\hbar} = \int dx \, \psi_B^*(x,t)\psi_A(x,t) \,. \qquad (10)$$

The sum on the left is over all paths from A at $t = 0$ to B at $t = T$. The amplitude $\psi_A(x,t)$ is the sum of $\exp\{iS[x(t)]\}$ over all paths from A at $t = 0$ to x at a time t between 0 and T. The amplitude $\psi_B(x,t)$ is similarly constructed from the paths between x at t to B at T.

The wave function $\psi_A(x,t)$ defines a state on constant time surfaces. Unitary evolution by the Schrödinger equation follows from its path integral construction.[7] The inner product between states defining a Hilbert space is specified by (10). In this way, the familiar 3+1 formulation of quantum mechanics is recovered from its spacetime form.

The equivalence represented in (10) relies on several special assumptions about the nature of spacetime and the fine-grained histories. In particular, it requires[8]:

- A fixed Lorentzian spacetime geometry to define timelike and spacelike directions.
- A foliating family of spacelike surfaces through which states can evolve.
- Fine-grained histories that are single-valued in the time labeling the spacelike surfaces in the foliating family.

As an illustrative example where the equivalence does not hold, consider quantum field theory in a fixed background spacetime with closed timelike curves (CTCs) such as those that can occur in wormhole spacetimes [39]. The fine-grained histories are four-dimensional field configurations that are single-valued on spacetime. But there is no foliating family of spacelike surfaces with which to define the Hamiltonian evolution of a quantum state. Thus, there is no usual 3+1 formulation of the quantum mechanics of fields in spacetimes with CTCs.

However, there is a four-dimensional sum-over-histories formulation of field theory in spacetimes with CTCs [40–42]. The resulting theory has some unattractive properties such as acausality and non-unitarity. But it does illustrate how closely usual quantum theory incorporates particular assumptions about spacetime, and also how these requirements can be relaxed in a suitable generalization of the usual theory.

---

[7]Reduction of the state vector (3) also follows from the path integral construction [37] when histories are coarse-grained by intervals of position at various times.

[8]The usual 3+1 formulation is also restricted to coarse-grained histories specified by alternatives at definite moments of time. More general spacetime coarse-grainings that are defined by quantities that extend over time can be used in the spacetime formulation. (See, *e.g.* [38] and references therein.) Spacetime alternatives are the only ones available in a diffeomorphism invariant quantum graviity.

2.1.7   *Generalized Quantum Theory*

In generalizing usual quantum mechanics to deal with quantum spacetime, some of its features will have to be left behind and others retained. What are the minimal essential features that characterize a quantum mechanical theory? The generalized quantum theory framework [8, 30, 10] provides one answer to this question. Just three elements abstracted from the quantum mechanics of closed systems in Section 4 define a generalized quantum theory.

- *Fine-grained Histories*: The sets of alternative fine-grained histories of the closed system which are the most refined descriptions of it physically possible.
- *Coarse-grained Histories*: These are partitions of a set of fine-grained histories into an exhaustive set of exclusive classes $\{c_\alpha\}, \alpha = 1, 2 \cdots$. Each class is a coarse-grained history.
- *Decoherence Functional*: A measure of quantum interference $D(\alpha, \alpha')$ between pairs of histories in a coarse-grained set, meeting the following conditions:

    i. Hermiticity: $D(\alpha, \alpha') = D^*(\alpha', \alpha)$
    ii. Positivity: $D(\alpha, \alpha) \geq 0$
    iii. Normalization: $\Sigma_{\alpha\alpha'} D(\alpha, \alpha') = 1$
    iv. Principle of superposition: If $\{\bar{c}_{\bar{\alpha}}\}$ is a further coarse-graining of $\{c_\alpha\}$, then

$$\bar{D}(\bar{\alpha}, \bar{\alpha}') = \sum_{\substack{\alpha \in \bar{\alpha} \\ \alpha' \in \bar{\alpha}'}} D(\alpha, \alpha')$$

Probabilities $p(\alpha)$ are assigned to sets of coarse-grained histories when they decohere according to the basic relation

$$D(\alpha, \alpha') \approx \delta_{\alpha\alpha'} \, p(\alpha) \, . \tag{11}$$

These $p(\alpha)$ satisfy the basic requirements for probabilities as a consequence of i)–iv) above. In particular, they satisfy the sum rule

$$p(\bar{\alpha}) = \sum_{\alpha \in \bar{\alpha}} p(\alpha) \tag{12}$$

as a consequence of i)–iv) and decoherence. For instance, the probabilities of an exhaustive set of alternatives always sum to 1.

The sum-over-histories formulation of usual quantum mechanics given in Section 4 is a particular example of a generalized quantum theory. The decoherence functional (4) satisfies the requirements i)–iv). But its particular form is not the only way of constructing a decoherence functional. Therein lies the possibility of generalization.

2.1.8 *A Quantum Theory of Spacetime Geometry*

The low energy, effective theory of quantum gravity is a quantum version of general relativity with a spacetime metric $g_{\alpha\beta}(x)$ coupled to matter fields. Of course, the divergences of this effective theory have to be regulated to extract predictions from it.[9]. These predictions can therefore be expected to be accurate only for limited coarse-grainings and certain states. But this effective theory does supply an instructive model for generalizations of quantum theory that can accommodate quantum spacetime. This generalization is sketched in this section.

The key idea is that the fine-grained histories do not have to represent evolution *in* spacetime. Rather they can be histories *of* spacetime. For this discussion we take these histories to be spatially closed cosmological four-geometries represented by metrics $g_{\alpha\beta}(x)$ on a fixed manifold $M = \boldsymbol{R} \times M^3$ where M^3 is a closed 3-manifold. For simplicity, we restrict attention to a single scalar matter field $\phi(x)$.

The three ingredients of a generalized quantum theory for spacetime geometry are then as follows:

- *Fine-grained Histories*: A fine-grained history is defined by a four-dimensional metric and matter field configuration on M.
- *Coarse-grainings*: The allowed coarse-grainings are partitions of the metrics and matter fields into four-dimensional *diffeomorphism invariant* classes $\{c_\alpha\}$.
- *Decoherence Functional*: A decoherence functional constructed on sum-over-history principles analogous to that described for usual quantum theory in Section 4. Schematically, branch state vectors $|\Psi_\alpha\rangle$ can be constructed for each coarse-grained history by summing over the metrics and fields in the corresponding class c_α of fine-grained histories, *viz.*

$$|\Psi_\alpha\rangle = \int_{c_\alpha} \delta g \delta \phi \, \exp\{iS[g,\phi]/\hbar\} \, |\Psi\rangle \, . \tag{13}$$

A decoherence functional satisfying the requirements of Section 6 is

$$D(\alpha',\alpha) = \langle \Psi_{\alpha'} | \Psi_\alpha \rangle \, . \tag{14}$$

Here, $S[g,\phi]$ is the action for general relativity coupled to the field $\phi(x)$, and $|\Psi\rangle$ is the initial cosmological state. The construction is only schematic because we did not spell out how the functional integrals are defined or regulated, nor did we specify the product between states that is implicit in both (13) and (14). These details can be made specific in models [9, 45, 46], but they will not be needed for the subsequent discussion.

A few remarks about the coarse-grained histories may be helpful. To every physical assertion that can be made about the geometry of the universe and the fields within, there corresponds a diffeomorphism invariant partition of the fine-grained histories into the class where the assertion is true and the class where it is

---

[9]Perhaps, most naturally by discrete approximations to geometry such as the Regge calculus (see, *e.g.* [43, 44])

false. The notion of coarse-grained history described above therefore supplies the most general notion of alternative describable in spacetime form. Among these we do not expect to find local alternatives because there is no diffeomorphism invariant notion of locality. In particular, we do not expect to find alternatives specified at a moment of time. We do expect to find alternatives referring to the kind of relational observables discussed in [47] and the references therein. We also expect to find observables referring to global properties of the universe such as the maximum size achieved over the history of its expansion.

This generalized quantum mechanics of spacetime geometry is in fully spacetime form with alternatives described by partitions of four-dimensional histories and a decoherence functional defined by sums over those histories. It is analogous to the spacetime formulation of usual quantum theory reviewed in Section 4.

However, unlike the theory in Section 4, we cannot expect an equivalent 3+1 formulation, of the kind described in Section 5, expressed in terms of states on spacelike surfaces and their unitary evolution between these surfaces. The fine-grained histories are not 'single-valued' in any geometrically defined variable labeling a spacelike surface. They therefore cannot be factored across a spacelike surface as in (10). More precisely, there is no geometrical variable that picks out a unique spacelike surface in all geometries.[10]

Even without a unitary evolution of states the generalized quantum theory is fully predictive because it assigns probabilities to the most general sets of coarse-grained alternative histories described in spacetime terms when these are decoherent.

How then is usual quantum theory used every day, with its unitarily evolving states, connected to this generalized quantum theory that is free from them? The answer is that usual quantum theory is an approximation to the more general framework that is appropriate for those coarse-grainings and initial state $|\Psi\rangle$ for which spacetime behaves classically. One equation will show the origin of this relation. Suppose we have a coarse-graining that distinguishes between fine-grained geometries only by their behavior on scales well above the Planck scale. Then, for suitable states $|\Psi\rangle$ we expect that the integral over metrics in (14) can be well approximated semiclassically by the method of steepest descents. Suppose further for simplicity that only a single classical geometry with metric $\hat{g}_{\alpha\beta}$ dominates the semiclassical approximation. Then, (14) becomes

$$|\Psi_\alpha\rangle \approx \int_{\hat{c}_\alpha} \delta\phi \, \exp\{iS[\hat{g}, \phi]/\hbar\} \, |\Psi\rangle \qquad (15)$$

where $\hat{c}_\alpha$ is the coarse-graining of $\phi(x)$ arising from c_α and the restriction of $g_{\alpha\beta}(x)$ to $\hat{g}_{\alpha\beta}(x)$. Eq. (15) effectively defines a quantum theory of the field $\phi(x)$ in the

---

[10]Spacelike surfaces labeled by the trace of the extrinsic curvature K foliate certain classes of classical spacetimes obeying the Einstein equation [48]. However, there is no reason to require that non-classical histories be foliable in this way. It is easy to construct geometries where surfaces of a given K occur arbitrarily often.

fixed background spacetime with the geometry specified by $\hat{g}_{\alpha\beta}(x)$. This is familiar territory. Field histories are single valued on spacetime. Sums-over-fields can thus be factored across spacelike surfaces in the geometry $\hat{g}$ as in (10) to define field states on spacelike surfaces, their unitary evolution, and their Hilbert space product. Usual quantum theory is thus recovered when spacetime behaves classically and provides the fixed spacetime geometry on which usual quantum theory relies.

From this perspective, familiar quantum theory and its unitary evolution of states is an effective approximation to a more general sum-over-histories formulation of quantum theory. The approximation is appropriate for those coarse-grainings and initial states in which spacetime geometry behaves classically.

2.1.9 *Beyond Spacetime*

The generalized quantum theory of spacetime sketched in the previous section assumed that geometry was a fundamental variable — part of the description of the fine-grained histories. But on almost every frontier in quantum gravity one finds the idea that continuum geometry is not fundamental, but will be replaced by something more fundamental. This is true for string theory [49], loop quantum gravity [50], and the causal set program [51, 52] although space does not permit a review of these speculations.

Can generalized quantum theory serve as a framework for theories where spacetime is emergent rather than fundamental? Certainly we cannot expect to have a notion of 'history'. But we can expect some fine-grained description, or a family of equivalent ones, and that is enough. A generalized quantum theory needs:

- The possible fine-grained descriptions of the system.
- The coarse-grained descriptions constructed from the fine-grained ones.
- A measure of quantum interference between different coarse-grained descriptions respecting conditions i)–iv) in Section VI.

Generalized quantum theory requires neither space nor time and can therefore serve as the basis for a quantum theory in which spacetime is emergent.

2.1.10 *Emergence/Excess Baggage*

The word 'emergent' appears in a number of places in the previous discussion. It probably has many meanings. This section aims at a more precise understanding of what is meant by the term in this essay.

Suppose we have a quantum theory defined by certain sets of fine-grained histories, coarse-grainings, and a decoherence functional. Let's call this the *fundamental* theory. It may happen that the decoherence and probabilities of limited kinds of sets of coarse-grained histories are given approximately by a second, *effective* theory. The two theories are related in the following way:

- Every fine-grained history of the effective theory is a coarse-grained history of the fundamental theory.
- The decoherence functionals approximately agree on a limited class of sets of coarse-grained histories.

$$D^{\text{fund}}(\alpha', \alpha) \approx D^{\text{eff}}(\alpha', \alpha). \qquad (16)$$

On the right, α' and α refer to the fine-grained histories of the effective theory. On the left, they refer to the corresponding coarse-grained histories of the fundamental theory.

When two theories are related in this way we can say that the effective theory is *emergent* from the fundamental theory. Loosely we can say that the restrictions, and the concepts that characterize them, are emergent. It should be emphasized that an approximate equality like (16) can be expected to hold, not just as a consequence of the particular dynamics incorporated into decoherence functionals, but also only for particular states.

Several examples of emergence in this sense have been considered in this essay: There is the possible emergence of a generalized quantum theory of spacetime geometry from a theory in which spacetime is not fundamental. There is the emergence of a 3+1 quantum theory of fields in a fixed background geometry from a four-dimensional generalized quantum theory in which geometry is a quantum variable. There is the emergence of the approximate quantum mechanics of measured subsystems (textbook quantum theory) from the quantum mechanics of the universe. And there is the emergence of classical physics from quantum physics.

Instead of looking at an effective theory as a restriction of a more fundamental one, we may look at the fundamental theory as a generalization of the effective one. That perspective is important because generalization is a way of searching for more comprehensive theories of nature. In passing from the specific to the more general some ideas have to be discarded. They are often ideas that were once perceived to be general because of our special place in the universe and the limited range of our experience. But, in fact, they arise from special situations in a more general theory. They are 'excess baggage' that has to be discarded to reach a more comprehensive theory [53]. Emergence and excess baggage are two ways of looking at the same thing.

Physics is replete with examples of emergence and excess baggage ranging from Earth-centered theories of the solar system to quantum electrodynamics. The chart on the next page helps understand the stages of emergence and generalization in quantum mechanics discussed in this essay provided it is not taken too rigidly or without qualification.

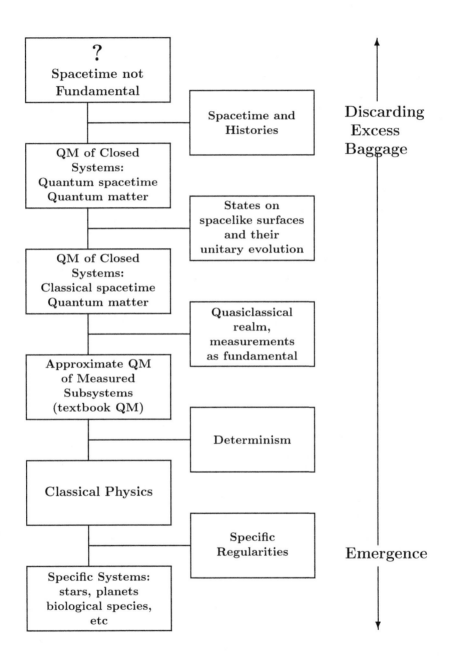

The chart can be read in two ways. Reading from the bottom up, the boxes on the left describe a path of generalization — from the specific to the general. Starting from the regularities of specific systems such as the planetary orbits, we move up to the general laws of classical physics, to textbook quantum theory, through various stages of assumptions about spacetime, to a yet unknown theory where spacetime is not fundamental. The excess baggage that must be jettisoned at each stage to reach a more general perspective is indicated in the middle tower of boxes.

Reading from the top down the chart tells a story of emergence. Each box on the left stands in the relation of an effective theory to the one before it. The middle boxes now describe phenomena = that are emergent at each stage.

2.1.11 *Emergence of Signature*

Classical spacetime has Lorentz signature. At each point it is possible to choose one timelike direction and three orthogonal spacelike ones. There are no physical spacetimes with zero timelike directions or with *two* timelike directions. But is such a seemingly basic property fundamental, or is it rather, emergent from a quantum theory of spacetime which allows for all possible signatures? This section sketches a simple model where that happens.

Classical behavior requires particular states [54]. Let's consider the possible classical behaviors of cosmological geometry assuming the 'no-boundary' quantum state of the universe [55] in a theory with only gravity and a cosmological constant Λ. The no-boundary wave function is given by a sum-over-geometries of the schematic form

$$\Psi[h] = \int_e \delta g \, e^{-I[g]/\hbar} . \tag{17}$$

For simplicity, we consider a = fixed manifold[11] M. The key requirement is that it be compact with one boundary for the argument of the wave function and no other boundary. The functional $I[g]$ is the Euclidean action for metric defining the geometry on M. The sum is over a complex contour $\mathcal{C}$ of g's that have finite action and match the three-metric h on the boundary that is the argument of Ψ.

Quantum theory predicts classical behavior when it predicts high probability for histories exhibiting the correlations in time implied by classical deterministic laws [58, 54]. The state Ψ is an input to the process of predicting those probabilities as described in Section 7. However, plausibly the output for the predicted classical spacetimes in this model are the extrema of the action in (17). We will assume this (see [9] for some justification). Further, to keep the discussion manageable, we will restrict it to the *real* extrema. These are the real tunneling geometries discussed in a much wider context in [59].

Let us ask for the semiclassical geometries which become large, *i.e.* contain

---

[11] Even the notion of manifold may be emergent in a more general theory of certain complexes [56, 57].

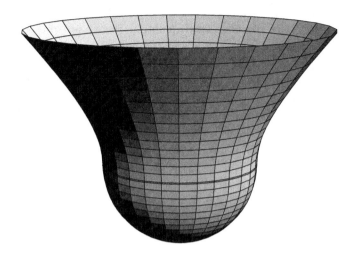

Fig. 2.4   The emergence of the Lorentz signature $(-,+,+,+)$ of spacetime. The semiclassical geometry describing a classical spacetime which becomes large according to the 'no-boundary' proposal for the universe's quantum state. The model is pure gravity and a cosmological constant. Purely Euclidean geometries $(+,+,+,+)$ or purely Lorentzian geometries are not allowed as described in the text. What is allowed is the real tunneling geometry illustrated above consisting of half a Euclidean four-sphere joined smoothly onto an expanding Lorentzian de Sitter space at the moment of maximum contraction. This can be described as the nucleation of classical Lorentz signatured spacetime. There is no similar nucleation of a classical geometry with signature $(-,-,+,+)$ because it could not match the Euclidean one across a spacelike surface.

symmetric three surfaces with size much larger than $(1/\Lambda)^{1/2}$. There are none with Euclidean signature. The purely Euclidean extremum is the round four-sphere with linear size $(1/\Lambda)^{1/2}$ and contains no symmetric three surfaces with larger size. There are none with purely Lorentzian signature either because these cannot be regular on M. There are, however, tunneling solutions of the kind illustrated in Figure 4 in which half of a Euclidean four-sphere is matched to expanding DeSitter space across a surface of vanishing extrinsic curvature.

Could a spacetime with two time and two space directions be nucleated in this way? The answer is 'no' because the geometry on a surface could not have the three spacelike directions necessary to match onto the half of a four-sphere.

Thus, in this very simple model, with many assumptions, if we live in a large universe it must have one time and three space dimensions. The Lorentzian signature of classical spacetime is an emergent property from an underlying theory not committed to this signature.

2.1.12   *Beyond Quantum Theory*

The path of generalization in the previous sections began with the textbook quantum mechanics of measurement outcomes in a fixed spacetime and ended in a quantum theory where neither measurements nor spacetime are fundamental. In this journey, the principles of generalized quantum theory are preserved, in particular the idea of quantum interference and the linearity inherent in the principle of superposition. But the end of this path is strikingly different from its beginning.

The founders of quantum theory thought that the indeterminacy of quantum theory "reflected the unavoidable interference in measurement dictated by the magnitude of the quantum of the action" (Bohr). But what then is the origin of quantum indeterminacy in a closed quantum universe which is never measured? Why enforce the principle of superposition in a framework for prediction of the universe which has but a single quantum state? In short, the endpoint of this journey of generalization forces us to ask John Wheeler's famous question, "How come the quantum?" [60].

Could quantum theory itself be an emergent effective theory? Many have thought so (Section 2). Extending quantum mechanics until it breaks could be one route to finding out. 'Traveler, there are no paths, paths are made by walking.'

2.1.13 *Conclusion*

Does quantum mechanics apply to spacetime? The answer is 'yes' provided that its familiar textbook formulation is suitably generalized. It must be generalized in two directions. First, to a quantum mechanics of closed systems, free from a fundamental role for measurements and observers and therefore applicable to cosmology. Second, it must be generalized so that it is free from any assumption of a fixed spacetime geometry and therefore applicable when spacetime geometry is a quantum variable.

Generalized quantum theory built on the pillars of fine-grained histories, coarse-graining, and decoherence provides a framework for investigating such generalizations. The fully, four-dimensional sum-over-histories effective quantum theory of spacetime geometry sketched in Section 7 is one example. In such fully four-dimensional generalizations of the usual theory, we cannot expect to recover an equivalent 3+1 formulation in terms of the unitary evolution of states on spacelike surfaces. There is no fixed notion of spacelike surface. Rather, the usual 3+1 formulation emerges as an effective approximation to the more general story for those coarse grainings and initial states in which spacetime geometry behaves classically.

If spacetime geometry is not fundamental, quantum mechanics will need further generalization and generalized quantum theory provides one framework for exploring that.

Acknowledgments

The author is grateful to Murray Gell-Mann and David Gross for delivering this paper at the Solvay meeting when he was unable to do so. Thanks are due to Murray Gell-Mann for discussions and collaboration on these issues over many years.

Bibliography

[1] Léon Rosenfeld, *Über die Gravitationswirkungen des Lichtes*, Z. Phys. **65**, 589–599 (1930).

[2] D.N. Page and C.D. Geilker, *Indirect Evidence for Quantum Gravity*, Phys. Rev. Lett. **47**, 979–982 (1981).

[3] Léon Rosenfeld, *On Quantization of Fields*, Nucl. Phys. **40**, 353–356 (1963).

[4] Léon Rosenfeld, *Quantentheorie und Gravitation* in *Entstehung, Entwicklung, und Perspektiven der Einsteinschen Gravitationstheorie*, Akademie-Verlag, Berlin, (1966) pp. 185–197. [English translation in *Selected Papers of Léon Rosenfeld*, ed. by R.S. Cohen and J. Stachel, D. Reidel, Dordrecht (1979).

[5] R.B. Griffiths, *Consistent Quantum Theory*, Cambridge University Press, Cambridge (2002).

[6] R. Omnès, *Interpretation of Quantum Mechanics*, Princeton University Press, Princeton (1994).

[7] M. Gell-Mann, *The Quark and the Jaguar*, W. Freeman San Francisco (1994).

[8] J.B. Hartle, *The Quantum Mechanics of Cosmology*, in *Quantum Cosmology and Baby Universes: Proceedings of the 1989 Jerusalem Winter School for Theoretical Physics*, ed. by S. Coleman, J.B. Hartle, T. Piran, and S. Weinberg, World Scientific, Singapore (1991), pp. 65-157.

[9] J.B. Hartle, *Spacetime Quantum Mechanics and the Quantum Mechanics of Spacetime* in *Gravitation and Quantizations*, Proceedings of the 1992 Les Houches Summer School, ed. by B. Julia and J. Zinn-Justin, Les Houches Summer School Proceedings Vol. LVII, North Holland, Amsterdam (1995); gr-qc/9304006.

[10] C.J. Isham, *Quantum Logic and the Histories Approach to Quantum Theory*, J. Math. Phys., **35**, 2157 (1994); C.J. Isham and N. Linden, *Quantum Temporal Logic in the Histories Approach to Generalized Quantum Theory*, J. Math. Phys., **35**, 5452 (1994).

[11] A.J. Leggett, *Testing the Limits of Quantum Mechanics: Motivation, State-of-Play, Prospects*, J. Phys. Cond. Matter **14**, R415 (2002).

[12] A. Tonomura, J. Endo, T. Matsuda, and T. Kawasaki, *Demonstration of Single-electron Build-up of an Interference Pattern*, Am. J. Phys. **57** 117–120 (1989).

[13] L. Hackermüller, *et al.*, *The Wave Nature of Biomolecules and Flurofullerenes*, Phys. Rev. Lett. **91**, 090408 (2003).

[14] C. van der Wal, *et al.*, *Quantum Superposition of Macroscopic Persistent-current States*, Science **290**, 773-777 (2000).

[15] I. Chiorescu, Y. Nakamura, C. Harmans, and J. Mooij, *Coherent Quantum Dynamics of a Superconducting Flux Orbit*, Science **299**, 1869 (2003).

[16] J. Friedman, *et al.*, *Quantum Superposition of Distinct Macroscopic States*, Nature **406**, 43–46 (2000).

[17] W.Marshall, C. Simon, R. Penrose, and D. Bouwmeester, *Quantum Superposition of a Mirror*, Phys. Rev. Lett. **91**, 13 (2003).

[18] S. Weinberg, *Dreams of a Final Theory*, Pantheon Books, New York (1992).

[19] G. Auletta, *Foundations and Interpretations of Quantum Mechanics*, World Scientific, Singapore (2000).

[20] D. Bohm and B.J. Hiley, *The Undivided Universe*, Routledge, London (1993).

[21] J.B. Hartle, *Bohmian Histories and Decoherent Histories*, Phys. Rev. A **69**, 042111 (2004); quant-ph/0308117.

[22] I. Percival, *Quantum State Diffusion*, Cambridge University Press, Cambridge, UK (1998).

[23] A. Bassi and G.C. Ghirardi, *Dynamical Reduction Models*, Phys. Rep. **379**, 257–426 (2003).

[24] F. Dowker and J. Henson, *A Spontaneous Collapse Model on a Lattice*, J. Stat. Phys., **115**, 1349 (2004); quant-ph/020951.

[25] S.L. Adler, *Quantum Theory as an Emergent Phenomenon*, Cambridge University Press, Cambridge, UK (2004).

[26] G. 't Hooft, *Determinism Beneath Quantum Mechanics*, in *Quo Vadis Quantum Mechanics*, ed. by A. Elitzur, S. Dolen, and N. Kolenda, Springer Verlag, Heidelburg, (2005); quant-ph/0212095, and his article in this volume (2006).

[27] R. Penrose, *Wavefunction Collapse as a Real Gravitational Effect*, in *Mathematical Physics 2000*, ed. by A. Fokas, T.W.B. Kibble, A. Grigourion, and B. Zegarlinski, Imperial College Press, London, pp. 266–282 (2000).

[28] R. Penrose, *The Road to Reality*, Jonathan Cape, London (2004) Chap. 30.

[29] K. Kuchař, *Time and Interpretations of Quantum Gravity*, in *Proceedings of the 4th Canadian Conference on General Relativity and Relativistic Astrophysics*, ed. by G. Kunstatter, D. Vincent, and J. Williams, World Scientific, Singapore, (1992); C. Isham, *Conceptual and Geometrical Problems in Quantum Gravity* in *Recent Aspects of Quantum Fields*, ed. by H. Mitter and H. Gausterer, Springer-Verlag, Berlin (1992); C. Isham, *Canonical Quantum Gravity and the Problem of Time* in *Integrable Systems, Quantum Groups, and Quantum Field Theories*, ed. by L.A. Ibort and M.A. Rodriguez, Kluwer Academic Publishers, London (1993); W. Unruh, *Time and Quantum Gravity*, in *Gravitation: A Banff Summer Institute*, ed. by R. Mann and P. Wesson, World Scientific, Singapore (1991).

[30] J.B. Hartle, *The Quantum Mechanics of Closed Systems*, in *Directions in General Relativity, Volume 1: A Symposium and Collection of Essays in honor of Professor Charles W. Misner's 60th Birthday*, ed. by B.-L. Hu, M.P. Ryan, and C.V. Vishveshwara, Cambridge University Press, Cambridge (1993); gr-qc/9210006.

[31] M. Gell-Mann and J.B. Hartle, *Alternative Decohering Histories in Quantum Mechanics*, in the *Proceedings of the 25th International Conference on High Energy Physics*, Singapore, August, 2-8, 1990, ed. by K.K. Phua and Y. Yamaguchi, South East Asia Theoretical Physics Association and Physical Society of Japan, distributed by World Scientific, Singapore (1990).

[32] M. Gell-Mann and J.B. Hartle, *Strong Decoherence*, in the *Proceedings of the 4th Drexel Symposium on Quantum Non-Integrability — The Quantum-Classical Correspondence*, Drexel University, September 8-11, 1994, ed. by D.-H. Feng and B.-L. Hu, International Press, Boston/Hong-Kong (1995); gr-qc/9509054.

[33] J.B. Hartle, *Linear Positivity and Virtual Probability*, Phys. Rev. A **70**, 022104 (2004); quant-ph/0401108.

[34] E. Joos and H.D. Zeh, *Emergence of Classical Properties through Interaction with the Environment*, Zeit. Phys. B **59**, 223 (1985).

[35] R.P. Feynman, *Space-time Approach to Non-Relativistic Quantum Mechanics*, Rev. Mod. Phys. **20**, 267 (1948).

[36] R.P. Feynman and A. Hibbs, *Quantum Mechanics and Path Integrals*, McGraw-Hill, New York (1965).

[37] C. Caves, *Quantum Mechanics and Measurements Distributed in Time I: A Path Integral Approach*, Phys. Rev. D **33**, 1643 (1986); *Quantum Mechanics and Measurements Distributed in Time II: Connections among Formalisms*, ibid. **35**, 1815 (1987).

[38] A.W. Bosse and J.B. Hartle, *Representations of Spacetime Alternatives and Their Classical Limits*, Phys. Rev. A **72**, 022105 (2005); quant-ph/0503182.

[39] M. Morris, K.S. Thorne, and U. Yurtsver, *Wormholes, Time Machines, and the Weak*

Energy Condition, Phys. Rev. Lett. **61**, 1446 (1988).

[40] J.B. Hartle, *Unitarity and Causality in Generalized Quantum Mechanics for Non-Chronal Spacetimes*, Phys. Rev. D **49**, 6543 (1994); quant-ph/9309012.

[41] J.L. Friedman, N.J. Papastamatiou, and J.Z. Simon, *Unitarity of Interacting Fields in Curved Spacetime*, Phys. Rev. D **46**, 4441 (1992); *Failure of Unitarity for Interacting Fields on Spacetimes with Closed Timelike Curves*, ibid, 4456 (1992).

[42] S. Rosenberg, *Testing Causality Violation on Spacetimes with Closed Timelike Curves*, Phys. Rev. D **57**, 3365 (1998).

[43] J.B. Hartle, *Simplicial Minisuperspace I. General Discussion*, J. Math. Phys. **26**, 804 (1985); *Simplicial Minisuperspace III: Integration Contours in a Five-Simplex Model*, ibid. **30**, 452 (1989).

[44] H. Hamber and R.M. Williams, *Non-perturbative Gravity and the Spin of the Lattice Gravition*, Phys. Rev. D **70**, 124007 (2004); hep-th/0407039.

[45] J.B. Hartle and D. Marolf, *Comparing Formulations of Generalized Quantum Mechanics for Reparametrization Invariant Systems*, Phys. Rev. D **56**, 6247-6257 (1997); gr-qc/9703021.

[46] D. Craig and J.B. Hartle, *Generalized Quantum Theories of Recollapsing, Homogeneous Cosmologies*, Phys. Rev. D **69**, 123525–123547 (2004); gr-qc/9703021.

[47] S. Giddings, D. Marolf, and J. Hartle, *Observables in Effective Gravity*; hep-th/0512200.

[48] J.E. Marsden and F. Tipler, *Maximal Hypersurfaces and Foliations of Constant Mean Curvature in General Relativity*, Physics Reports **66**, 109 (1980).

[49] N. Seiberg, *Emergent Spacetime*, this volume, hep-th/0601234.

[50] A. Ashtekar and J. Lewandowski, *Background Independent Gravity: A status report*, Class. Quant. Grav. **21**, R53 (2004).

[51] F. Dowker, *Causal Sets and the Deep Structure of Spacetime*, in *100 Years of Relativity*, ed. by A. Ashtekar, World Scientific, Singapore (2005); gr-qc/0508109.

[52] J. Henson, *The Causal Set Approach to Quantum Gravity*; gr-qc/0601121.

[53] J.B. Hartle, *Excess Baggage*, in *Elementary Particles and the Universe: Essays in Honor of Murray Gell-Mann* ed. by J. Schwarz, Cambridge University Press, Cambridge (1990); gr-qc/0508001.

[54] J.B. Hartle, *Quasiclassical Domains In A Quantum Universe*, in *Proceedings of the Cornelius Lanczos International Centenary Conference*, North Carolina State University, December 1992, ed. by J.D. Brown, M.T. Chu, D.C. Ellison, R.J. Plemmons, SIAM, Philadelphia, (1994); gr-qc/9404017.

[55] J.B. Hartle and S.W. Hawking, *Wave Function of the Universe*, Phys. Rev. D **28**, 2960 (1983).

[56] J.B. Hartle, *Unruly Topologies in Two Dimensional Quantum Gravity*, Class. & Quant. Grav., **2**, 707 (1985).

[57] K. Schleich and D. Witt, *Generalized Sums over Histories for Quantum Gravity: I. Smooth Conifolds*, Nucl. Phys. **402**, 411 (1993); *II. Simplicial Conifolds*, ibid. **402**, 469 (1993).

[58] M. Gell-Mann and J.B. Hartle, *Classical Equations for Quantum Systems*, Phys. Rev. D **47**, 3345 (1993); gr-qc/9210010.

[59] G.W. Gibbons and J.B. Hartle, *Real Tunneling Geometries and the Large-scale Topology of the Universe*, Phys. Rev. D **42**, 2458 (1990).

[60] J.A. Wheeler, *How Come the Quantum?*, in *New Techniques and Ideas in Quantum Measurement Theory*, ed. by D. Greenberger, Ann. N.Y. Acad. Sci **480**, 304–316 (1986).

2.2 Discussion

G. Gibbons If Hartle were here I would ask him the question that I always ask him. He lists the axioms of generalized quantum mechanics and one of them is that the decoherence functional should be complex valued and has to satisfy hermiticity. But it seems to me that the most vulnerable thing about quantum mechanics in quantum gravity is the idea that we have a complex Hilbert space with unitary evolution. We introduce the complex numbers precisely so that we have a first order equation of motion, and as Jim pointed out in his overheads, you do not have a unique notion of time in general relativity. So it seems to me that a good candidate for one of the things we should jettison in quantum mechanics is the complex structure of quantum mechanics.

D. Gross Do you mean we should go back to the real numbers?

G. Gibbons Basically, return to the real numbers and only get to the complex numbers in some approximation when we have a well defined notion of time.

S. Weinberg I have a very elementary question which goes back to Gell-Mann's talk. I agree completely that the textbook interpretation of quantum mechanics is absurd but I am worried whether the formalism of decoherent histories, that Hartle, Gell-Mann and others have developed, is a satisfactory resting place, or a satisfactory alternative. It has to do with the word probability, which still appears. Gell-Mann talked about the probabilities of different decoherent histories, or coarse grained histories. But what does the word probability mean? To me, it means what happens when an experimenter does an experiment a number of times. If half the time he gets the spin up and half the time he gets the spin down then we say that the probability is one half. Now, if it does not mean that, if Gell-Mann has some other meaning to the word probability, then there is a responsibility to relate his probability to the probability that is used in the textbooks. In other words, even if you replace the textbook interpretation, then you have to explain why the textbook interpretation works so well. That is a responsibility that has not, it seems to me, been met. The apparatus, and the observer, and the Physical Review journal in which these results are published are all described by a wave function. It is necessary, by using the deterministic evolution of the wave function, to explain how the observer, or the reader of the journal article, becomes convinced that the probability is one half, in the situation where it is one half. This is not a subject on which I am an expert, but it seems to me that Abner Shimony and Sidney Coleman have taken steps in this direction, even though the work is not completed. In other words, the work I am describing is to explain how, within a deterministic framework of the evolution of the wave function, observers who are also described by wave functions get convinced about probabilities having certain values. I would like to ask Gell-Mann whether he thinks that is in a satisfactory state or not.

M. Gell-Mann I think it is, but there is one direction in which it can be improved, and you are right in saying that Coleman and collaborators were pursuing that. The probabilities are a priori probabilities, we do not always deal with reproducible situations. When you make a personal decision of some kind, it is not usually a statistical sample that you are dealing with, you anticipate a certain outcome on the basis of the theory, and an individual case does not have to be statistical. You can then show that in a statistical situation the a priori probability becomes the statistical probability. In the course of doing that, there is a mathematical point that needs further elaboration, and that one was being worked on by Coleman and others. I am sure that some day it will be improved somewhat. The general idea, I think, is very simple: the a priori probability becomes a statistical one in a statistical situation.

A. Polyakov I think that, as far as we are talking about normal physics, the problem is there is no problem. All these things which I call many worlds interpretation of quantum mechanics are completely unnecessary, a single world is enough. But I am really worried, just as Weinberg, about the notion of probability which we have to discuss. The notion of probability is inevitably subjective. If we ask why probability theory describes the natural world, the answer in the classical world is very obvious. It is because when we throw a dice, it is described by some chaotic differential equation, so some small uncertainty develops. But what is this small uncertainty in the case of the Universe? Who decoheres the whole Universe as a closed system? I think that a possible answer to this could be that, if we view the Universe in the Euclidean signature and obtain physical results by analytic continuation, in the Euclidean signature the Feynman principle looks precisely as the Boltzmann-Gibbs principle. We know that the Gibbs distribution is not a fundamental concept, it is an approximation to underlying dynamics. So it is not unthinkable that we will need a similar more fundamental approach in the case of the Universe or spacetime: we have some differential equation with sensitive dependence on the initial conditions which eventually may or may not lead to the statistical description in terms of the Gibbs distribution. In this case the notion of probability itself does not arise. Basically my confusion is that I see probability as a self-referential notion, a subjective notion. I think the best definition of probability was given by Poincaré who said that it is the measure of our ignorance. What would be the objective counterpart of that, I do not know. Maybe we will hear something about it.

D. Gross It is fascinating that some of the discussions here could have been made 70 years ago.

A. Polyakov That is right. Actually I think that Einstein's point of view was that quantum mechanics was just a statistical approximation.

D. Gross What I meant was that it is interesting, and perhaps discouraging, that we are still engaged in these discussions.

A. Strominger I am not so interested in the probability or interpretation issue,

I like what Hartle said about it in private though maybe not in his talks: it is the word problem in physics. That is, it is not about the measurement that you do or the calculation that you do, but the words that you say while you are doing them. So that does not really seem so interesting to me, but what is interesting to me is what I view as the likely possibility that quantum mechanics is deeply wrong in some very fundamental way. For a long time it looked as though in the context of black holes there might be some problems with quantum mechanics. But now, though I do not think the nail is quite in the coffin, it seems that all the behavior of black holes, at least when we use results from string theory, is consistent with quantum mechanics. But I do not see any reason why quantum mechanics should not, maybe relatively soon, go the way of all our other cherished notions in physics, that is, need to undergo some basic renovation. One reason why we might believe this is what I view as a kind of white elephant standing in the room, which is the Big Bang. If you believe in unitary evolution, you can take your quantum state and evolve it forever. On the other hand we believe that the Universe had a beginning (of course there is the ekpyrotic or other kinds of infinitely existing universes), but that seems to me inconsistent with quantum mechanics which does not allow for a beginning of time. More generally, I just think that because the Schrödinger equation involves d/dt in such a preferred way is very much against the spirit of general relativity. So I think the more interesting question is: is it time now for quantum mechanics to be modified, or is that something that is still ahead of us in the future? I think we should be open to the possibility of a very deep modification of quantum mechanics.

D. Gross To some extent, I think that Hartle dealt with some of those issues, and I have not heard anything that refutes his statement in your discussion. You do not need to formulate quantum mechanics in terms of the Schrödinger equation with d/dt. Normally, the initial conditions can be separated from the kinematical framework.

J. Maldacena I like this hypothesis of radical conservatism, because we do not have anything better to replace quantum mechanics with. We just have to assume that it is correct and get as far as we can. I am not sure there is a problem with time, because as Hartle said, time could emerge. In the example of the Hartle-Hawking wave function, time is some kind of emergent property and you can describe this de Sitter universe in a perfectly consistent fashion. It looks like we need some framework which allows us to compute for example quantum corrections to that. I think we probably need to put that whole discussion in the string theory framework.

D. Gross The one problem I wonder about is: in the absence of spacetime, how do we recover a causal structure and in particular a sequence that would be the equivalent of histories without this underlying spacetime?

M. Gell-Mann You said it yourself a little while ago, and I am sure Hartle would have said the same thing: if time fades away a bit in describing spacetime as emergent from something different, we still have left this sequence which in the straightforward quantum mechanics of today is a sequence of projection operators in the histories. In the future it might be something slightly more subtle, but that gives the "nacheinander" quality of time, one thing after another, that we need. It can in many cases replace the role of time. In today's situation we have a dual role: we have the time and we have the succession. They are aligned with each other but if time fades a bit, we still have the succession, and that has to be kept. It is the answer to several questions that people have asked.

M. Douglas I have a different question which could not have been asked seventy or even twenty years ago. Many people are trying to build quantum computers, systems that would maintain what seem to be very complicated quantum superpositions, that actually do things that you might not be able to do, or do so quickly, with classical physics, like factoring. Should we regard these as interesting new experimental probes of quantum mechanics at this level? If one believes that quantum mechanics is not fundamental does that suggest in any way that such a computer might not be possible, or will a surprise be seen in these attempts? These are questions for all the speakers.

G. 't Hooft This is definitely a question which came to my mind of course and I think I have a rather precise answer. Maybe people will not like it so much, but that is the one prediction I can give from my theory. If indeed there is something more deterministic underlying quantum mechanics, you could call it a hidden variable theory or whatever, then it should be possible to mimic that on a quantum computer. So the conjecture I am making, which in principle can be falsified by people who construct really good quantum computers, is that no quantum computer will work in a way better than if you take a classical computer and you scale it up. Even if it is impossible in practice, in the imagination we could scale up the performance of a classical computer, say its bits and bytes are acting at the Planck scale. That classical computer should work better than any quantum computer anybody will ever make. This of course will not make the search for a quantum computer futile because nobody can make such a classical Planck scale thing. That is why quantum computers can probably do miracles that no other computer can do. The miracle will not be truly exponential, non-polynomial, but there will be a limit set by the Planck scale to what the quantum computer can perform. Now that is a prediction, at least one prediction I can make which can in principle be falsified.

F. Wilczek I would like to make a comment, it is not so much a question. The thing that has always bothered me about quantum mechanics is that it is not unified with the rest of physics. In the standard model and other applications of quantum mechanics you formulate some symmetry principles and then quantize, namely you separately postulate commutation relations. It is suspicious that the commutation relations take the same form as symmetry relations where you have commutators and Lie algebras. So I suspect that the separation of quantum mechanics and the rest of physics is something we will have to get beyond. Another sign of this is in line with Planck's units, which originally were c, G and $\hbar$. But they could have been c, G and e, the charge of the electron. Those are perfectly adequate to link the mass and the time, and if you take that attitude, you should be able to derive $\hbar$.

D. Gross I have often thought, and this ties in with what Strominger said, that we are probably headed toward a situation where kinematics and dynamics are not separated. We have quantum mechanics or quantum field theory on the one hand and specific dynamical models on the other hand. Such a separation seems bizarre. In string theory, in fact, we seemed to be headed towards a unification of the kinematical and dynamical schemes, since it appears that any consistent generally relativistic quantum theory is part of string theory. The fact that there might be no separation between kinematics and dynamics might very well tie into the issue of the initial conditions as well. In that sense our view of quantum mechanics might change but I doubt that it will become more deterministic. It might become as much a part of our total physical theory as the dynamical scheme, and as such might be less mysterious or disturbing.

E. Rabinovici We were told that you come to the Solvay conference to reexamine your prejudices. So one of my prejudices is against the anthropic principle. But as I hear the talks, it is not clear to me actually what the arguments are that quantum mechanics does not come out of an anthropic principle. It seems that, if I had been here seventy or eighty years ago and heard about the quantization levels and that the hydrogen atom is stable, I would have learned that quantization leads to it. So why should we not add to these things that we reexamine using the anthropic principle also quantum mechanics. I suggest we do that.

D. Gross We undoubtedly will get to the anthropic principle later, but I hope that we will not imagine that the laws of mathematics and logic and quantum mechanics are up for grabs anthropically as well.

S. Weinberg I will make two comments. First of all, quantum mechanics is deterministic. Until we begin worrying about the nature of time, it is the theory of the deterministic evolution of the wave function. The secret agenda behind my earlier question was the following: if by studying the linear evolution of the wave function as Coleman and others do we can understand why observers who are part of this wave function come to the conclusion that probabilities have certain values, then we do not need the probabilistic assumption as a separate assumption within quantum mechanics. I still think you need something about Hilbert space. Namely, following what I understand of what Coleman and Shimony have done, you need an idea of what it means for one state vector to be close to another state vector. That is where the Hilbert space norm comes in. All this business about particle trajectories and so on, none of that is real. What is real is the wave function. It evolves deterministically. There are no probabilities and we should get off that subject. The other point I wanted to make is an experimental point, which I thought might be refreshing. Although it is very hard to think of alternatives to quantum mechanics, there have been efforts to test the linearity of the evolution of the wave function in simple atomic systems. In particular, one prediction of the linear quality of the wave function is that the precession frequency of an atomic spin around a magnetic field does not depend on the angle that the spin makes with the magnetic field. That is a principle underlying atomic clocks and has been tested to much better than one part in ten to the twentieth.

B. Greene I think a lot of the discussion is about some nature of time in a quantum mechanical framework, but of course it has also to do with the nature of space. One of the developments the Solvay conferences have given part of the solution of, is the notion of non locality as a fundamental feature of quantum mechanics. You can argue that the interpretation of the measurement problem is just about words. But it is actually more than just about words because different interpretations of quantum mechanics do have different views on whether non-locality is an essential feature of space in a quantum mechanical context. I think that this is the issue that ultimately needs to be resolved. I know Gell-Mann has already raised his hand and has his views on this, which I do not really agree with. But I think that there really is something there, there is a real implication of this interpretation which is not just words.

L. Faddeev In relation with Wilczek's statement: I think that there is a great difference between h and the charge of the electron. h is a parameter of deformation of unstable degenerate classical mechanics to quantum mechanics. The same role is played by c and G. So these three parameters are certainly distinguished.

N. Seiberg I would really like to address Wilczek's point. The view that we have a classical system, which is later quantized, is the way we were taught physics. But we have many examples in string theory where this is not the case. The theory is intrinsically quantum mechanical and does not have a parameter like $\hbar$. For example, there are self-dual fields which do not come from the quantization of any system. Eleven dimensional supergravity does not have an $\hbar$. The conifold is another example where part of the system is intrinsically quantum mechanical. I think this goes a long way to show that quantum mechanics is really part of the story and not something that is an add-on to classical string theory.

F. Wilczek If you can calculate the fine structure constant, which is what this amounts to, then I will be impressed.

M. Gell-Mann I would like to make a couple of brief comments. One is that people like Coleman and many others, who were unhappy about a probability postulate in quantum mechanics, have imagined, for purposes of argument, that the universe is reproduced in many copies independent of one another. They then look at the quantum mechanical behavior of each system and they show that the statistical weight in this hypothetical set of universes, the statistical weight of a given situation, is what is called the probability. In other words, the calculation of the two things is the same. Therefore you do not have to worry. You do not have to agonize about this independent probability postulate. It just follows from identifying probability with the result of this artificially created statistical situation. The second comment I would like to make is about the so-called non-locality of quantum mechanics. I think it is simply a misnomer. What people have noticed is that, if you try for example in the Einstein-Rosen-Podolsky-Bohm experiment as performed by many people in the laboratory, to interpret the results classically, then you would need either non-locality or negative probabilities. But you do not interpret them classically. You interpret them in quantum mechanics and in that case there is no non-locality of any kind whatsoever.

D. Gross Especially since we use local quantum field theory to describe these experiments.

Session 3

Singularities

Chair: *Gary T. Horowitz*, UCSB, USA
Rapporteur: *Gary W. Gibbons*, DAMPT, Cambridge, UK
Scientific secretary: *Jan Troost* (Ecole Normale Supérieure, Paris, France)

3.1 Rapporteur talk: Singularities, by Gary W. Gibbons

3.1.1 *Introduction*

Einstein's General Relativity is incomplete because

- It predicts that **gravitational collapse** , both at the Big Bang and inside black holes, brings about spacetime singularities as at which the theory breaks down
- It gives no account of 'matter 'as opposed to geometry , and in particular the nature of classical 'particles '
- It is incompatible with quantum mechanics

I have been asked to review the first problem.
I will cover the following topics.

- Singularity Theorems
- Cosmic Censorship
- Classical Boundary conditions and stabiity
- Higher dimensional resolutions
- Singularities at the end of Hawking Evaporation
- Maldacena's conjecture

3.1.2 *Singularity Theorems*

First discovered in Friedmann-Lemaitre models, it was shown by Roger Penrose [19] that these arise if closed trapped surfaces occur during gravitational collapse and work by Geroch , Hawking and Penrose [20] showed that as long as matter satisfies various **positive energy conditions**, then spacetime singularities are inevitable in the future of certain types of Cauchy data.

Thus unlike classical Yang-Mills theory [1] and scalar fields theories with renormalisable potentials, **Leibniz-Laplace Determinism** breaks down for General Relativity. It can at best be an effective theory.

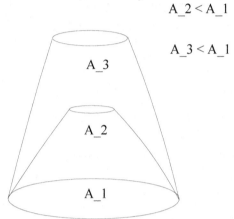

$$A\_2 < A\_1$$

$$A\_3 < A\_1$$

The area of a closed trapped 2-surface decreases in both the inward *and* the outward directions if pushed to the future along its two lightlike normals.

The singularity theorem give very little information about the nature of the singularities, in effect they deny the existence of timelike or null geodesically complete spacetimes. The reason for the incompleteness is not predicted.

It is widely believed that incompleteness is due to divergences of curvature invariants [2], or the components of the curvature in certain privileged, example parallelly propagated frames.

The singularity theorems also fail to predict the scale at which singularities arise. Indeed classical general relativity has no in built scale,

$$g_{\mu\nu} \to \lambda^2 g_{\mu\nu} \tag{1}$$

with λ constant is a symmetry of the theory.

Many people believe that the resolution of the problem of singularities will come from modifications of the Einstein equations due to Quantum Gravity at the Planck scale, but this is by no means obvious. The necessary modification could, in principle, have nothing to do with quantum mechanics.

It might for example entail the introduction of higher curvature terms [3] . However examples tend to show that some singularities still remain, e.g. those in singular pp-waves which are in effect solutions of almost all theories of gravity. In addition many, but not all admit ghosts.

Well motivated modifications of Einstein's theory include supergravity theories and the low energy limits of String theories.

---

[1] or possibly Born-Infeld theory

[2] but these may all vanish, e.g. for singular pp waves

[3] For example Born-Infeld Gravity [26]

However, the singularity theorems also apply to classical supergravity theories in all relevant dimensions since the matter fields satisfy the energy conditions. If supermatter is added, then only if potentials for scalars are positive (which cannot happen for pure supergravity) could singularities conceivably be avoided. However one may also truncate to the pure gravity sector and we are back to the same problem.

The same problem arises in String Theory in the zero slope limit. Only higher curvature terms could could conceivably evade the problem.

3.1.2.1 *The Strong Energy Condition*

$$T_{\hat{0}\hat{0}} + \sum_i T_{\hat{i}\hat{i}} \geq 0$$ (2)

is the most important energy condition ('Gravity is attractive').

It can only fail if potentials for scalars are positive.

As an aside , one should note that unless it fails, cosmic acceleration (e.g. a positive cosmological constant, $\Lambda > 0$, is impossible [27]).

Thus there can be no inflation in pure supergravity theories, or the zero slope limit of String theory.[4]

If one takes the view that the problem of singularities will be resolved quantum mechanically, one might be tempted to argue that no particular classical spacetime is of particular significance, and that classical or semi-classical studies of singularities are misguided.

At a fundamental level that is probably correct, and is a certainly a valid criticism of much current speculation on the final outcome of Hawking evaporation for example, but as a practical matter almost all of the large scale universe appears to be essentially classical. Astrophysicists should not need quantum gravity to understand X-ray sources or the black hole at the centre of our galaxy.

Thus we need understand better classical singularities.

3.1.3 *Cosmic Censorship*

The singularities that arise from localised gravitational collapse are associated with black holes. Intuitively, Penrose's **Cosmic Censorship Hypothesis** [21] postulates that all singularities are hidden inside the event horizon, i.e. inside $\dot{I}^-(\mathcal{I}^+)$, the boundary of the past of future null infinity, $\mathcal{I}^+$, the latter is usually assume to be complete.

Investigating this problem is extremely challenging mathematically. At present one is limited to looking at spherically symmetric spacetimes coupled to matter, e.g. massless scalar field, or, in the non-spherical case, to numerical simulations.

---

[4]except in models with time-dependent extra dimensions which have other problems

The work of Christodoulou on spherically symmetric massless scalar collapse [40][41][42][43] shows that Cosmic Censorship in the strict sense fails, because of transient singularities associated with a very special choice of initial data.

These singularities are associated with a discrete self-similar behaviour, referred to as critical behaviour, first uncovered numerically by Choptuik [44].

It is rather doubtful that this behaviour will survive quantum corrections.

The strategy of Christodoulou is essentially to reduce the problem to a 1+1 dimensional non-linear wave problem.

This technique has been exploited recently by Dafermos [35][36][34] [33][32] who is able to treat charged gravitational collapse and establish a form of cosmic censorship in the presence of Cauchy horizons assuming the existence of closed trapped surfaces and also to justify the assumption that $\mathcal{I}^+$ is complete.

Dafermos's methods also extend to higher dimensions [46] where, following numerical numerical work by Bizon, Chmaj and Schmidt in 4+1 [30] and 8+1 [31] Dafermos and Holzegel [29] were recently able to extend some of these results vacuum gravitational collapse in 4+ 1 .

There has been a great deal of work on homogeneous solutions of Einstein's equations, particularly near singularities. This can be partially extended to inhomogeneous solutions provided one assumes spatial derivatives are small (velocity dominated approximation).

On this basis, Belinsky, Lifshitz and Khalatnikov proposed that generically, singularities are of chaotic, oscillatory, Mixmaster type, first seen in Bianchi IX models.

Recent mathematically rigorous work tends to confirm that this may happen, at least for an open set of Cauchy data.

The story in higher dimensions will be the subject of a report by T Damour.

3.1.4   *Classical Boundary Conditions and Stability*

The basic problem raised by spacetime by singularities is what boundary conditions are to be posed in their presence? Cosmic Censorship is an attempt to evade that problem as long as one is outside the event horizon. Even if it is true, what happens inside the horizon?.

The choice of boundary conditions may affect questions of genericity or stability.

In some cases, despite singularities the choice of boundary condition may be unique.[5]

Typically in gravitational situations however, this is not the case and choices must be made.

For example, in a recent paper, Gibbons, Hartnoll and Ishibashi [5] showed that there is a choice of boundary conditions such that even negative mass Schwarzschild

---

[5] for example the unique self-adjoint extension for the Hydrogen Atom

is stable against linear perturbations[6].

Another case of non-uniqueness concerns quantum fields near cosmic strings and orbifolds singularities.

3.1.5 *Boundary Conditions in Cosmology*

The issue of boundary conditions is particularly important in cosmology. For example one typically thinks of Minkowski spacetime as being stable.However this is manifestly not the case if one considers perturbations which do not die off at large distances.

3.1.5.1 *Instability of Flat space*

Consider Kasner solutions of the vacuum Einstein equations,

$$ds^2 = -dt^2 + t^{2p_1}dx^2 + t^{2p_2}dy^2 + t^{2p_3}dz^2 \,, \tag{3}$$

where p_1, p_2, p_3 are constants such that

$$p_1 + p_2 + p_3 = 1 = p_1^2 + p_2^2 + p_3^2 \,. \tag{4}$$

Unless one of the p_i is equal to 1, these metrics have a singularity at $t = 0$. If we set $t = 1 - t'$, then the metric near $t = 1$ starts out looking like a small deformation of the flat metric, with a small homogeneous mode growing linearly with t'. Ultimately, however, non-linear effects take over and the universe ends in a Big Crunch at $t' = 1$, *i.e.* $t = 0$.

This instability is universal in gravity theories, and is closely related to the modulus problem in theories with extra dimensions.

Consider, for example, the exact ten-dimensional Ricci-flat metric

$$ds^2 = t^{1/2}(-dt^2 + d\mathbf{x}^2) + t^{1/2}g_{mn}(y)dy^m dy^n, \tag{5}$$

where $g_{mn}(y)$ is a six-dimensional metric on a Calabi-Yau space K. This starts off at $t = 1$ looking like $E^{3,1} \times K$ with a small perturbation growing linearly in $t' = 1 - t$. However by the time it reaches $t' = 1$, $t = 0$, the solution has evolved to give a spacetime singularity. From the point of view of the four-dimensional reduced theory, the logarithm of the volume of the Calabi-Yau behaves like a massless scalar field – the modulus field which is sometimes thought of as a kind of Goldstone mode for a spontaneously-broken global scaling symmetry. This causes an isotropic expansion or contraction of the three spatial dimensions, with the scale factor $a(\tau)$ going like $\tau^{\frac{1}{3}}$, which is what one expects for a fluid whose energy density equals its pressure.

In recent work, Chenm Gibbons, Hu and Pope[9] [10] have shown that a wide variety of BPS brane configurations, including the Horawa-Witten model are cosmologically unstable. .

---

[6]non-linear stability remains problematic

3.1.5.2   *Hořava-Witten solution from Heterotic M-Theory*

The equations of motion for the metric and ϕ may be consistently obtained from the Lagrangian

$$\mathcal{L}_5 = \sqrt{-g}\left(R - \frac{1}{2}(\partial\phi)^2 - m^2\, e^{2\phi}\right). \tag{6}$$

The scalar field ϕ characterises the size of the internal Calabi-Yau space.

3.1.5.3   *Exact static supersymmetric solution*

$$ds_5^2 = \widetilde{H}\left(-dt^2 + d\mathbf{x}^2\right) + \widetilde{H}^4\, d\tilde{y}^2\,,$$

$$\widetilde{H} = 1 + \tilde{k}\,|\tilde{y}|\,, \tag{7}$$

$$\phi = -3\log\widetilde{H}\,,$$

$\tilde{k}$ is a constant In the Hořava-Witten picture a second domain wall is introduce, at $y = L$, by taking y to be periodic with period $2L$, such that $y = L$ is identified with $y = -L$. Furthermore, one makes the Z_2 identification $y \leftrightarrow -y$.

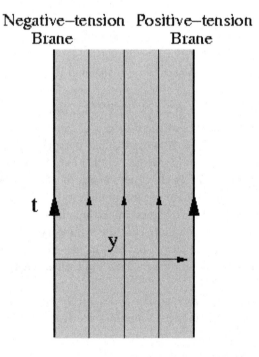

Negative–tension Brane   Positive–tension Brane

3.1.5.4 *Time-dependent solutions*

$$ds_5^2 = H^{1/2}\left(-dt^2 + d\mathbf{x}^2\right) + H\,dy^2\,,$$

$$H = h\,t + k\,|y|\,, \tag{8}$$

$$\phi = -\frac{3}{2}\log H\,,$$

$k^2 = 8m^2/3$, and h is an arbitrary constant.

If we turn off the time dependence (by setting $h = 0$), the relation to the previous static solution is seen by making a coordinate transformation of the form $y = \tilde{y}^2$. If we set the parameter m in the Lagrangian to zero, the solution describes a **Kasner universe**.

When we lift the solution back to $D = 11$, the metric becomes

$$ds_{11}^2 = H^{-1/2}(-dt^2 + d\mathbf{x}^2) + H^{1/2}\,ds_{CY_6}^2 + dy^2\,. \tag{9}$$

The static solution, in the orbifold limit, can be viewed as an intersection of three equal-charge M5-branes. Turning off the brane charge, the time-dependent metric describes a direct product of a ten-dimensional Kasner universe and a line segment.

3.1.5.5 *Local Static Form: Chamblin-Reall picture*

If we temporarily drop the modulus sign around y in (8), then the coordinate transformation from t and y to $\tilde{t}$ and r given by

$$dt = d\tilde{t} - \frac{hr^{1/2}}{k^2 f}\,dr\,,$$

$$H = r\,, \tag{10}$$

$$f = 1 - \frac{h^2\,r^{1/2}}{k^2}$$

transforms the solution into the static form [25]

$$ds_5^2 = r^{1/2}\left(-f\,d\tilde{t}^2 + d\mathbf{x}^2 + r^{1/2}\frac{dr^2}{k^2 f}\right),$$

$$\phi = -\frac{3}{2}\log r\,. \tag{11}$$

This can be recognised as a black 3-brane, with an horizon at $f = 0$.

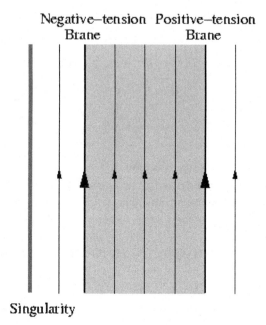

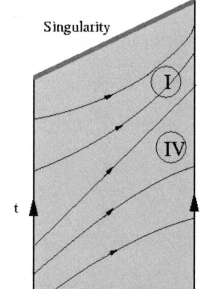

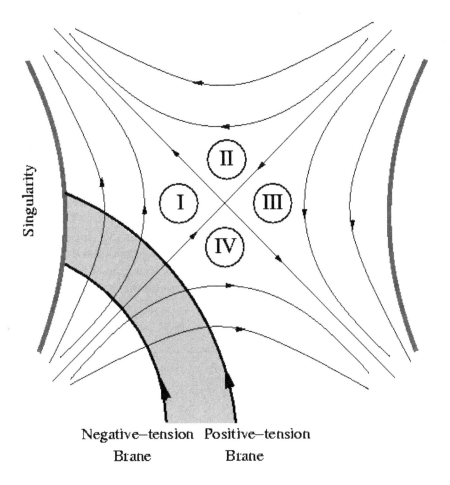

Negative–tension Positive–tension
Brane Brane

3.1.6 *Higher dimensional resolutions*

Serini, Einstein, Pauli and Lichnerowicz were able to show that there are no static or stationary soliton like solutions of the vacuum Einstein equations without horizons (see [45] for a review of this type of result). The presence of an horizon implies a singularity. These results extend straightforwardly [45] to include the sort of matter encountered in ungauged supergravity theories and Klauza-Klein theory[7] or the zero slope limits of String Theory. They follow essentially because these theories do not admit a length scale: rigid dilation

$$g_{\mu\nu} \to \lambda^2 g_{\mu\nu} \qquad \lambda \ \text{constant}\,, \tag{12}$$

is a symmetry of the equations of motion.

---

[7]with the proviso that the fields in four dimensions are regular, see later

The situation improves if one considers the Kaluza-Klein monopoles of Gross Perry [3] and Sorkin [4].

$$ds^2 = dt^2 + V^{-1}(dx^5 + \omega_i dx^i)^2 + V d\mathbf{x}^2 , \tag{13}$$

$$\operatorname{grad} V = \operatorname{curl} \omega \tag{14}$$

$$V = 1 + \sum_a \frac{M_a}{|\mathbf{x} - \mathbf{x}_a|} \tag{15}$$

Periodicity of x^5 imposes a quantisation condition on the Klauza-Klein charges and on the magnetic monopole moment.

In addition, the singularities of the four-dimensional metric receive a **Higher Dimensional Resolution**: they are mere coordinate artifacts in five dimensions. In this way, they evade the Pauli-Einstein theorem.

Gibbons, Horowitz and Townsend [2] have shown that higher dimensional resolutions are quite common. However the problem of singularities and the ultimate outcome of gravitational collapse and Hawking evaporation cannot be solved in this this way. Moreover, the solution is unstable in the sense that

$$ds^2 = dt^2 + \tilde{V}^{-1}(dx^5 + \omega_i dx^i)^2 + \tilde{V} d\mathbf{x}^2 , \tag{16}$$

$$\operatorname{grad} V = \operatorname{curl} \omega \tag{17}$$

$$\tilde{V} = -ht + \sum_a \frac{M_a}{|\mathbf{x} - \mathbf{x}_a|} \tag{18}$$

is an exact, time-dependent solution.

This is a particular example of the general tendency of higher dimensions to undergo gravitational collapse [17], which is fatal to the dimensional reduction programme unless some means can be found to stabilize the various 'moduli' fields.

These examples also underscore the need for **a theory of initial conditions** in order to understand cosmology and the initial singularity or big bang. As emphasised by Penrose among others, elementary thermodynamic arguments indicate that the Universe began in a very special state and even proponents of eternal inflation have had to concede, following Borde, Guth and Vilenkin [28] , that eternity is past incomplete. In other words if inflation is past eternal then spacetime is geodesically incomplete. **Penrose's Weyl Curvature Hypothesis** postulates a connection between gravitational entropy and Weyl curvature[8] and hence demands of the universe that the initial singularity has vanishing, or possibly finite, Weyl curvature, as in F-L-R-W. models.In general such singularities are called isotropic and Tod and co-workers [22] [23] [24] have proven existence and uniqueness results for the associated Cauchy problem. However there is as yet no derivation of this condition from something deeper and it remains a purely classical viewpoint.

Hartle and Hawking's No Boundary Proposal achieves a similar purpose but at the expense of leaving the realm of Lorentzian metrics. In principle this is

---

[8]despite the example of de-Sitter spacetime

a complete quantum mechanical answer to the problem of singularities but so far has only been explored at the semi-classical level. Probably all the resources of String/M-theory will be required for a full treatment.

3.1.7 *Singularities at the end of Hawking Evaporation*

The ultimate fate of the singularities inside black holes is inextricably mixed up with the ultimate end of Hawking Evaporation. It may be shown (e.g. Kodama [1]) that the well known classical spacetime model incoporating back re-action must contain a (transient) naked singularity.

However it is by no means clear that the semi-classical approximation applies. One must surely have to take into account the **quantum interference of space-times**.

But how to do this?

3.1.8 *Maldacena's Conjecture*

One way in which this might be achieved is to consider Hawking Evaporation in AdS. String theory in the bulk is supposedly dual to conformal field theory on the conformal boundary. The latter is believed to be unitary and non-singular, hence so must the former.

Much work has been done relating black holes in AdS_5 and $\mathcal{N} = 4$ SU(N) SUSYM. A more tractable case is AdS_3. Using it, Maldacena has suggested [6] a deep connection between unitarity and the **ergodic properties of quantum fields** and this has recently been taken up by Barbon and Rabinovici [14] [15], [16] and by Hawking himself [7][8].

However, presumably the case of greatest physical interest AdS_4 which has received much less attention. Little is known about the CFT.

Klebanov and Polyakov [13] have proposed a correspondence valid at weak coupling but this invovles a bulk theory containing **infinitely many spins**. Apart from some work of Hartnoll and Kumar [11] and Warnick[12], little detailed work has been done on this case.

Hertog, Horowitz and Maeda have argued [39][38] that cosmic scensorship is easier to violate in AdS backgrounds, but the remain uncertainties about the details [35][37].

Bibliography

[1] H. Kodama, Prog. Theor. Phys. **62** (1979) 1434.
[2] G. W. Gibbons, G. T. Horowitz and P. K. Townsend, Class. Quant. Grav. **12** (1995) 297 [arXiv:hep-th/9410073].
[3] D. J. Gross and M. J. Perry, Nucl. Phys. B **226**, 29 (1983).
[4] R. d. Sorkin, Phys. Rev. Lett. **51**, 87 (1983).

[5] G. W. Gibbons, S. A. Hartnoll and A. Ishibashi, Prog. Theor. Phys. **113** (2005) 963 [arXiv:hep-th/0409307].

[6] J. M. Maldacena, JHEP **0304** (2003) 021 [arXiv:hep-th/0106112].

[7] S. W. Hawking, Phys. Rev. D **72**, 084013 (2005) [arXiv:hep-th/0507171].

[8] S. Hawking, SPIRES entry *Prepared for GR17: 17th International Conference on General Relativity and Gravitation, Dublin, Ireland, 18-24 Jul 2004*

[9] W. Chen, Z. W. Chong, G. W. Gibbons, H. Lu and C. N. Pope, Nucl. Phys. B **732**, 118 (2006) [arXiv:hep-th/0502077].

[10] G. W. Gibbons, H. Lu and C. N. Pope, Phys. Rev. Lett. **94**, 131602 (2005) [arXiv:hep-th/0501117].

[11] S. A. Hartnoll and S. P. Kumar, JHEP **0506**, 012 (2005) [arXiv:hep-th/0503238].

[12] C. Warnick, arXiv:hep-th/0602127.

[13] I. R. Klebanov and A. M. Polyakov, Phys. Lett. B **550**, 213 (2002) [arXiv:hep-th/0210114].

[14] J. L. F. Barbon and E. Rabinovici, arXiv:hep-th/0503144.

[15] J. L. F. Barbon and E. Rabinovici, Fortsch. Phys. **52**, 642 (2004) [arXiv:hep-th/0403268].

[16] J. L. F. Barbon and E. Rabinovici, JHEP **0311**, 047 (2003) [arXiv:hep-th/0308063].

[17] R. Penrose, *Prepared for Workshop on Conference on the Future of Theoretical Physics and Cosmology in Honor of Steven Hawking's 60th Birthday, Cambridge, England, 7-10 Jan 2002.*

[18] R. Penrose, *Prepared for Workshop on Conference on the Future of Theoretical Physics and Cosmology in Honor of Steven Hawking's 60th Birthday, Cambridge, England, 7-10 Jan 2002.*

[19] R. Penrose, Phys. Rev. Lett. **14**, 57 (1965).

[20] S. W. Hawking and R. Penrose, Proc. Roy. Soc. Lond. A **314**, 529 (1970).

[21] R. Penrose, Riv. Nuovo Cim. **1**, 252 (1969) [Gen. Rel. Grav. **34**, 1141 (2002)].

[22] K. P. Tod, Class. Quant. Grav. **20**, 521 (2003) [arXiv:gr-qc/0209071].

[23] K. Anguige and K. P. Tod, Annals Phys. **276**, 294 (1999) [arXiv:gr-qc/9903009].

[24] K. Anguige and K. P. Tod, Annals Phys. **276**, 257 (1999) [arXiv:gr-qc/9903008].

[25] H. A. Chamblin and H. S. Reall, Nucl. Phys. B **562**, 133 (1999) [arXiv:hep-th/9903225].

[26] S. Deser and G. W. Gibbons, Class. Quant. Grav. **15** (1998) L35 [arXiv:hep-th/9803049].

[27] G. W. Gibbons, Print-84-0265 (CAMBRIDGE) *Lectures given at 4th Silarg Symposium, Caracas.*

[28] A. Borde, A. H. Guth and A. Vilenkin, Phys. Rev. Lett. **90**, 151301 (2003) [arXiv:gr-qc/0110012].

[29] M. Dafermos and G. Holzegel, arXiv:gr-qc/0510051.

[30] P. Bizon, T. Chmaj and B. G. Schmidt, Phys. Rev. Lett. **95**, 071102 (2005) [arXiv:gr-qc/0506074].

[31] P. Bizon, T. Chmaj, A. Rostworowski, B. G. Schmidt and Z. Tabor, Phys. Rev. D **72**, 121502 (2005) [arXiv:gr-qc/0511064].

[32] M. Dafermos, arXiv:gr-qc/0209052.

[33] M. Dafermos, arXiv:gr-qc/0307013.

[34] M. Dafermos, arXiv:gr-qc/0310040.

[35] M. Dafermos, arXiv:gr-qc/0403033.

[36] M. Dafermos, Class. Quant. Grav. **22**, 2221 (2005) [arXiv:gr-qc/0403032].

[37] T. Hertog, G. T. Horowitz and K. Maeda, arXiv:gr-qc/0405050.

[38] T. Hertog and G. T. Horowitz, JHEP **0407** (2004) 073 [arXiv:hep-th/0406134].

[39] T. Hertog, G. T. Horowitz and K. Maeda, Phys. Rev. Lett. **92** (2004) 131101 [arXiv:gr-qc/0307102].

[40] D. Christodoulou, Commun. Math. Phys. **109**, 613 (1987).

[41] D. Christodoulou, Commun. Math. Phys. **109**, 591 (1987).

[42] D. Christodoulou, Commun. Math. Phys. **106**, 587 (1986).

[43] D. Christodoulou, Commun. Math. Phys. **105**, 337 (1986).

[44] M. W. Choptuik, Phys. Rev. Lett. **70**, 9 (1993).

[45] P. Breitenlohner, D. Maison and G. W. Gibbons, Commun. Math. Phys. **120**, 295 (1988).

[46] Gibbons G. W. (2005), Talk given in September 2005, during the Programme Global Problems in Mathematical Relativity held in Newton Institute.
http://www.newton.cam.ac.uk/webseminars/pg+ws/2005/gmr/0905/gibbons/

3.2 Discussion

A. Linde Eternal inflation indeed requires initial conditions and is not future geodesically complete. However, one should recall the difference between being
past eternal and past complete. For any point, any geodesic starting in it has
finite distance, however, for any geodesic, there is a longer one. So for a particular observer, the universe begins and ends, but not the universe as a whole.
This rectifies a common misconception.

G. Gibbons I agree, there is no big bang in that model.

A. Strominger Cosmic censorship proposed a solution to the singularity problems, namely that they always lie behind horizons. However, Hawking radiation
shows that black holes become smaller over time, and that presumably Planck-
ian effects take over the dynamics at some point in time. So, cosmic censorship
would not anymore seem to make general relativity complete in any way. Why
is it then so close to being true?

G. Gibbons Recall that the cosmic censorship idea came from a practical question
in X-ray cosmology. Did general relativity need to be changed at the scale these
experimenters were interested in? The conclusion they came to was that there
was no obvious reason why general relativity should break down at macroscopic
scales. Still, cosmic censorship is an important property for classical general
relativity, and it is still important to know whether it is satisfied.

A. Strominger If it was not, we could see naked singularities in the sky.

G. Gibbons Which was always Penrose's viewpoint. The counter was put forward
by more conservative opponents.

S. Weinberg If only on aesthetic grounds, it seems obvious that at short distance
scales additional terms in the gravitational action will become important. General relativity is an effective theory as is the theory of soft pions. Obviously, it is
not the whole answer. Original singularity theorems are important in showing
that higher order terms will become relevant at some point.

G. Gibbons Yes, however, we distinguish two possibilities: they can become important already classically, or at the quantum level. The latter is more plausible.

S. Weinberg There is no problem with ghosts and the higher derivative terms.
They arise only when misusing perturbation theory.

D. Gross Indeed, there is no problem. But only if you have a sensible theory to
do perturbation theory in, which is not a priori given.

3.3 Prepared Comments

3.3.1 *Gary Horowitz: Singularities in String Theory*

I would like to give a brief overview of singularities in string theory. Many different types of singularities have been shown to be harmless in this theory. They are resolved by a variety of different mechanisms using different aspects of the theory. Some rely on the existence of other extended objects called branes, while others are resolved just in perturbative string theory. However it is known that not all singularities are resolved, and as we will review below, this necessary for the theory to have a stable ground state. There are very few general results in this subject. In particular, there is nothing like the singularity theorems of general relativity which give general conditions under which singularities form. So far, singularities have been studied on a case by case basis.

The starting point for our discussion is the fact that strings sense spacetime differently than point particles. Perturbative strings feel the metric through a two dimensional field theory called a sigma model. This means that two spacetimes which give rise to equivalent sigma models are indistinguishable in string theory. Apparently trivial changes to the sigma model can result in dramatically different spacetimes. Let me give two examples:

1) T-duality: If the spacetime metric is independent of a periodic coordinate x, then a change of variables in the sigma model describes strings on a new spacetime with $g_{xx} \to 1/g_{xx}$ [1, 2].

2) Mirror symmetry: In string theory, we often consider spacetimes of the form $M_4 \times K$ where M_4 is four dimensional Minkowski spacetime and K is a Calabi-Yau space [3], i.e. a compact six dimensional Ricci flat space. One can show that changing a sign in the (supersymmetric) sigma model changes the spacetime from $M_4 \times K$ to $M_4 \times K'$ where K' is topologically different Calabi-Yau space [4].

Using these facts it is easy to show that spacetimes which are singular in general relativity can be nonsingular in string theory. A simple example is the quotient of Euclidean space by a discrete subgroup of the rotation group. The resulting space, called an orbifold, has a conical singularity at the origin. Even though this leads to geodesic incompleteness in general relativity, it is completely harmless in string theory [5]. This is essentially because strings are extended objects.

The orbifold has a very mild singularity, but even curvature singularities can be harmless in string theory. A simple example follows from applying T-duality to rotations in the plane. This results in the metric $ds^2 = dr^2 + (1/r^2)d\phi^2$ which has a curvature singularity at the origin. However strings on this space are completely equivalent to strings in flat space.

As mentioned above, string theory has exact solutions which are the product of four dimensional Minkowski space and a compact Calabi-Yau space. A given Calabi-Yau manifold usually admits a whole family of Ricci flat metrics. So one can construct a solution in which the four large dimensions stay approximately flat and the geometry of the Calabi-Yau manifold changes slowly from one Ricci flat metric to another. In this process the Calabi-Yau space can develop a curvature singularity. In many cases, this is the result of a topologically nontrivial sphere S^2 or S^3 being shrunk down to zero area. It has been shown that when this happens, string theory remains completely well defined. The evolution continues through the geometrical singularity to a nonsingular Calabi-Yau space on the other side.

There are two qualitatively different ways in which this can happen. In one case, an S^2 collapses to zero size and then re-expands as a topologically different S^2. This is known as a flop transition. It was shown in [6] that the mirror description of this is completely nonsingular. Under mirror symmetry, this transition corresponds to evolution through nonsingular metrics. In the second case, an S^3 collapses down to zero size and re-expands as an S^2. This is called a conifold singularity. This transition is nonsingular if you include branes wrapped around the S^3 [7]. As long as the area of the surface is nonzero, these degrees of freedom are massive, and it is consistent to ignore them. However when the surface shrinks to zero volume these degrees of freedom become massless, and one must include them in the analysis. When this is done, the theory is nonsingular. These examples show that topology can change in a nonsingular way in string theory.

I will divide the remaining examples of singularity resolution into three classes depending on whether the singularities are timelike, null, or spacelike. Some space-times with timelike singularities can be replaced by entirely smooth solutions. In some cases this involves replacing the singularity with a source consisting of a smooth distribution of branes as in the "enhancon" [8]. Other cases can be done purely geometrically and do not need a source [9]. In this case, the smooth solution has less symmetry than the singular one. Although there is no argument here that strings in the singular space are equivalent to strings in the nonsingular space, there are arguments that the nonsingular description is the correct description of the physical situation.

Branes carry charges which source higher rank generalizations of a Maxwell field called RR fields. The gravitational field produced by a collection of branes wrapped around cycles often contain null singularities. In some cases, one can find nonsingular geometries with the same charge but no brane sources. This is possible since they contain nontrivial topology which supports nonzero RR flux. Many examples of this have been found for solutions involving two charges [10]. This phenomena of branes being replaced by fluxes is generally called geometric transitions.

Under certain conditions, string theory has tachyons, i.e. states with $m^2 < 0$. In the past, these tachyons were mysterious, but recently they have been understood as just indicating an instability of the space. In fact, tachyons can be very useful in avoiding black hole and cosmological singularities. There are situations in which a tachyon arises in the evolution toward a spacelike singularity. The evolution past this point is then governed by the dynamics of the tachyon and no longer agrees with general relativity [11].

Despite all these examples, it is simply not true that all singularities are removed in string theory. Nonlinear gravitational plane waves are not only vacuum solutions to general relativity, but also exact solutions to string theory [12]. These solutions contain arbitrary functions describing the amplitude of each polarization of the wave. If one of the amplitudes diverge at a finite point, then the plane wave is singular. One can study string propagation in this background and show that in some cases, the string does not have well behaved propagation through this curvature singularity [13]. The divergent tidal forces cause the string to become infinitely excited.

It is a good thing that string theory does not resolve all singularities. Consider the Schwarzschild solution with $M < 0$. This describes a negative mass solution with a naked singularity at the origin. If this singularity was resolved, there would be states with arbitrarily negative energy. String theory would not have a ground state. This argument, of course, is not restricted to string theory but applies to any candidate quantum theory of gravity.

One of the main goals of quantum gravity is to provide a better understanding of the big bang or big crunch singularities of cosmology. Perhaps the most fundamental question is whether they provide a true beginning or end of time, or whether there is a bounce. Hertog and I have recently studied this question using the AdS/CFT correspondence [14], which states that string theory in spacetimes which are asymptotically anti de Sitter (AdS) is equivalent to a conformal field theory (CFT). We found supergravity solutions in which asymptotically AdS initial data evolve to big crunch singularities [15]. The dual description involves a CFT with a potential unbounded from below. In the large N limit, the expectation value of some CFT operators diverge in finite time. A minisuperspace approximation leads to a bounce, but there are arguments that this is not possible in the full CFT. Although more work is still needed to completely understand the dual description, this suggests that a big crunch is not a big bounce [15].

Acknowledgment: Preparation of this comment was supported in part by NSF grant PHY-0244764.

Bibliography

[1] T. H. Buscher, "A Symmetry Of The String Background Field Equations," Phys. Lett. B **194** (1987) 59; M. Rocek and E. Verlinde, "Duality, quotients, and currents," Nucl. Phys. B **373** (1992) 630 [arXiv:hep-th/9110053].

[2] A. Giveon, M. Porrati and E. Rabinovici, "Target space duality in string theory," Phys. Rept. **244** (1994) 77 [arXiv:hep-th/9401139].

[3] P. Candelas, G. T. Horowitz, A. Strominger and E. Witten, "Vacuum Configurations For Superstrings," Nucl. Phys. B **258** (1985) 46.

[4] B. R. Greene and M. R. Plesser, "Duality In Calabi-Yau Moduli Space," Nucl. Phys. B **338** (1990) 15; P. Candelas, M. Lynker and R. Schimmrigk, "Calabi-Yau Manifolds In Weighted P(4)," Nucl. Phys. B **341** (1990) 383.

[5] L. J. Dixon, J. A. Harvey, C. Vafa and E. Witten, "Strings On Orbifolds," Nucl. Phys. B **261** (1985) 678; "Strings On Orbifolds. 2," Nucl. Phys. B **274** (1986) 285.

[6] P. S. Aspinwall, B. R. Greene and D. R. Morrison, "Calabi-Yau moduli space, mirror manifolds and spacetime topology change in string theory," Nucl. Phys. B **416** (1994) 414 [arXiv:hep-th/9309097].

[7] A. Strominger, "Massless black holes and conifolds in string theory," Nucl. Phys. B **451** (1995) 96 [arXiv:hep-th/9504090]; B. R. Greene, D. R. Morrison and A. Strominger, "Black hole condensation and the unification of string vacua," Nucl. Phys. B **451** (1995) 109 [arXiv:hep-th/9504145].

[8] C. V. Johnson, A. W. Peet and J. Polchinski, "Gauge theory and the excision of repulson singularities," Phys. Rev. D **61**, 086001 (2000) [arXiv:hep-th/9911161].

[9] I. R. Klebanov and M. J. Strassler, "Supergravity and a confining gauge theory: Duality cascades and chiSB-resolution of naked singularities," JHEP **0008**, 052 (2000) [arXiv:hep-th/0007191].

[10] O. Lunin, S. D. Mathur and A. Saxena, "What is the gravity dual of a chiral primary?," Nucl. Phys. B **655**, 185 (2003) [arXiv:hep-th/0211292]; O. Lunin, J. Maldacena and L. Maoz, "Gravity solutions for the D1-D5 system with angular momentum," arXiv:hep-th/0212210.

[11] J. McGreevy and E. Silverstein, "The tachyon at the end of the universe," JHEP **0508**, 090 (2005) [arXiv:hep-th/0506130]; E. Silverstein, contribution in this volume.

[12] D. Amati and C. Klimcik, "Nonperturbative Computation Of The Weyl Anomaly For A Class Of Nontrivial Backgrounds," Phys. Lett. B **219** (1989) 443.

[13] G. T. Horowitz and A. R. Steif, "Space-Time Singularities In String Theory," Phys. Rev. Lett. **64** (1990) 260.

[14] J. M. Maldacena, "The large N limit of superconformal field theories and supergravity," Adv. Theor. Math. Phys. **2** (1998) 231 [Int. J. Theor. Phys. **38** (1999) 1113] [arXiv:hep-th/9711200]; O. Aharony, S. S. Gubser, J. M. Maldacena, H. Ooguri and Y. Oz, "Large N field theories, string theory and gravity," Phys. Rept. **323** (2000) 183 [arXiv:hep-th/9905111].

[15] T. Hertog and G. T. Horowitz, "Holographic description of AdS cosmologies," JHEP **0504**, 005 (2005) [arXiv:hep-th/0503071]; "Towards a big crunch dual," JHEP **0407**, 073 (2004) [arXiv:hep-th/0406134].

3.4 Discussion

Prepared comment by S. Shenker.

D. Gross In the context of resolving the information paradox, it has been suggested by J. Maldacena and S. Hawking that puzzles disappear when we take into account different semi-classical space-times. When resolving singularities, might it not be the case that taking seriously superpositions of geometries would similarly resolve puzzles?

S. Shenker We have defined in pedestrian fashion an observable with a pole (on the second sheet) and we can ask what happens to that pole. What happens to the observable might be related to many geometries contributing . A first signal of that would be large $g_s \propto 1/N$ corrections of the sort $g_s^2 (\frac{1}{t-t_c})^x$ for some power x. The closer the geodesic gets to the singularity at $t = t_c$ the larger the quantum effects are. That could be interpretable as many geometries becoming important. Or as branes becoming light. Imagine further that we have a power series $g_s^n (\frac{1}{t-t_c})^{xn}$ that we resum, and then taking a double scaling limit. That might lead to an understanding of non-perturbative physics at the singularity.

A. Ashtekar In how far do your conclusions depend on analyticity?

S. Shenker They heavily do. Analyticity is a crucial property of quantum field theory and it would be bad to lose it.

G. Gibbons You have these two disjoint components to the boundary and a field theory on each. How, within field theory, do they talk to one another?

S. Shenker As J. Maldacena observed, the state in which you evaluate the correlation functions is a correlated state, the Hartle-Hawking state.

G. Gibbons So, you are not thinking of the quantum field theories as disjoint?

S. Shenker Well, you cannot build these states from non-singular data (if that is your question).

G. Gibbons I am more concerned with the quantum mechanics on the boundary. I thought the idea was to reduce the problem to standard quantum mechanics?

S. Shenker There is a corollary to that which is that if you have two Hilbert spaces in this case, one for each boundary.

G. Gibbons Is this a true generalization of quantum mechanics?

S. Shenker I am not going to agree with that. Perhaps somebody else would like to.

3.5 Prepared Comments

3.5.1 *Eva Silverstein: Singularities: Closed String Tachyons and Singularities*

3.5.1.1 *Singularities and Winding Modes*

A basic problem in gravitational physics is the resolution of spacetime singularities where general relativity breaks down. The simplest such singularities are conical singularities arising from orbifold identifications of flat space, and the most challenging are spacelike singularities inside black holes (and in cosmology). Topology changing processes also require evolution through classically singular spacetimes. In this contribution I will briefly review how a phase of closed string tachyon condensate replaces, and helps to resolve, basic singularities of each of these types. Finally I will discuss some interesting features of singularities arising in the small volume limit of compact negatively curved spaces.

In the framework of string theory, several types of general relativistic singularities are replaced by a phase of closed string tachyon condensate. The simplest class of examples involves spacetimes containing 1-cycles with antiperiodic Fermion boundary conditions. This class includes spacetimes which are globally stable, such as backgrounds with late-time long-distance supersymmetry and/or AdS boundary conditions.

In the presence of such a circle, the spectrum of strings includes winding modes around the circle. The Casimir energy on the worldsheet of the string contributes a negative contribution to the mass squared, which is of the form

$$M^2 = -\frac{1}{l_s}^2 + \frac{L^2}{l_s^4} \tag{1}$$

where L is the circle radius and $1/l_s^2$ is the string tension scale. For small L, the winding state develops a negative mass squared and condenses, deforming the system away from the $L < l_s$ extrapolation of general relativity. This statement is under control as long as L is static or shrinking very slowly as it crosses the string scale.

Examples include the following. Generic orbifold singularities have twisted sector tachyons, *i.e.* tachyons from strings wound around the angular direction of the cone. The result of their condensation is that the cone smooths out [1], as seen in calculations of D-brane probes, worldsheet RG, and time dependent GR in their regimes of applicability (see [2] for reviews). Topology changing transitions in which a Riemann surface target space loses a handle or factorizes into separate surfaces are also mediated by winding tachyon condensation [3]. Tachyon condensation replaces certain spacelike singularities of a cosmological type in which some number of circles shrinks homogeneously in the far past (or future) [4].

Finally, tachyons condensing *quasilocally* over a spacelike surface appear in black hole problems and in a new set of examples sharing some of their features [5][6].

One interesting new example is an AdS/CFT dual pair in which an infrared mass gap (confinement) arises at late times in a system which starts out in an unconfined phase out on its approximate Coulomb branch. As an example, consider the $\mathcal{N} = 4$ SYM theory on a (time dependent) Scherk-Schwarz circle, with scalar VEVs turned on putting it out on its Coulomb branch. As the circle shrinks to a finite size and the scalars roll back toward the origin, the infrared physics of the gauge theory becomes dominated by a three dimensional confining theory. The gravity-side description of this is via a shell of D3-branes which enclose a finite region with a shrinking Scherk-Schwarz cylinder. When the cylinder's radius shrinks below the string scale, a winding tachyon turns on. At the level of bulk spacetime gravity, a candidate dual for the confining theory exists [7]; it is a type of "bubble of nothing" in which the geometry smoothly caps off in the region corresponding to the infrared limit of the gauge theory. This arises in the time dependent problem via the tachyon condensate phase replacing the region of the geometry corresponding to the deep IR limit of the field theory.

For all these reasons, it is important to understand the physics of the tachyon condensate phase. The tachyon condensation process renders the background time-dependent; the linearized solution to the tachyon equation of motion yields an exponentially growing solution $T \propto \mu e^{\kappa X^0}$. As such there is no a priori preferred vacuum state. The simplest state to control is a state $|out >$ obtained by a Euclidean continuation in the target space, and describes a state in which nothing is excited in the far future when the tachyon dominates. This is a perturbative analogue of the Hartle-Hawking choice of state. At the worldsheet level (whose self-consistency we must check in each background to which we apply it), the tachyon condensation shifts the semiclassical action appearing in the path integrand. String amplitudes are given by

$$< \prod \int V >\sim \int DX^0 D\vec{X} e^{-S_E} \prod \int V \qquad (2)$$

where I work in conformal gauge and suppress the fermions and ghosts. Here X^μ are the embedding coordinates of the string in the target space and $\int V$ are the integrated vertex operators corresponding to the bulk asymptotic string states appearing in the amplitude. The semiclassical action in the Euclidean theory is

$$S_E = S_0 + \int d^2\sigma \mu^2 e^{2\kappa X^0} \hat{T}(\vec{X}) \qquad (3)$$

with S_0 the action without tachyon condensation and $\hat{T}(\vec{X})$ a winding (sine-Gordon) operator on the worldsheet. These amplitudes compute the components of the state $|out >$ in a basis of multiple free string states arising in the far past bulk spacetime when the tachyon is absent. The tachyon term behaves like a worldsheet potential energy term, suppressing contributions from otherwise singular regions of the path integration.

Before moving to summarize the full calculation of basic amplitudes, let me note two heuristic indications that the tachyon condensation effectively masses up degrees of freedom of the system. First, the tachyon term in (3) behaves like a spacetime dependent mass squared term in the analogue of this action arising in the case of a first quantized worldline action for a relativistic particle [8]. Second, the dependence of the tachyon term on the spatial variables $\vec{X}$ is via a relevant operator, dressed by worldsheet gravity (which in conformal gauge is encapsulated in the fluctuations of the timelike embedding coordinate X^0). The worldsheet renormalization group evolution with scale is different from the time dependent evolution, since fluctuations of X^0 contribute. However in some cases, such as localized tachyon condensates and highly supercritical systems, the two processes yield similar endpoints. In any case, as a heuristic indicator of the effect of tachyon condensation, the worldsheet RG suggests a massing up of degrees of freedom at the level of the worldsheet theory as time evolves forward.

Fortunately we do not need to rely too heavily on these heuristics, as the methods of Liouville field theory enable us to calculate basic physical quantities in the problem. In the Euclidean state defined by the above path integral, regulating the bulk contribution by cutting off X^0 in the far past at $ln\mu_*$, one finds a partition function Z with real part

$$Re(Z) = -\frac{ln(\mu/\mu_*)}{\kappa}\hat{Z}_{free} \tag{4}$$

This is to be compared with the result from non-tachyonic flat space $Z_0 = \delta(0)\hat{Z}_{free}$ [4], where $\delta(0)$ is the infinite volume of time, and $\hat{Z}_{free}$ is the rest of the partition function. In the tachyonic background (4), the first factor is replaced by a truncated temporal volume which ends when the tachyon turns on. A similar calculation of the two point function yields the Bogoliubov coefficients corresponding to a pure state in the bulk with thermal occupation numbers of particles, with temperature proportional to κ. This technique was first suggested in [8], where it was applied to bulk tachyons for which $\kappa \sim 1/l_s$ and the resulting total energy density blows up. In the examples of interest for singularities, the tachyon arises from a winding mode for which $\kappa \ll 1/l_s$, and the method [8] yields a self-consistently small energy density [4]. In the case of an initial singularity, this gives a perturbative string mechanism for the Hartle/Hawking idea of starting time from nothing. This timelike Liouville theory provides a perturbative example of "emergent time", in the same sense that spatial Liouville theory provides a worldsheet notion of "emergent space".[9]

So far this analysis applied to a particular vacuum. It is important to understand the status of other states of the system. In particular, the worldsheet path integral has a saddle point describing a single free string sitting in the tachyon phase. Do putative states such as this with extra excitations above the tachyon condensate

---

[9]This was also noted by M. Douglas in the discussion period in the session on emergent spacetime, in which G. Horowitz also noted existing examples. As explained by the speakers in that session, no complete *non-perturbative* formulation involving emergent time exists, in contrast to the situation with spatial dimensions where matrix models and AdS/CFT provide examples (but see [13] for an interesting example of a null singularity with a proposed non-perturbative description in terms of matrix theory).

constitute consistent states? This question is important for the problem of unitarity in black hole physics and in more general backgrounds where a tachyon condenses quasilocally, excising regions of ordinary spacetime. If nontrivial states persist in the tachyon phase in such systems, this would be tantamount to the existence of hidden remnants destroying bulk spacetime unitarity.

In fact, we find significant indications that the state where a string sits in the tachyon phase does not survive as a consistent state in the interacting theory [9][6]. The saddle point solution has the property that the embedding coordinate X^0 goes to infinity in finite worldsheet time τ. This corresponds to a hole in the worldsheet, which is generically not BRST invariant by itself. If mapped unitarily to another hole in the worldsheet obtained from a correlated negative frequency particle impinging on the singularity, worldsheet unitarity may be restored. This prescription is a version of the Horowitz/Maldacena proposal of a black hole final state [10]; the tachyon condensate seems to provide a microphysical basis for this suggestion.

A more dynamical effect which evacuates the tachyon region also arises in this system. A particle in danger of getting stuck in the tachyon phase drags fields (for example the dilaton and graviton) along with it. The heuristic model of the tachyon condensate as an effective mass for these modes [8] suggests that the fields themselves are getting heavy. The resulting total energy of the configuration, computed in [6] for a particle of initial mass m_0 coupled with strength λ to a field whose mass also grows at late times like $M(x^0)$, is

$$E = m_0^2 \lambda^2 M(x^0) \cos^2 \left(\int^{x^0} M(t') dt' \right) F(R) \tag{5}$$

This is proportional to a function $F(R)$ which increases with greater penetration distance R of the particle into the tachyon phase. Hence we expect a force on any configuration left in the tachyon phase which sources fields (including higher components of the string field). This does not mean every particle classically gets forced out of the tachyonic sector: for example in black hole physics, the partners of Hawking particles which fall inside the black hole provide negative frequency modes that correlate with the matter forming the black hole.

The analysis of this dynamical effect in generic states relies on the field-theoretic (worldsheet minisuperspace) model for tachyon dynamics. It is of interest to develop complete worldsheet techniques to analyze other putative vacua beyond the Euclidean vacuum. In the case of the Euclidean vacuum, the worldsheet analyses [8][4] reproduce the behavior expected from the heuristic model, so we have tentatively taken it as a reasonable guide to the physics in more general states as well.

The string-theoretic tachyon mode which drives the system away from the GR singularities is clearly accessible perturbatively. But it is important to understand whether the whole background has a self-consistent perturbative string description. In the Euclidean vacuum, this seems to be the case: the worldsheet amplitudes are shut off in the tachyon phase in a way similar to that obtained in spatial Liouville theory. In other states, it is not a priori clear how far the perturbative treatment extends. One indication for continued perturbativity is that according to the simple field theory model, every state gets heavy in the tachyon phase, including fluctuations of the dilaton, which may therefore be stuck at its bulk weak coupling value. It could be useful to employ AdS/CFT methods [11] to help decide this point.

3.5.1.2 *Discussion and Zoology*

Many timelike singularities are resolved in a way that involves new *light* degrees of freedom appearing at the singularity. In the examples reviewed in section 1, ordinary spacetime ends where the tachyon background becomes important. The tachyon at first constitutes a new light mode in the system, but its condensation replaces the would-be short-distance singularity with a phase where degrees of freedom ultimately become *heavy*. However, there are strong indications that there is a whole zoo of possible behaviors at cosmological spacelike singularities, including examples in which the GR singularity is replaced by a phase with *more* light degrees of freedom [12] (see [13] for an interesting null singularity where a similar behavior obtains).

Consider a spacetime with compact negative curvature spatial slices, for example a Riemann surface. The corresponding nonlinear sigma model is strongly coupled in the UV, and requires a completion containing more degrees of freedom. In supercritical string theory, the dilaton beta function has a term proportional to $D - D_{crit}$. The corresponding contribution in a Riemann surface compactification is $(2h - 2)/V \sim 1/R^2$ where V is the volume of the surface in string units, h the genus and R the curvature radius in string units. This suggests that there are effectively $(2h - 2)/V$ extra (supercritical) degrees of freedom in the Riemann surface case. Interestingly, this count of extra degrees of freedom arises from the states supported by the fundamental group of the Riemann surface.[10] For simplicity one can work at constant curvature and obtain the Riemann surface as an orbifold of Euclidean AdS_2, and apply the Selberg trace formula to obtain the asymptotic number density of periodic geodesics (as reviewed for example in [14]). This yields a density of states from a sum over the ground states in the winding sectors proportional to $e^{ml_s\sqrt{2h/V}}$ where m is proportional to the mass of the string state. Another check arises by modular invariance which relates the high energy behavior of the partition function to the lowest lying state: the system contains a light normalizable volume mode whose mass scales the right way to account for the modular transform of this density of states.

At large radius, the system is clearly two dimensional to a good approximation, and the 2d oscillator modes are entropically favored at high energy. It is interesting to contemplate possibility of cases where the winding states persist to the limit $V \to 1$, in which case the density of states from this sector becomes that of a $2h$

---

[10]I thank A. Maloney, J. McGreevy, and others for discussions on these points.

dimensional theory and the system crosses over to a very supercritical theory in which these states become part of the oscillator spectrum. In particular, states formed from the string wrapping generating cycles of the fundamental group in arbitrary orders (up to a small number of relations) constitute a $2h$-dimensional lattice random walk. At large volume, the lattice spacing is much greater than the string scale and the system is far from its continuum limit, so these states are a small effect. But at small Riemann surface volume it is an interesting possibility that these states cross over to the high energy spectrum of oscillator modes in $2h$ dimensions [12].

Of course as emphasized in [12], there are many possible behaviors at early times, including ones where the above states do not persist to small radius [3] and ones where they do persist but are part of a still larger system. One simple way to complete this sigma model is to extend it to a linear sigma model (containing more degrees of freedom) which flows to the Riemann surface model in the IR. Coupling this system to worldsheet gravity yields in general a complicated time dependent evolution, whose late time behavior is well described by the nonlinear sigma model on an expanding Riemann surface. If one couples this system to a large supercritical spectator sector, the time dependent evolution approaches the RG flow of the linear sigma model, which yields a controlled regime in which it is clear that at earlier times the system had more degrees of freedom.

Clearly a priori this can happen in many ways. In addition to the landscape of metastable vacua of string theory I believe the conservative expectation is that there will be a zoo of possible cosmological histories with similar late time behavior; indeed inflationary cosmology already has this feature. While it may be tempting to reject this possibility out of hand in hopes of a unique prediction for cosmological singularities, this seems to me much more speculative. However there are various indications that gravity may simplify in large dimensions (see e.g. [15]) and it would be interesting to try to obtain from this an organizing principle or measure applying to the plethora of cosmological singularities of this type.[11]

In any case, the singularities discussed in section 1, which are replaced by a phase of tachyon condensate, are simpler, appear more constrained [10][6], and apply more directly to black hole physics. It would be interesting to understand if there is any relation between black hole singularities and cosmological singularities.

Acknowledgements

I would like to thank the organizers for a very interesting conference and for the invitation to present these results. On these topics I have many people to thank, including my collaborators O. Aharony, A. Adams, G. Horowitz, X. Liu, A. Maloney, J. McGreevy, J. Polchinski, A. Saltman, and A. Strominger.

---

[11] as mentioned for example in J. Polchinski's talk in the cosmology session.

Bibliography

[1] A. Adams, J. Polchinski and E. Silverstein, JHEP **0110**, 029 (2001) [arXiv:hep-th/0108075].

[2] M. Headrick, S. Minwalla and T. Takayanagi, Class. Quant. Grav. **21**, S1539 (2004) [arXiv:hep-th/0405064]; E. J. Martinec, arXiv:hep-th/0305148.

[3] A. Adams, X. Liu, J. McGreevy, A. Saltman and E. Silverstein, JHEP **0510**, 033 (2005) [arXiv:hep-th/0502021].

[4] J. McGreevy and E. Silverstein, JHEP **0508**, 090 (2005) [arXiv:hep-th/0506130].

[5] G. T. Horowitz, JHEP **0508** (2005) 091 [arXiv:hep-th/0506166]. S. F. Ross, JHEP **0510** (2005) 112 [arXiv:hep-th/0509066].

[6] G. T. Horowitz and E. Silverstein, arXiv:hep-th/0601032.

[7] G. T. Horowitz and R. C. Myers, Phys. Rev. D **59** (1999) 026005 [arXiv:hep-th/9808079].

[8] A. Strominger and T. Takayanagi, Adv. Theor. Math. Phys. **7** (2003) 369 [arXiv:hep-th/0303221].

[9] S. Fredenhagen and V. Schomerus, JHEP **0312**, 003 (2003) [arXiv:hep-th/0308205].

[10] G. T. Horowitz and J. Maldacena, JHEP **0402**, 008 (2004) [arXiv:hep-th/0310281].

[11] L. Fidkowski, V. Hubeny, M. Kleban and S. Shenker, JHEP **0402** (2004) 014 [arXiv:hep-th/0306170].

[12] E. Silverstein, arXiv:hep-th/0510044.

[13] B. Craps, S. Sethi and E. P. Verlinde, JHEP **0510** (2005) 005 [arXiv:hep-th/0506180].

[14] N. L. Balazs and A. Voros, Phys. Rept. **143** (1986) 109.

[15] A. Strominger, "The Inverse Dimensional Expansion In Quantum Gravity," Phys. Rev. D **24**, 3082 (1981). A. Maloney, E. Silverstein and A. Strominger, arXiv:hep-th/0205316; Published in *Cambridge 2002, The future of theoretical physics and cosmology* 570-591 J. Polchinski, private discussions and rapporteur talk at this conference.

3.6   Discussion

T. Banks Generically, in your models you seem to have few initial states and many final states. What does unitarity mean in a context like that?

E. Silverstein Which model are you referring to specifically?

T. Banks The model where you use the Hartle-Hawking initial state.

E. Silverstein In that model, we merely determine what that state looks like in the bulk.

T. Banks I was under the impression that you were endorsing the fact that a Hartle-Hawking prescription picks out one initial state.

E. Silverstein I am not claiming that consistency picks out this one state, merely that perturbative time-dependent string theory with this initial state is well-defined.

T. Banks So there might be an infinite Hilbert space of consistent initial states?

E. Silverstein That is what I discussed in another part of my talk. There might only be a subset allowed by consistency.

3.7   Prepared Comments

3.7.1   *Thibault Damour: Cosmological Singularities and E_{10}*

Near spacelike singularity limit and a $SUGRA/[E_{10}/K(E_{10})]$ correspondence

The consideration of the *near horizon limit* (Maldacena) of certain black D-branes has greatly enriched our comprehension of string theory. In this limit, there emerges a *correspondence* between two seemingly different theories: 10-dimensional string theory in ADS spacetime on one side, and a lower-dimensional CFT on the other side. It is believed that this correspondence maps two different descriptions of the (continuation of the) same physics.

In recent years, the consideration of the *near spacelike singularity limit* of generic (classical) solutions of 10-dimensional string theories, or 11-dimensional supergravity, has suggested the existence of a *correspondence* between (say) 11-dimensional supergravity and a *one-dimensional $E_{10}/K(E_{10})$* nonlinear σ model [1]. If this correspondence were confirmed, it might provide both the basis of a new definition of M-theory, and a description of the 'de-emergence' of space near a cosmological singularity (where the 10-dimensional spatial extension would be replaced by the infinite number of coordinates of the $E_{10}/K(E_{10})$ coset space).

Cosmological billiards

The first hint of a correspondence $SUGRA/[E_{10}/K(E_{10})]$ (or, for short, $SUGRA/E_{10}$) emerged through the study, à la Belinskii-Khalatnikov-Lifshitz (BKL), of the structure of cosmological singularities in string theory and $SUGRA_{11}$ [2]. Belinskii, Khalatnikov and Lifshitz [3] introduced an approximate way of discussing the structure of generic cosmological singularities. The basic idea is that, near a spacelike singularity, the time derivatives are expected to dominate over spatial derivatives. More precisely, BKL found that spatial derivatives introduce terms in the equations of motion for the metric which are similar to the "walls" of a billiard table [3]. In an Hamiltonian formulation [4], [5] where one takes as basic gravitational variables the *logarithms* of the diagonal components of the metric, say β^a, these walls are Toda-like potential walls, i.e. exponentials of linear combinations of the β's, say $\exp - 2w(\beta)$, where $w(\beta) = \sum_a w_a \beta^a$. To each wall is therefore associated a certain *linear form* in β space, $w(\beta) = \sum_a w_a \beta^a$, and also a corresponding hyperplane $\sum_a w_a \beta^a = 0$. Ref. [2] found that the set of leading walls $w_i(\beta)$ entering the cosmological dynamics of $SUGRA_{11}$ or type-II string theories could be identified with the *Weyl chamber* of the hyperbolic Kac-Moody algebra E_{10} [6], i.e. the set of hyperplanes defined by the *simple roots* $\alpha_i(h)$ of E_{10}. Here h parametrizes a generic element of a Cartan subalgebra (CSA) of E_{10}, and the index i labels both the leading walls and the simple roots. [i takes r values, where r denotes the rank of the considered Lie algebra. For E_{10}, $r = 10$.] Let us also note that, for Heterotic and type-I string theories, the cosmological billiard is the Weyl chamber of another rank-10 hyperbolic Kac-Moody algebra, namely BE_{10}.

The appearance of E_{10} in the BKL behaviour of $SUGRA_{11}$ revived an old suggestion of B. Julia [7] about the possible role of E_{10} in a *one-dimensional reduction* of $SUGRA_{11}$. A posteriori, one can see the BKL behaviour as a kind of spontaneous reduction to one dimension (time) of a multidimensional theory. Note, however, that it is essential to consider a generic *inhomogeneous* solution (instead of a naively one-dimensionally reduced one) because the wall structure comes from the sub-leading ($\partial_x \ll \partial_t$) spatial derivatives.

Gradient expansion versus height expansion of the $E_{10}/K(E_{10})$ coset model

Refs. [1, 8] went beyond the leading BKL analysis of Ref. [2] by including the first three "layers" of spatial gradients modifying the zeroth-order *free billiard* dynamics defined by keeping only the time derivatives of the (diagonal) metric. This *gradient expansion* [5] can be graded by counting how many leading wall forms $w_i(\beta)$ are contained in the exponents of the sub-leading potential walls associated to these higher-order spatial gradients. As further discussed below, it was then found that this counting could be related (up to height 29 included) to the grading defined by the *height* of the roots entering the Toda-like Hamiltonian walls of the dynamics defined by the motion of a massless particle on the coset space $E_{10}/K(E_{10})$, with action

$$S_{E_{10}/K(E_{10})} = \int \frac{dt}{n(t)} (v^{\text{sym}} | v^{\text{sym}}). \tag{1}$$

Here, $v^{\text{sym}} \equiv \frac{1}{2}(v + v^T)(\equiv P$ in [6]) is the symmetric part of the "velocity" $v \equiv (dg/dt)g^{-1}$ of a group element $g(t)$ running over E_{10}. The transpose operation

T is the negative of the Chevalley involution ω, so that the elements of the Lie algebra of $K(E_{10})$ are "T-antisymmetric", $k^T = -k$ (which is equivalent to them being fixed under ω: $\omega(k) = +k$).

Current tests of the $SUGRA/E_{10}$ correspondence

An E_{10} group element $g(t)$ is parametrized by an infinite number of coordinates. When decomposing (the Lie algebra of) E_{10} with respect to (the Lie algebra of) the $GL(10)$ subgroup defined by the horizontal line in the Dynkin diagram of E_{10}, the various components of $g(t)$ can be graded by their $GL(10)$ level ℓ. At the $\ell = 0$ level $g(t)$ is parametrized by a $GL(10)$ matrix k^i_j, to which is associated (in the coset space $GL(10)/SO(10)$) a symmetric matrix $g^{ij} = (e^k)^i_s (e^k)^j_s$. [The indices $i, j, \cdots$ take ten values.] At the level $\ell = 1$, one finds a 3-form A_{ijk}. At the level $\ell = 2$, a 6-form A_{ijklmn}, and at the level $\ell = 3$ a 9-index object A_9 with Young-tableau symmetry $\{8, 1\}$.

The coset action (1) then defines a coupled set of equations of motion for $g_{ij}(t), A_{ijk}(t), A_{i_1 i_2 \cdots i_6}(t), A_{i_1 i_2 \cdots i_9}(t), \cdots$. By explicit calculations, it was shown that these coupled equations of motion could be identified, modulo terms which correspond to potential walls of height at least 30, to the $SUGRA_{11}$ equations of motion. This identification between the coset dynamics and the $SUGRA_{11}$ one is obtained by means of a *dictionary* which maps: (1) $g_{ij}(t)$ to the spatial components of the 11-dimensional metric $G_{ij}(t, \mathbf{x}_0)$ in a certain coframe (Ndt, θ^i), (2) $\dot{A}_{ijk}(t)$ to the mixed temporal-spatial ('electric') components of the 11-dimensional field strength $\mathcal{F} = d\mathcal{A}$ in the same coframe, (3) the conjugate momentum of $A_{i_1 i_2 \cdots i_6}(t)$ to the dual (using $\epsilon^{i_1 i_2 \cdots i_{10}}$) of the spatial ('magnetic') frame components of $\mathcal{F} = d\mathcal{A}$, and (4) the conjugate momentum of $A_{i_1 i_2 \cdots i_9}(t)$ to the ϵ^{10} dual (on jk) of the structure constants C^i_{jk} of the coframe θ, i.e. $d\theta^i = \frac{1}{2} C^i_{jk} \theta^j \wedge \theta^k$. Here all the SUGRA field variables are considered at some fixed (but arbitrary) spatial point $\mathbf{x}_0$.

The fact that at levels $\ell = 2$ (A_6), and $\ell = 3$ (A_9) the dictionary between SUGRA and coset variables is such that the first spatial gradients of the SUGRA variables $G, \mathcal{A}$ are mapped onto (time derivatives of) coset variables suggested the conjecture that the infinite tower of coset variables could fully encode all the spatial derivatives of the SUGRA variables, thereby explaining how a one-dimensional coset dynamics could correspond to an 11-dimensional one. Some evidence for this conjecture comes from the fact that among the infinite number of generators of E_{10} there do exist towers of generators that have the appropriate $GL(10)$ index structure for representing the infinite sequence of spatial gradients of the various SUGRA variables.

It is not known how to extend this dictionary beyond the level $\ell = 3$ (corresponding to A_9). The difficulty in extending the dictionary might be due (similarly to what happens in the ADS/CFT case) to the non-existence of a common domain of validity for the two descriptions.

However, Ref. [9] found evidence for a nice compatibility between some high-level contributions in the coset action, corresponding to *imaginary* roots ($(\alpha, \alpha) < 0$, by contrast to the roots that entered the checks [1, 8] which were all "real" with $(\alpha, \alpha) = +2$), and M-theory one-loop corrections to $SUGRA_{11}$, notably the terms quartic in the curvature tensor. This finding suggests new ways of testing the conjecture by looking at the structure of higher loop terms. [See also [10] for a different approach to the possible role of the imaginary roots of E_{10}.]

Two recent studies of the fermionic sector of $SUGRA_{11}$ have also found a nice compatibility between $SUGRA_{11}$ and the extension of the (bosonic) massless particle action (1) to an action describing the (supersymmetric) dynamics of a *massless spinning particle* on $E_{10}/K(E_{10})$ [11, 12]. In this extension $K(E_{10})$ plays the role of a generalized 'R symmetry'.

Conclusion

Much work, and probably new tools, are needed to establish the conjectured correspondence between $SUGRA_{11}$, or hopefully M-theory, and the dynamics of a (quantum) massless spinning particle on the coset space $E_{10}/K(E_{10})$. It is, however, interesting to speculate that, as one approaches a cosmological singularity, space 'de-emerges' in the sense that the 11-dimensional description of $SUGRA_{11}/M$-theory gets replaced (roughly when the curvature exceeds the – 11-dimensional– Planck scale) by a 1-dimensional $E_{10}/K(E_{10})$ coset model (where the only remaining dimension is timelike).

Acknowledgments: It is a pleasure to thank my dear friends and collaborators Marc Henneaux and Hermann Nicolai for exciting interactions over several years.

Bibliography

[1] T. Damour, M. Henneaux and H. Nicolai, "E_{10} and a small tension expansion of M theory," Phys. Rev. Lett. **89**, 221601 (2002) [arXiv:hep-th/0207267].

[2] T. Damour and M. Henneaux, "E(10), BE(10) and arithmetical chaos in superstring cosmology," Phys. Rev. Lett. **86**, 4749 (2001) [arXiv:hep-th/0012172].

[3] V.A. Belinskii, I.M. Khalatnikov and E.M. Lifshitz, "Oscillatory approach to a singular point in the relativistic cosmology," Adv. Phys. **19**, 525 (1970).

[4] C. W. Misner, "Quantum Cosmology. 1," Phys. Rev. **186**, 1319 (1969); "Minisuperspace," in: J R Klauder, Magic Without Magic, San Francisco 1972, 441-473.

[5] T. Damour, M. Henneaux and H. Nicolai, "Cosmological Billiards", Class. Quant. Grav. **20** (2003) R145–R200 [arXiv: hep-th/0212256].

[6] For an introduction to E_{10}, and its maximally compact subgroup $K(E_{10})$, see, in these proceedings, the prepared comment of H. Nicolai in the *Mathematical Structures* session.

[7] B. Julia, in: Lectures in Applied Mathematics, Vol. 21 (1985), AMS-SIAM, p. 335; preprint LPTENS 80/16.

[8] T. Damour and H. Nicolai, "Eleven dimensional supergravity and the $E_{10}/ K(E_{10})$ σ-model at low A_9 levels", in: Group Theoretical Methods in Physics, Institute of Physics Conference Series No. 185, IoP Publishing, 2005 [arXiv: hep-th/0410245].

[9] T. Damour and H. Nicolai, "Higher order M theory corrections and the Kac-Moody algebra E_{10}", Class. Quant. Grav. **22** (2005) 2849 [arXiv: hep-th/0504153].

[10] J. Brown, O. J. Ganor and C. Helfgott, "M-theory and E(10): Billiards, branes, and imaginary roots," JHEP **0408** (2004) 063 [arXiv:hep-th/0401053].

[11] T. Damour, A. Kleinschmidt and H. Nicolai, "Hidden symmetries and the fermionic sector of eleven-dimensional supergravity," Phys. Lett. B 634 (2006) 319–324 [arXiv:hep-th/0512163].

[12] S. de Buyl, M. Henneaux and L. Paulot, "Extended E(8) invariance of 11-dimensional supergravity," [arXiv:hep-th/0512292].

3.8   Discussion

J. Harvey There is a five-dimensional supergravity formally similar to eleven-dimensional supergravity, with a one-form potential with Chern-Simons term. Do you know whether there is a similar algebraic structure associated to that supergravity in this particular limit?

T. Damour Yes, Kac-Moody algebras can generically be associated to supergravities, and even to ordinary gravitational theories in various dimensions.

A. Ashtekar If I consider the coset space dynamics as being fundamental, where is $\hbar$?

T. Damour It is encoded in the eleven-dimensional Planck length.

A. Ashtekar It appears in the metric?

T. Damour: Yes, for instance R^4 terms in the action are made dimensionless using the Planck length.

F. Englert I am wondering whether your theory is not simply valid near a space-like singularity for different reasons, namely that near a space-like singularity you essentially only have the time-direction. After all, if you dimensionally reduce gravity to one dimension you expect E_{10} as a symmetry. Near a singularity this is verified because the problem becomes essentially one-dimensional.

T. Damour The singularity is used here as a tool to reveal a symmetry structure which exists independently. Recall that here indeed one find E_{10} starting from any dimension. One never sees E_7, or other duality groups appear.

Prepared comment by N. Turok.

E. Silverstein Wound strings might not get blue-shifted at the singularity, but other modes will?

N. Turok Other modes will be non-perturbative states like black holes, etc. If perturbative gravity goes through the singularity, I would already be delighted. Let us deal with the non-perturbative states later. If you work out are going to be created at the crunch, the creation rate is proportional to θ_0 which can be taken small.

E. Silverstein They will not be created, but what if you send one in?

N. Turok If you send a black hole in ...

E. Silverstein Just send a perturbative mode in.

Answer by N. Turok An initial particle? A pragmatic answer is that the density of particles going in will be negligible. Still one should worry.

A. Polyakov There is a comment I wanted to make, concerning singularities. When we try to treat the Big Bang, Big Crunch singularities in string theory, you write down a sigma-model and then try a one-loop approximation and you find the Einstein equation with Friedman type solution. From the point of view of the sigma-model, the singularity in such a solution is nothing but the Landau-pole, which follows from the renormalization group flow at one loop. In many cases, the singularity can be resolved as e.g. in the $O(3)$ sigma-model (which becomes massive). The Big Bang would then not be so big after all.

3.9 Prepared Comments

3.9.1 *Abhay Ashtekar: Singularities: quantum nature of the big bang in loop quantum gravity*

3.9.1.1 *Introduction*

A central feature of general relativity is that gravity is encoded in the very geometry of space-time. Loop quantum gravity is a non-perturbative approach to unifying general relativity with quantum physics which retains this interplay between geometry and gravity [1, 2]. There is no background space-time; matter as well as geometry are 'quantum from birth'. Effects of quantum geometry are negligible under ordinary circumstances but they dominate near singularities. There, quantum space-time is dramatically different from the smooth continuum of general relativity. In particular, quantum geometry effects have led to a natural resolution of space-like singularities in a number of mini and midi-superspaces. These encompass both black hole and cosmological contexts.

In the cosmological setting, there are several long-standing questions that have been relegated to quantum gravity. Examples are:

• How close to the big-bang does a smooth space-time of general relativity make sense? In particular, can one show from first principles that this approximation is valid at the onset of inflation?
• Is the Big-Bang singularity naturally resolved by quantum gravity? Or, is some external input such as a new principle or a boundary condition at the Big Bang essential?
• Is the quantum evolution across the 'singularity' deterministic? One needs a fully non-perturbative framework to answer this question in the affirmative. (In the Ekpyrotic and Pre-Big-Bang scenarios, for example, the answer is in the negative.)
• If the singularity is resolved, what is on the 'other side'? Is there just a *quantum foam* far removed from any classical space-time, or, is there another large, classical universe?

Using loop quantum gravity, these and related questions have been answered in detail in several models by combining analytic and numerical methods.

3.9.1.2 *Novel features of loop quantum cosmology*

Quantum cosmology is an old subject. It was studied extensively in the framework of geometrodynamics where quantum states are taken to be functions of 3-geometries and matter fields. In the cosmological context, the wave functions $\Psi(a, \phi)$ depend on the scale factor a and the matter field ϕ. They are subject to a quantum constraint, called the Wheeler-DeWitt equation. Initially, it was hoped that the quantum evolution dictated by this equation would resolve classical singularities. Unfortunately, this hope was not realized. For example in the simplest of homogeneous isotropic models, if one begins with a semi-classical state at late times and evolves it back via Wheeler DeWitt equation, one finds that it just follows the classical trajectory into the big bang singularity.

Loop quantum gravity is based on spin-connections rather than metrics and is thus closer in spirit to gauge theories. The basic dynamical variables are holonomies h of a gravitational spin-connection A and electric fields E canonically conjugate to these connections. However, the E's now have a dual, geometrical interpretation: they represent orthonormal triads which determine the Riemannian geometry. Thanks to the contributions from 2 dozen or so groups since the mid-nineties, the subject has reached a high degree of mathematical precision [1]. In particular, it has been shown that the fundamental quantum algebra based on h's and E's admits a *unique* diffeomorphism covariant representation [3]. From the perspective of Minkowskian field theories, this result is surprising and brings out the powerful role of the requirement of diffeomorphism covariance (i.e., background independence). In this representation, there are well-defined holonomy operators $\hat{h}$ but *there is no operator $\hat{A}$ corresponding to the connection itself.* The second key feature is that Riemannian geometry is now quantized: there are well-defined operators corresponding to, say, lengths, areas and volumes, and all their eigenvalues are discrete.

In quantum cosmology, one deals with symmetry reduced models. However, in loop quantum cosmology, quantization is carried out by closely mimicking the procedure used in the full theory, and the resulting theory turns out to be qualitatively different from the Wheeler DeWitt theory. Specifically, because only the holonomy operators are well-defined and there is no operator corresponding to the connection itself, the von-Neumann uniqueness theorem is by-passed. A new representation of the algebra generated by holonomies and triads becomes available. *We have new quantum mechanics.* In the resulting theory, the Wheeler-DeWitt differential equation is replaced by a *difference* equation (Eq (1) below), the size of the step being dictated by the first non-zero area eigenvalue —i.e., the 'area gap'— in quantum geometry. Qualitative differences from the Wheeler-DeWitt theory emerge precisely near the big-bang singularity. Specifically, the evolution does not follow the classical trajectory. Because of quantum geometry effects, gravity becomes *repulsive* near the singularity and there is a quantum bounce.

3.9.1.3 *A Simple model*

I will now illustrate these general features through a simple model: Homogeneous, isotropic $k = 0$ cosmologies with a zero rest mass scalar field. Since there is no potential in this model, the big-bang singularity is inevitable in the classical theory. The momentum p_ϕ of the scalar field is a constant of motion and for each value of

p_ϕ, there are two trajectories: one starting out at the Big Bang and expanding and the other contracting in to a Big Crunch, each with a singularity.

Classical dynamics suggests that here, as well as in the closed models, one can take the scalar field as an *internal clock* defined by the system itself —unrelated to any choice of coordinates or a background space-time. This idea can be successfully transported to quantum theory because the Hamiltonian constraint equation

$$\frac{\partial^2 \Psi}{\partial \phi^2} = C^+(v)\Psi(v+4,\phi) + C^o(v)\Psi(v,\phi) + C^-(v)\Psi(v-4,\phi) \tag{1}$$

'evolves' the wave functions $\Psi(v,\phi)$ with respect to the internal time ϕ. (Here v is the oriented volume of a fixed fiducial cell in Planck units, so $v \sim \pm(\text{scale factor})^3$, and $C^\pm, C^o$ are simple algebraic functions on v.) The detailed theory is fully compatible with this interpretation. Thus, this simple model provides a concrete realization of the *emergent time* scenario, discussed in another session of this conference.

A standard ('group averaging') procedure enables one to introduce a natural Hilbert space structure on the space of solutions to the Hamiltonian constraint. There are complete sets of Dirac observables using which one can rigorously construct semi-classical states and follow their evolution. Since we do not want to prejudice the issue by stating at the outset what the wave function should do at the singularity, let us specify the wave function at late time —say *now*— and take it to be sharply peaked at a point on the expanding branch. Let us use the Hamiltonian constraint to evolve the state backwards towards the classical singularity. Computer simulations show that the state remains sharply peaked on the classical trajectory till very early times, when the density becomes comparable to the Planck density. The fluctuations are all under control and we can say that the the continuum space-time of general relativity is an excellent approximation till this very early epoch. In particular, space-time can be taken to be classical at the onset of standard inflation. But in the Planck regime the fluctuations are significant and there is no unambiguous classical trajectory. This is to be expected. But then something unexpected happens. The state re-emerges on the other side again as a semi-classical state, now peaked on a contracting branch. Thus, in the Planck regime, although there are significant quantum fluctuations, we do not have a quantum foam on the other side. Rather, there is a quantum bounce. *Quantum geometry in the Planck regime serves as a bridge between two large classical universes.* The fact that the state is again semi-classical in the past was unforeseen and emerged from detailed numerical simulations [4]. However, knowing that this occurs, one can derive an effective modification of the Friedmann equation: $(\dot{a}/a)^2 = (8\pi G/3)\,\rho\,[1 - \rho/\rho_\star] + \text{higher order terms}$, where ρ is the matter density and $\rho_\star$, the critical density, is given by $\rho_\star = \text{const}\,(1/8\pi G\Delta)$, Δ being the smallest non-zero eigenvalue of the area operator. The key feature is that, without any extra input, the quantum geometry correction naturally comes with a *negative* sign making gravity repulsive in the Planck regime, giving rise to the bounce. The correction is completely negligible when the matter density is very small compared to the Planck density, i.e., when the universe is large. Finally, a key consequence of (1) is that the quantum evolution is deterministic across the 'quantum bridge'; no new input was required to 'join' the two branches. This is because, thanks to quantum geometry, one can treat the Planck regime fully non-perturbatively, without any need of a classical background geometry.

The singularity resolution feature is robust for the mini and midi-superspace models we have studied so far *provided* we use background independent description and quantum geometry. For example, in the anisotropic case, the evolution is again non-singular if we treat the full model non-perturbatively, using quantum geometry. But if one treats anisotropies as perturbations using the standard, Wheeler-DeWitt type Hilbert spaces, the perturbations blow up and the singularity is not resolved. Finally, the Schwarzschild singularity has also been resolved. This resolution suggests a paradigm for the black hole evaporation process which can explain why there is no information loss in the setting of the physical, Lorentzian space-times [5].

To summarize, quantum geometry effects have led to a resolution of a number of space-like singularities showing that quantum space-times can be significantly larger than their classical counterparts. These results have direct physical and conceptual ramifications. I should emphasize however that so far the work has been restricted to mini and midi superspaces and a systematic analysis of generic singularities of the full theory is still to be undertaken.

Bibliography

[1] A. Ashtekar and J. Lewandowski, *Background independent quantum gravity: A status report*, Class. Quant. Grav. **21** (2004) R53-R152, gr-qc/0404018.
[2] C. Rovelli *Quantum Gravity*, (CUP, Cambridge, 2004).
[3] J. Lewandowski, A. Okolow, H. Sahlmann and T. Thiemann, *Uniqueness of diffeomorphism invariant states on holonomy flux algebras*, gr-qc/0504147.
[4] A. Ashtekar, T. Pawlowski and P. Singh, *Quantum nature of the big bang* (IGPG preprint); *Quantum nature of the big bang: An analytical and numerical investigation, I and II* (IGPG pre-prints).
[5] A. Ashtekar and M. Bojowald, *Black hole evaporation: A paradigm*, Class. Quant. Grav. **22** (2005) 3349-3362, gr-qc/0504029; *Quantum geometry and the Schwarzschild singularity*, Class. Quant. Grav. **2** (2006) 391-411, gr-qc/0509075.

3.10   Discussion

S. Kachru Through much of your discussion, you worked with a truncated Hilbert
space. What are your reasons for believing that approximation in the region
where the curvature is large?

A. Ashtekar In some models that we have completely worked through, one can
indeed see that one can go through the singularity, but, yes, in general your
question is very relevant.

T. Banks The loop quantum gravity equation shares with the Wheeler-De Witt
equation its hyperbolic nature. The set of solutions then does not allow for a
positive definite metric and yet you talk of Hilbert spaces. Could you comment?

A. Ashtekar: In this simple model, there is a Hilbert space and the analogue of
the Wheeler-De Witt equation looks like a Klein-Gordon equation. One treats
this theory then similarly to the Klein-Gordon equation, and one can define as
in that case, a positive definite inner product.

T. Banks Is the evolution in your model unitary?

A. Ashtekar In the initial region ϕ can be chosen as time and the evolution is
unitary in time. In the intermediate (crunch) region time is not well-defined,
but one can still choose the value of ϕ as denoting time. If one does this, the
evolution can still be considered unitary.

General Discussion on Singularities starts.

M. Douglas Two questions I would like to pose are: 1. For infinite time scenarios,
why does entropy not increase eternally? and 2. Does the bounce not give less
predictivity than an initial state?

G. Veneziano On question 1: the entropy after the Big Bang satisfies the holographic upper bound. A pre-Big Bang seems to be what is necessary to satisfy
the bound, and otherwise, this amount of entropy would be difficult to understand. Initial entropy would not seem to be a problem for bouncing cosmologies.

A. Strominger A closely related issue is degrees of freedom. We would like to
think that there is one degree of freedom per Planck volume and not an infinite
number as in field theory. A Big Bang could be thought of as a point before
which there are no degrees of freedom. A Big Bounce seems to say that there
are only few degrees of freedom finally. Having a universe of microscopic size
would not change that conclusion much. So what do people advancing a Big
Bounce have to say about having a large number of degrees of freedom today?

G. Horowitz You are assuming a closed universe when referring to it as being
small?

A. Strominger Yes, I guess one could allow for an infinite volume bounce.

T. Banks I do not understand the issue of having a small number of degrees of
freedom at the Big Bang. One can have a unitary evolution that stops at a
given time if one has a time-dependent Hamiltonian. The issue with the Big
Bang is that there is a particle horizon and that only a small number of degrees

of freedom are correlated and interact with each other, but one can allow for many more, non-interacting degrees of freedom. I do believe there is an issue (raised by R. Penrose) with why the initial state had little entropy and why the thermodynamical and cosmological arrow of time coincide.

Question by S. Weinberg (asking for a prediction of the ekpyrotic universe)

N. Turok A prediction of our [ekpyrotic] model (compared to inflation) is that non-gaussianities are very highly suppressed.

S. Weinberg So, non-gaussian perturbations would rule out the model?

N. Turok Yes, as would tensor perturbations.

P. Steinhardt The tensor perturbations would not be precisely zero, just exponentially suppressed. The reason is that the Hubble constant at the time the perturbations are generated is exponentially small compared to today. This is possible because the fluctuations are not caused by rapid expansion or contraction, as they are for inflation, but, instead, by a different instability that occurs in a modulus field as it rolls down its exponentially steep potential.

S. Weinberg That confuses me. In the usual inflationary scenario the Hubble constant sets the scale of the perturbations we see today and we believe the primordial non-gaussianity is small because we see small perturbations now and therefore the Hubble constant in Planck units at the time of horizon exit in the usual picture is about 10^{-5}. So, in your scenario, are not the perturbations themselves very small?

P. Steinhardt No, because they are not produced by gravity itself, but by the potential term in the model.

S. Weinberg That is clear.

A. Ashtekar Is it possible to have a heuristic picture of which singularities are resolved in string theory?

G. Horowitz There are time-like, space-like and null singularities which have been resolved in string theory, with different mechanisms. I do not have a criterium to say which singularities will be resolved and which not.

A. Ashtekar Is there a heuristic, or an intuition? For example, is it the case that in the resolved cases the total energy was always positive?

G. Horowitz Yes, in the usual asymptotically flat context that is so.

A. Linde From my perspective, the theory of a pre-Big Bang is a kind of inflation, but not sufficiently good to solve flatness and horizon problems. Perhaps it works, but not by itself, but perhaps by adding on usual inflation afterwards. But then, who cares about the bounce? On the ekpyrotic scenario, I would like to say that we studied it. It may be possible to have a tachyonic instability before the singularity to produce fluctuations, but producing fluctuations is not the main difficulty. The main problem is how to make the universe isotropic, homogeneous, flat, ... In order to achieve this, the ekpyrotic scenario also uses a long stage of exponential expansion, as in inflation. What is different in inflation is that inflationary theory protects one from having to think about the singularity, whether or not there was a bounce.

P. Steinhardt Your description of the situation is a misconception. The accelerated expansion that occurs in the cyclic model is not necessary to make the universe smooth and flat. In fact, it may be that there are only a few e-folds of dark energy domination in our future before the contraction begins. The key in our scenario is the slowly contracting phase with very high pressure, which, it turns, smoothes and flattens the universe as well. So, we have learned from these studies that there are two ways of obtaining a flat, homogeneous and isotropic universe: a rapidly expanding phase with w near -1, conventional inflation, or a slowly contracting phase with $w > 1$.

W. Fischler It is always said that inflation solves the homogeneity and flatness problems, however, that is put in by hand. At the moment where inflation starts, homogeneity of the inflaton at scales larger than the causal region at that point in time, is fed into the model.

Session 4

Mathematical Structures

Chair: *Hiroshi Ooguri*, Caltech, USA
Rapporteur: *Robbert Dijkgraaf*, University of Amsterdam, the Netherlands
Scientific secretaries: *Philippe Spindel* (Université de Mons-Hainaut) and *Laurent Houart* (Université Libre de Bruxelles)

4.1 Rapporteur talk: Mathematical Structures, by Robbert Dijkgraaf

4.1.1 *Abstract*

The search for a quantum theory of gravity has stimulated many developments in mathematics. String theory in particular has had a profound impact, generating many new structures and concepts that extend classical geometry and give indications of what a full theory of quantum gravity should entail. I will try to put some of these ingredients in a broader mathematical context.

4.1.2 *Quantum Theory and Mathematics*

Over the years the search for a theory of quantum gravity has both depended on and enriched many fields of mathematics. String theory [1] in particular has had an enormous impact in mathematical thinking. Subjects like algebraic and differential geometry, topology, representation theory, infinite dimensional analysis and many others have been stimulated by new concepts such as mirror symmetry [2], [3], quantum cohomology [4] and conformal field theory [5]. In fact, one can argue that this influence in mathematics will be a lasting and rewarding impact of string theory, whatever its final role in fundamental physics. String theory seem to be the most complex and richest mathematical object that has so far appeared in physics and the inspiring dialogue between mathematics and physics that it has triggered is blooming and spreading in wider and wider circles.

This synergy between physics and mathematics that is driving so many developments in modern theoretical physics, in particularly in the field of quantum geometry, is definitely not a new phenomenon. Mathematics has a long history of drawing inspiration from the physical sciences, going back to astrology, architecture and land measurements in Babylonian and Egyptian times. Certainly this reached a high point in the 16th and 17th centuries with the development of what we now call classical mechanics. One of its leading architects, Galileo, has given us the famous image of the "Book of Nature" in *Il Saggiatore*, waiting to be decoded by scientists

> *Philosophy is written in this grand book, the universe, which stands continually open to our gaze. But the book cannot be understood unless one first learns to comprehend the language and read the characters in which it is written. It is written in the language of mathematics, and its characters are triangles, circles, and other geometric figures without which it is humanly impossible to understand a single word of it; without these one is wandering in a dark labyrinth.*

This deep respect for mathematics didn't disappear after the 17th century. Again in the beginning of the last century we saw again a wonderful intellectual union of physics and mathematics when the great theories of general relativity and quantum mechanics were developed. In all the centers of the mathematical world this was closely watched and mathematicians actively participated. If anywhere this was so in Göttingen, where Hilbert, Minkowksi, Weyl, Von Neumann and many other mathematicians made important contributions to physics.

Theoretical physics have always been fascinated by the beauty of their equations. Here we can even quote Feynman, who was certainly not known as a fine connoisseur of higher abstract mathematics[1]

> *To those who do not know mathematics it is difficult to get across a real feeling as to the beauty, the deepest beauty, of nature ... If you want to learn about nature, to appreciate nature, it is necessary to understand the language that she speaks in.*

But despite the warm feelings of Feynman, the paths of fundamental physics and mathematics started to diverge dramatically in the 1950s and 1960s. In the struggle with all the new subatomic particles physicists were close to giving up the hope of a beautiful underlying mathematical structure. On the other hand mathematicians were very much in an introspective mode these years. Because the fields were standing back to back, Dyson famously stated in his Gibbs Lecture in 1972:

> *I am acutely aware of the fact that the marriage between mathematics and physics, which was so enormously fruitful in past centuries, has recently ended in divorce.*

But this was a premature remark, since just at time the Standard Model was being born. This brought geometry in the form of non-abelian gauge fields, spinors

---

[1]But then Feynman also said "If all mathematics disappeared today, physics would be set back exactly one week." One mathematician's answer to this remark was: "This was the week God created the world."

and topology back to forefront. Indeed, it is remarkable fact, that all the ingredients of the standard model have a completely natural mathematical interpretation in terms of connections, vector bundles and Clifford algebras. Soon mathematicians and physicists started to build this dictionary and through the work of Atiyah, Singer, 't Hooft, Polyakov and many others a new period of fruitful interactions between mathematics and physics was born.

Remarkably, the recent influx of ideas from quantum theory has also led to many new developments in pure mathematics. In this regards one can paraphrase Wigner [6] and speak of the "unreasonable effectiveness of quantum physics in mathematics." There is a one immediate reason why quantum theory is so effective. Mathematics studies abstract patterns and structures. As such it has a hierarchical view of the world, where things are first put in broadly defined categories and then are more and more refined and distinguished. In topology one studies spaces in a very crude fashion, whereas in geometry the actual shape of a space matters. For example, two-dimensional (closed, connected, oriented) surfaces are topologically completely determined by their genus or number of handles $g = 0, 1, 2, \ldots$ So we have one simple topological invariant g that associates to each surface a non-negative number

$$g : \{Surfaces\} \rightarrow \mathbb{Z}_{\geq 0}.$$

More complicated examples are the knot invariants that distinguish embeddings of a circle in $\mathbb{R}^3$ up to isotopy. In that case there are an infinite number of such invariants

$$Z : \{Knots\} \rightarrow \mathbb{C}.$$

But in general such invariants are very hard to come by – the first knot invariant was discovered by J.W. Alexander in 1923, the second one sixty years later by V. Jones.

Quantum physics, in particular particle and string theory, has proven to be a remarkable fruitful source of inspiration for new topological invariants of knots and manifolds. With hindsight this should perhaps not come as a complete surprise. Roughly one can say that quantum theory takes a geometric object (a manifold, a knot, a map) and associates to it a (complex) number, that represents the probability amplitude for a certain physical process represented by the object. For example, a knot in $\mathbb{R}^3$ can stand for the world-line of a particular particle and a manifold for a particular space-time geometry. So the rules of quantum theory are perfectly set up to provide invariants.

Once we have associated concrete numbers to geometric objects one can operate on them with various algebraic operations. In knot theory one has the concept of relating knots through recursion relations (skein relations) or even differentiation (Vassiliev invariants). In this very general way quantization can be thought of as a map (functor)

$$Geometry \rightarrow Algebra.$$

that brings objects out the world of geometry into the real of algebra. This often gives powerful new perspectives, as we will see in a few examples later.

4.1.2.1   *String theory and mathematics*

First of all, it must be said that the subject of *Quantum Field Theory* (QFT) is already a powerful source for mathematical inspiration. There are important challenges in constructive and algebraic QFT: for example, the rigorous construction of four-dimensional asymptotically free non-abelian gauge theories and the establishment of a mass gap (one of the seven Millennium Prize problems of the Clay Mathematics Institute [7]).

Even in perturbative QFT there remain many beautiful mathematical structures to be discovered. Recently, a surprisingly rich algebraic structures has been discovered in the combinatorics of Feynman diagrams by Connes, Kreimer and others, relating Hopf algebras, multiple zeta-functions, and various notions from number theory [8]. Also the reinvigorated program of the twistor reformulation of (self-dual) Yang-Mills and gravity theories should be mentioned [9]. This development relates directly to the special properties of so-called MHV (maximal helicity violating) amplitudes and many other hidden mathematical structures in perturbative gauge theory [10].

In fact, a much deeper conceptual question seems to underlie the formulation of QFT. Modern developments have stressed the importance of quantum dualities, special symmetries of the quantum system that are not present in the classical system. These dualities can relate gauge theories of different gauge groups (*e.g.* Langlands dual gauge groups in the $\mathcal{N} = 4$ supersymmetric Yang-Mills theory [11]) and even different matter representations (Seiberg duality [12]). All of this points to the conclusion that a QFT is more than simply the quantization of a classical (gauge) field and that even the path-integral formulation is at best one particular, duality-dependent choice of parametrization. This makes one wonder whether a formulation of QFT exist that is manifest duality invariant.

Although the mathematical aspects of quantum field theory are far from exhausted, it is fair to say, I believe, that the renewed bond between mathematics and physics has been greatly further stimulated with the advent of string theory. There is quite a history of developing and applying of new mathematical concepts in the "old days" of string theory, leading among others to representations of Kac-Moody and Virasoro algebras, vertex operators and supersymmetry. But since the seminal work of Green and Schwarz in 1984 on anomaly cancellations, these interactions have truly exploded. In particular with the discovery of Calabi-Yau manifolds as compactifications of the heterotic strings with promising phenomenological prospectives by the pioneering work of Witten and others, many techniques of algebraic geometry entered the field.

Most of these developments have been based on the perturbative formulation of string theory, either in the Lagrangian formalism in terms of maps of Riemann surfaces into manifolds or in terms of the quantization of loop spaces. This perturbative approach is however only an approximate description that appears for small values of the quantization parameter.

Recently there has been much progress in understanding a more fundamental description of string theory that is sometimes described as M-theory. It seems to unify three great ideas of twentieth century theoretical physics and their related mathematical fields:

- General relativity; the idea that gravity can be described by the Riemannian geometry of space-time. The corresponding mathematical fields are topology, differential and algebraic geometry, global analysis.
- Gauge theory; the description of forces between elementary particles using connections on vector bundles. In mathematics this involves the notions of K-theory and index theorems and more generally non-commutative algebra.
- Strings, or more generally extended objects (branes) as a natural generalization of point particles. Mathematically this means that we study spaces primarily through their (quantized) loop spaces. This relates naturally to infinite-dimensional analysis and representation theory.

At present it seems that these three independent ideas are closely related, and perhaps essentially equivalent. To some extent physics is trying to build a dictionary between geometry, gauge theory and strings. From a mathematical perspective it is extremely interesting that such diverse fields are intimately related. It makes one wonder what the overarching structure will be.

It must be said that in all developments there have been two further ingredients that are absolutely crucial. The first is quantum mechanics – the description of physical reality in terms of operator algebras acting on Hilbert spaces. In most attempts to understand string theory quantum mechanics has been the foundation, and there is little indication that this is going to change.

The second ingredient is supersymmetry – the unification of matter and forces. In mathematical terms supersymmetry is closely related to De Rham complexes and algebraic topology. In some way much of the miraculous interconnections in string theory only work if supersymmetry is present. Since we are essentially working with a complex, it should not come to a surprise to mathematicians that there are various 'topological' indices that are stable under perturbation and can be computed exactly in an appropriate limit. Indeed it is the existence of these topological quantities, that are not sensitive to the full theory, that make it possible to make precise mathematical predictions, even though the final theory is far from complete.

4.1.2.2 *What is quantum geometry?*

Physical intuition tells us that the traditional pseudo-Riemannian geometry of space-time cannot be a definite description of physical reality. Quantum corrections will change this picture at short-distances on the order of the Planck scale $\ell_P \sim 10^{-35}$ m.

Several ideas seem to be necessary ingredients of any complete quantum gravity theory.

- *Correspondence principle.* Whatever quantum geometry is, it should reduce to the classical space-times of general relativity in the limit $\ell_P \to 0$.
- *Space-time non-commutativity.* The space-time coordinates x^μ are no longer real numbers, but most likely should become eigenvalues of quantum operators. These operators should no longer commute, but instead obey relations of the form

$$[x^\mu, x^\nu] \sim \ell_P^2.$$

  In particular, space-like and time-like coordinates should no longer commute. There is a well-known simple physical argument for this: precise short-distance spatial measurements Δx require such high energy waves compressed in such a small volume, that a microscopic black hole can be formed. Due to Hawking evaporation, such a black hole is only meta-stable, and it will have a typical decay time $\Delta t \sim \ell_P^2/\Delta x$.
- *Quantum foam.* In some sense one should be able to interpret quantum geometry as a path-integral over fluctuating space-time histories. Short-distance space-time geometries should be therefore be subject to quantum corrections that have arbitrary complicated topologies. This induces some quantized, discrete structures. The sizes of the topologically non-trivial cycles (handles, loops, "holes", ...) should be quantized in units of ℓ_P. Together with the idea of non-commutativity of the space-time coordinates, this reminds one of semi-classical Bohr-Sommerfeld quantization.
- *Holography.* As is discussed in much greater details in the rapporteur talk of Seiberg [13], the ideas of holography in black hole physics [14] suggest that space-time geometry should be an *emergent* concept. It should arise in the limit $N \to \infty$, where N is some measure of the total degrees of freedom of the quantum system. In this context the analogy with the emergence of the laws of thermodynamics out of the properties of a statistical mechanical system has often been mentioned. In fact, both thermodynamics and general relativity were discover first as macroscopic theories, before the corresponding microscopic formulations were found. They are also in a precise sense universal theories: in the suitable macroscopic limit any system is subject to the laws of thermodynamics and any gravity theory will produce Einsteinian gravity.
- *Probe dependence.* Experience of string theory has taught us that the measured geometry will depend on the object that one uses to probe the system. Roughly, the metric $g_{\mu\nu}(x)$ will appear as an effective coupling constant in the world-volume theory of a particle, string or brane that is used as probe. With Σ the worldvolume of the probe, one has

$$S_{probe}[g_{\mu\nu}] = \int_\Sigma g_{\mu\nu}(x)dx^\mu \wedge *dx^\nu + \dots$$

As such the space-time metric $g_{\mu\nu}$ is not an invariant concept, but dependent on the "duality frame." For example, certain singularities can appear from the perspective of one kind a brane, but not from another one where the geometry seems perfectly smooth. So, even to *define* a space-time we have to split the total degrees of freedom in a large source system, that produces an emergent geometry, and a small probe system, that measures that effective geometry.

- *Alternative variables.* It is in no way obvious (and most likely simply wrong) that a suitable theory of quantum gravity can be obtained as a (non-perturbative) quantization of the metric tensor field $g_{\mu\nu}(x)$. The ultimate quantum degrees of freedom are probably not directly related to the usual quantities of classical geometry. One possible direction, as suggested for example in the case of three-dimensional geometry [15] and the Ashtekar program of loop quantum gravity [16], is that some form of gauge fields could be appropriate, possible a p-form generalization of that [17]. However, in view of the strong physical arguments for holography, it is likely that this change of dynamical variables should entail more that simply replacing the metric field with another space-time quantum field.

4.1.3 The quantum geometry of string theory

Let us now put the mathematical structures in some perspective. For pedagogical purposes we will consider string theory as a two parameter family of deformations of "classical" Riemannian geometry. Let us introduce these two parameters heuristically. (We will give a more precise explanation later.)

First, in perturbative string theory we study the loops in a space-time manifold. These loops can be thought to have an intrinsic length ℓ_s, the *string length*. Because of the finite extent of a string, the geometry is necessarily "fuzzy." At least at an intuitive level it is clear that in the limit $\ell_s \to 0$ the string degenerates to a point, a constant loop, and the classical geometry is recovered. The parameter ℓ_s controls the "stringyness" of the model. We will see how the quantity $\ell_s^2 = \alpha'$ plays the role of Planck's constant on the worldsheet of the string. That is, it controls the quantum correction of the two-dimensional field theory on the world-sheet of the string.

A second deformation of classical geometry has to do with the fact that strings can split and join, sweeping out a surface Σ of general topology in space-time. According to the general rules of quantum mechanics we have to include a sum over all topologies. Such a sum over topologies can be regulated if we can introduce a formal parameter g_s, the *string coupling*, such that a surface of genus h gets weighted by a factor g_s^{2h-2}. Higher genus topologies can be interpreted as virtual processes wherein strings split and join — a typical quantum phenomenon. Therefore the parameter g_s controls the quantum corrections. In fact we can equate g_s^2 with Planck's constant in space-time. Only for small values of g_s can string theory be described in terms of loop spaces and sums over surfaces.

In theories of four-dimensional gravity, the Planck scale is determined as

$$\ell_P = g_s \ell_s.$$

This is the scale at which we expect to find the effects of quantum geometry, such as non-commutativity and space-time foam. So, in a perturbative regime, where g_s is by definition small, the Planck scale will be much smaller than the string scale $\ell_P \ll \ell_s$ and we will typical have 3 regimes of geometry, depending on which length scale we will probe the space-time: a classical regime at large scales, a "stringy" regime where we study the loop space for scales around ℓ_s, finally and a truly quantum regime for scales around ℓ_P.

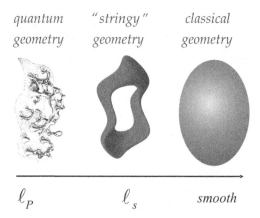

For large values of g_s this picture changes drastically. In the case of particles we know that for large $\hbar$ it is better to think in terms of waves, or more precisely quantum fields. So one could expect that for large g_s and ℓ_s the right framework is string field theory [18]. This is partly true, but it is in general difficult to analyze (closed) string field theory in all its generality.

Summarizing we can distinguish two kinds of deformations: *stringy* effects parameterized by ℓ_s or α', and *quantum* effects parameterized by g_s. This situation can be described with the following diagram

α' large	conformal field theory	M-theory
	strings	*string fields, branes*
$\alpha' \approx 0$	quantum mechanics	quantum field theory
	particles	*fields*
	$g_s \approx 0$	g_s large

4.1.3.1 *Quantum mechanics and point particles*

As a warm-up let us start by briefly reviewing the quantum mechanics of point particles in more abstract mathematical terms.

In classical mechanics we describe point particles on a Riemannian manifold X that we think of as a (Euclidean) space-time. Pedantically speaking we look at X through maps

$$x : pt \to X$$

of an abstract point into X. Quantum mechanics associates to the classical configuration space X the Hilbert space $\mathcal{H} = L^2(X)$ of square-integrable wavefunctions. We want to think of this Hilbert space as associated to a point

$$\mathcal{H} = \mathcal{H}_{pt}.$$

For a supersymmetric point particle, we have bosonic coordinates x^μ and fermionic variables θ^μ satisfying

$$\theta^\mu \theta^\nu = -\theta^\nu \theta^\mu.$$

We can think of these fermionic variables geometrically as one-forms $\theta^\mu = dx^\mu$. So, the supersymmetric wavefunction $\Psi(x, \theta)$ can be interpreted as a linear superposition of differential forms on X

$$\Psi(x, \theta) = \sum_n \Psi_{\mu_1 \ldots \mu_n} dx^{\mu_1} \wedge \ldots \wedge dx^{\mu_n}.$$

So, in this case the Hilbert space is given by the space of (square-integrable) de Rham differential forms $\mathcal{H} = \Omega^*(X)$.

Classically a particle can go in a time t from point x to point y along some preferred path, typically a geodesic. Quantum mechanically we instead have a linear evolution operator

$$\Phi_t : \mathcal{H} \to \mathcal{H}.$$

that describes the time evolution. Through the Feynman path-integral this operator is associated to maps of the line interval of length t into X. More precisely, the kernel $\Phi_t(x, y)$ of the operator Φ_t, that gives the probability amplitude of a particle situated at x to arrive at position y in time t, is given by the path-integral

$$\Phi_t(x, y) = \int_{x(\tau)} \mathcal{D}x \, \exp\left[-\int_0^t d\tau |\dot{x}|^2\right]$$

over all paths $x(\tau)$ with $x(0) = x$ and $x(t) = y$. Φ_t is a famous mathematical object — the integral kernel of the heat equation

$$\frac{d}{dt}\Phi_t = \Delta\Phi_t, \qquad \Phi_0(x, y) = \delta(x - y).$$

These path-integrals have a natural gluing property: if we first evolve over a time t_1 and then over a time t_2 this should be equivalent to evolving over time $t_1 + t_2$. That is, we have the composition property of the corresponding linear maps

$$\Phi_{t_1} \circ \Phi_{t_2} = \Phi_{t_1 + t_2}. \tag{1}$$

This allows us to write

$$\Phi_t = e^{-tH}$$

with H the Hamiltonian. In the case of a particle on X the Hamiltonian is of course simply given by (minus) the Laplacian $H = -\Delta$. The composition property (1) is a general property of quantum field theories. It leads us to Segal's functorial view of quantum field theory, as a functor between the categories of manifolds (with bordisms) to vector spaces (with linear maps) [19].

In the supersymmetric case the Hamiltonian can be written as

$$H = -\Delta = -(dd^* + d^*d)$$

Here the differentials d, d^* play the role of the supercharges

$$d = \psi^\mu \frac{\partial}{\partial x^\mu}, \qquad d^* = g^{\mu\nu} \frac{\partial^2}{\partial \psi^\mu \partial x^\nu}.$$

The ground states of the supersymmetric quantum mechanics satisfy $H\Psi = 0$ and are therefore harmonic forms

$$d\Psi = 0, \qquad d^*\Psi = 0.$$

Therefore they are in 1-to-1 correspondence with the de Rham cohomology group of the space-time manifold

$$\Psi \in \mathrm{Harm}^*(X) \cong H^*(X).$$

We want to make two additional remarks. First we can consider also a closed 1-manifold, namely a circle S^1 of length t. Since a circle is obtained by identifying two ends of an interval we can write

$$Z_{S^1} = \mathrm{Tr}_{\mathcal{H}} e^{-tH}.$$

Here the partition function Z_{S^1} is a number associated to the circle S^1 that encodes the spectrum of the operator Δ. We can also compute the supersymmetric partition function by using the fermion number F (defined as the degree of the corresponding differential form). It computes the Euler number

$$\mathrm{Tr}_{\mathcal{H}} \left((-1)^F e^{-tH} \right) = \dim H^{even}(X) - \dim H^{odd}(X) = \chi(X).$$

4.1.3.2   *Conformal field theory and strings*

We will now introduce our first deformation parameter α' and generalize from point particles and quantum mechanics to strings and conformal field theory.

A string can be considered as a parameterized loop. So, in this case we study the manifold X through maps

$$x : S^1 \rightarrow X,$$

that is, through the free loop space $\mathcal{L}X$.

Quantization will associate a Hilbert space to this loop space. Roughly one can think of this Hilbert space as $L^2(\mathcal{L}X)$, but it is better to think of it as a quantization of an infinitesimal thickening of the locus of constant loops $X \subset \mathcal{L}X$. These constant loops are the fixed points under the obvious S^1 action on the loop space. The normal bundle to X in $\mathcal{L}X$ decomposes into eigenspaces under this S^1 action, and this gives a description (valid for large volume of X) of the Hilbert space $\mathcal{H}_{S^1}$ associated to the circle as the normalizable sections of an infinite Fock space bundle over X.

$$\mathcal{H}_{S^1} = L^2(X, \mathcal{F}_+ \otimes \mathcal{F}_-)$$

where the Fock bundle is defined as

$$\mathcal{F} = \bigotimes_{n \geq 1} S_{q^n}(TX) = \mathbb{C} \oplus qTX \oplus \cdots$$

Here we use the formal variable q to indicate the $\mathbb{Z}$-grading of $\mathcal{F}$ and we use the standard notation

$$S_q V = \bigoplus_{N \geq 0} q^N S^N V$$

for the generating function of symmetric products of a vector space V.

When a string moves in time it sweeps out a surface Σ. For a free string Σ has the topology of $S^1 \times I$, but we can also consider at no extra cost interacting strings that join and split. In that case Σ will be a oriented surface of arbitrary topology. So in the Lagrangian formalism one is let to consider maps

$$x : \Sigma \to X.$$

There is a natural action for such a sigma model if we pick a Hogde star or conformal structure on Σ (together with of course a Riemannian metric g on X)

$$S(x) = \int_\Sigma g_{\mu\nu}(x) dx^\mu \wedge *dx^\nu$$

The critical points of $S(x)$ are the harmonic maps. In the Lagrangian quantization formalism one considers the formal path-integral over all maps $x : \Sigma \to X$

$$\Phi_\Sigma = \int_{x:\ \Sigma \to X} e^{-S/\alpha'}.$$

Here the constant α' plays the role of Planck's constant on the string worldsheet Σ. It can be absorbed in the volume of the target X by rescaling the metric as $g \to \alpha' \cdot g$. The semi-classical limit $\alpha' \to 0$ is therefore equivalent to the limit $vol(X) \to \infty$.

4.1.3.3 *Functorial description*

In the functorial description of conformal field theory the maps Φ_Σ are abstracted away from the concrete sigma model definition. Starting point is now an arbitrary (closed, oriented) Riemann surface Σ with boundary. This boundary consists of a collections of oriented circles. One declares these circles in-coming or out-going depending on whether their orientation matches that of the surface Σ or not. To a surface Σ with m in-coming and n out-going boundaries one associates a linear map

$$\Phi_\Sigma : \mathcal{H}_{S^1}^{\otimes n} \to \mathcal{H}_{S^1}^{\otimes n}.$$

These maps are not independent but satisfy gluing axioms that generalize the simple composition law (1)

$$\Phi_{\Sigma_1} \circ \Phi_{\Sigma_2} = \Phi_\Sigma, \tag{2}$$

where Σ is obtained by gluing Σ_1 and Σ_2 on their out-going and incoming boundaries respectively.

  In this way we obtain what is known as a modular functor. It has a rich algebraic structure. For instance, the sphere with three holes

gives rise to a product

$$\Phi : \mathcal{H}_{S^1} \otimes \mathcal{H}_{S^1} \to \mathcal{H}_{S^1}.$$

Using the fact that a sphere with four holes can be glued together from two copies of the three-holed sphere one shows that this product is essentially commutative and associative

Once translated in terms of transition amplitudes, these relation lead to non-trivial differential equations and integrable hierarchies. For more details see *e.g.* [4, 20].

4.1.3.4 *Stringy geometry and T-duality*

Two-dimensional sigma models give a natural one-parameter deformation of classical geometry. The deformation parameter is Planck's constant α'. In the limit $\alpha' \to 0$ we localize on constant loops and recover quantum mechanics or point particle theory. For non-zero α' the non-constant loops contribute.

In fact we can picture the moduli space of CFT's roughly as follows.

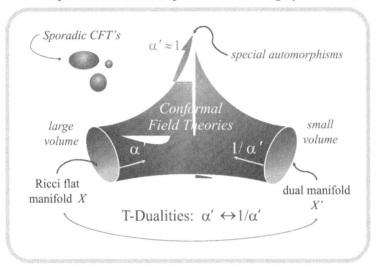

The moduli space of conformal field theories.

It will have components that can be described in terms of a target spaces X. For these models the moduli parameterize Ricci-flat metrics plus a choice of B-field. These components have a boundary 'at infinity' which describe the large volume manifolds. We can use the parameter α' as local transverse coordinate on the collar around this boundary. If we move away from this boundary stringy corrections set in. In the middle of the moduli space exotic phenomena can take place. For example, the automorphism group of the CFT can jump, which gives rise to orbifold singularities at enhanced symmetry points.

The most striking phenomena that the moduli space can have another boundary that allows again for a semi-classical interpretation in terms of a second classical geometry $\hat{X}$. These points look like quantum or small volume in terms of the original variables on X but can also be interpreted as large volume in terms of a set of dual variables on a dual or mirror manifold $\hat{X}$. In this case we speak of a T-duality. In this way two manifold X and $\hat{X}$ are related since they give rise to the same CFT.

The most simple example of such a T-duality occurs for toroidal compactification. If $X = \mathbb{T}$ is an torus, the CFT's on $\mathbb{T}$ and its dual $\mathbb{T}^*$ are isomorphic. We will explain this now in more detail. These kind of T-dualities have led to spectacular mathematical application in mirror symmetry, as we will review after that.

Let us consider a particle or a string on a space-time that is given by a n-dimensional torus, written as the quotient

$$\mathbb{T} = \mathbb{R}^n / L$$

with L a rank n lattice. States of a quantum mechanical point particle on $\mathbb{T}$ are conveniently labeled by their momentum

$$p \in L^*.$$

The wavefunctions $\Psi(x) = e^{ipx}$ form a basis of $\mathcal{H} = L^2(\mathbb{T})$ that diagonalizes the Hamiltonian $H = -\Delta = p^2$. So we can decompose the Hilbert space as

$$\mathcal{H} = \bigoplus_{p \in L^*} \mathcal{H}_p,$$

where the graded pieces $\mathcal{H}_p$ are all one-dimensional. There is a natural action of the symmetry group

$$G = SL(n, \mathbb{Z}) = \mathrm{Aut}(L)$$

on the lattice $\Gamma = L$ and the Hilbert space $\mathcal{H}$.

In the case of a string moving on the torus $\mathbb{T}$ states are labeled by a second quantum number: their winding number

$$w \in L,$$

which is simply the class in $\pi_1 \mathbb{T}$ of the corresponding classical configuration. The winding number simply distinguishes the various connected components of the loop space $\mathcal{L}\mathbb{T}$, since $\pi_0 \mathcal{L}\mathbb{T} = \pi_1 \mathbb{T} \cong L$. We therefore see a natural occurrence of the so-called Narain lattice $\Gamma^{n,n}$, which is the set of momenta $p \in L^*$ and winding numbers $w \in L$

$$\Gamma^{n,n} = L \oplus L^*$$

This is an even self-dual lattice of signature (n, n) with inner product

$$p^2 = 2w \cdot k, \qquad p = (w, k) \in \Gamma^{n,n}.$$

It turns out that all the symmetries of the lattice $\Gamma^{n,n}$ lift to symmetries of the full conformal field theory built up by quantizing the loop space. The elements of the symmetry group of the Narain lattice

$$SO(n, n, \mathbb{Z}) = \mathrm{Aut}\,(\Gamma^{n,n})$$

are examples of $\mathbb{T}$-dualities. A particular example is the interchange of the torus with its dual

$$\mathbb{T} \leftrightarrow \mathbb{T}^*.$$

T-dualities that interchange a torus with its dual can be also applied fiberwise. If the manifold X allows for a fibration $X \to B$ whose fibers are tori, then we can produce a dual fibration where we dualize all the fibers. This gives a new manifold $\widehat{X} \to B$. Under suitable circumstances this produces an equivalent supersymmetric sigma model. The symmetry that interchanges these two manifolds

$$X \leftrightarrow \widehat{X}$$

is called mirror symmetry [2], [3].

4.1.3.5 *Topological string theory*

In the case of point particles it was instructive to consider the supersymmetric extension since we naturally produced differential form on the target space. These differential forms are able, through the De Rahm complex, to capture the topology of the manifold. In fact, reducing the theory to the ground states, we obtained exactly the harmonic forms that are unique representatives of the cohomology groups. In this way we made the step from functional analysis and operator theory to topology.

In a similar fashion there is a formulation of string theory that is able to capture the topology of string configurations. This is called topological string theory. This is quite a technical subject, that is impossible to do justice to within the confines of this survey, but I will sketch the essential features. For more details see *e.g.* [3].

Topological string theory is important for several reasons

- It is a "toy model" of string theory that allows many exact computations. In this sense, its relation to the full superstring theory is a bit like topology versus Riemannian geometry.
- It is the main connection between string theory and various fields in mathematics.
- Topological strings compute so-called BPS or supersymmetric amplitudes in the full-fledged superstring and therefore also capture exact physical information.

Roughly, the idea is the following. First, just as in the point particle case, one introduces fermion fields θ^μ. Now these are considered as spinors on the two-dimensional world-sheet and they have two components $\theta_L^\mu, \theta_R^\mu$. One furthermore assumes that the target space X is (almost) complex so that one can use holomorphic local coordinates $x^i, \bar{x}^{\bar{i}}$ with a similar decomposition for the fermions. When complemented with the appropriate higher order terms this gives a sigma model that has $\mathcal{N} = (2,2)$ supersymmetry.

One now changes the spins of the fermionic fields to produce the topological string. This can be done in two inequivalent ways called the A-model and the B-model. Depending on the nature of this topological twisting the path-integral of the sigma model localizes to a finite-dimensional space.

The A-model restricts to holomorphic maps

$$\frac{\partial x^i}{\partial \bar{z}} = 0$$

This reduces the full path-integral over all maps from Σ into X to a finite-dimensional integral over the moduli space $\mathcal{M}$ of *holomorphic maps*. More precisely, it is the moduli space of pairs (Σ, f) where Σ is a Riemann surface and f is a holomorphic map $f : \Sigma \to X$. The A-model only depends on the Kähler class

$$t = [\omega] \in H^2(X)$$

of the manifold X.

A-model topological strings give an important example of a typical stringy generalization of a classical geometric structure. Quantum cohomology [4] is a deformation of the De Rahm cohomology ring $H^*(X)$ of a manifold. Classically this ring captures the intersection properties of submanifolds. More precisely, if we have three cohomology classes

$$\alpha, \beta, \gamma \in H^*(X)$$

that are Poincaré dual to three subvarieties $A, B, C \subset X$, the quantity

$$I(\alpha, \beta, \gamma) = \int_X \alpha \wedge \beta \wedge \gamma$$

computes the intersection of the three classes A, B, and C. That is, it counts (with signs) the number of points in $A \cap B \cap C$.

In the case of the A-model we have to assume that X is a Kähler manifold or at least a symplectic manifold with symplectic form ω. Now the "stringy" intersection product is related to the three-string vertex. Mathematically it defined as

$$I_{qu}(\alpha, \beta, \gamma) = \sum_d q^d \int_{\overline{\mathcal{M}}_0(X, d)} \alpha \wedge \beta \wedge \gamma,$$

where we integrate our differential forms now over the moduli space of pseudo-holomorphic maps of degree d of a sphere into the manifold X. These maps are weighted by the classical instanton action

$$q^d = \exp\left[-\frac{1}{\alpha'} \int_{S^2} \omega\right] = e^{-d \cdot t/\alpha'}.$$

Clearly in the limit $\alpha' \to 0$ only the holomorphic maps of degree zero contribute. But these maps are necessarily constant and so we recover the classical definition of the intersection product by means of an integral over the space X. Geometrically, we can think of the quantum intersection product as follows: it counts the pseudo-holomorphic spheres inside X that intersect each of the three cycles A, B and C. So, in the quantum case these cycles do no longer need to actually intersect. It is enough if there is a pseudo-holomorphic sphere with points $a, b, c \in \mathbb{P}^1$ such that $a \in A$, $b \in B$ and $c \in C$, *i.e.*, if there is a string world-sheet that connect the three cycles.

For example, for the projective space $\mathbb{P}^n$ the classical cohomology ring is given by

$$H^*(\mathbb{P}^n) = \mathbb{C}[x]/(x^{n+1}).$$

The quantum ring takes the form

$$QH^*(\mathbb{P}^n) = \mathbb{C}[x]/(x^{n+1} = q),$$

with $q = e^{-t/\alpha'}$.

In the B-model one can reduce to (almost) constant maps. This model only depends on the complex structure moduli of X. It most important feature is that mirror symmetry will interchange the A-model with the B-model. A famous example of the power of mirror symmetry is the original computation of Candelas et. al. [21] of the quintic Calabi-Yau manifold given by the equation

$$X : \; x_1^5 + x_2^5 + x_3^5 + x_4^5 + x_5^5 = 0$$

in $\mathbb{P}^4$. In the case the A-model computation leads to an expression of the form

$$F_0(q) = \sum_d n_d \, q^d$$

where n_d computes the number of rational curves in X of degree d. These numbers are notoriously difficult to compute. The number $n_1 = 2875$ of lines is a classical result from the 19th century. The next one $n_2 = 609250$ counts the different conics in the quintic and was only computed around 1980. Finally the number of twisted cubics $n_3 = 317206375$ was the result of a complicated computer program. However, now we know all these numbers and many more thanks to string theory. Here are the first ten

d	n_d
1	2875
2	6 09250
3	3172 06375
4	24 24675 30000
5	22930 59999 87625
6	248 24974 21180 22000
7	2 95091 05057 08456 59250
8	3756 32160 93747 66035 50000
9	50 38405 10416 98524 36451 06250
10	70428 81649 78454 68611 34882 49750

How are physicists able to compute these numbers? Mirror symmetry does the job. It relates the "stringy" invariants coming from the A-model on the manifold X to the classical invariants of the B-model on the mirror manifold $\widehat{X}$. In particular this leads to a so-called Fuchsian differential equation for the function $F_0(q)$. Solving this equation one reads off the integers n_d.

4.1.4 *Non-perturbative string theory and branes*

We have seen how CFT gives rise to a rich structure in terms of the modular geometry as formulated in terms of the maps Φ_Σ. To go from CFT to string theory we have to make two more steps.

4.1.4.1   *Summing over string topologies*

First, we want to generalize to the situation where the maps Φ_Σ are not just functions on the moduli space $\mathcal{M}_{g,n}$ of Riemann surfaces but more general differential forms. In fact, we are particular interested in the case where they are volume forms since then we can define the so-called string amplitudes as

$$F_g = \int_{\mathcal{M}_g} \Phi_\Sigma$$

This is also the general definition of Gromov-Witten invariants [4] as we will come to later. Although we suppress the dependence on the CFT moduli, we should realize that the amplitudes A_g (now associated to a *topological* surface of genus g) still have (among others) α' dependence.

Secondly, it is not enough to consider a string amplitude of a given topology. Just as in field theory one sums over all possible Feynman graphs, in string theory we have to sum over all topologies of the string world-sheet. In fact, we have to ensemble these amplitudes into a generating function.

$$F(g_s) \approx \sum_{g \geq 0} g_s{}^{2g-2} F_g.$$

Here we introduce the string coupling constant g_s. Unfortunately, in general this generating function can be at best an asymptotic series expansion of an analytical function $F(g_s)$. A rough estimate of the volume of $\mathcal{M}_g$ shows that typically

$$F_g \sim 2g!$$

so the sum over string topologies will not converge. Indeed, general physics arguments tell us that the *non-perturbative* amplitudes $F(g_s)$ have corrections of the form

$$F(g_s) = \sum_{g \geq 0} g_s{}^{2g-2} F_g + \mathcal{O}(e^{-1/g_s})$$

Clearly to approach the proper definition of the string amplitudes these non-perturbative corrections have to be understood.

As will be reviewed at much greater length in other lectures, the last years have seen remarkable progress in the direction of developing such a non-perturbative formulation. Remarkable, it has brought very different kind of mathematics into the game. It involves some remarkable new ideas.

- *Branes.* String theory is not a theory of strings. It is simply not enough to consider loop spaces and their quantization. We should also include other extended objects, collectively known as branes. One can try to think of these objects as associated to more general maps $Y \to X$ where Y is a higher-dimensional space. But the problem is that there is not a consistent quantization starting from 'small' branes along the lines of string theory, that is, an expansion where we control the size of Y (through α') and the topology (through g_s). However, through the formalism of D-branes [22] these can be analyzed exactly in string perturbation theory. D-branes give contribution that are of order

$$e^{-1/g_s}$$

and therefore complement the asymptotic string perturbation series.

- *Gauge theory.* These D-branes are described by non-abelian gauge theories and therefore by definition non-commutative structures. This suggests that an alternative formulation of string theory makes use of non-commutative variables. These gauge-gravity dualities are the driving force of all recent progress in string theory [23].

- *Extra dimensions.* As we stressed, the full quantum amplitudes F depend on many parameters or moduli. Apart from the string coupling g_s all other moduli have a geometric interpretation, in terms of the metric and B-field on X. The second new ingredient is the insight that string theory on X with string coupling g_s can be given a fully geometric realization in terms of a new theory called M-theory on the manifold $X \times S^1$, where the length of the circle S^1 is g_s [24].

Summarizing, the moduli space of string theory solutions has a structure that in many aspects resembles the structures that described moduli of CFT's. In this case there are S-dualities that relate various perturbative regimes.

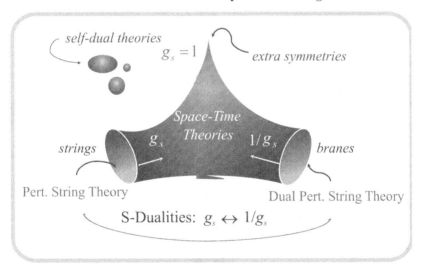

The moduli space of string theory vacua.

4.1.4.2 *Topological strings and quantum crystals*

The way in which quantum geometry can emerge from a non-perturbative completion of a perturbative string theory can be nicely illustrated by a topological string example.

In general the topological string partition function (of the A model) takes the form

$$Z_{top} = \exp \sum_{g \geq 0} g_s^{2g-2} F_g(t),$$

where the genus g contribution F_g can be expended as a sum over degree d maps

$$F_g(t) = \sum_d GW_{g,d}\, e^{-dt/\alpha'}.$$

Here $GW_{g,d} \in \mathbb{Q}$ denotes the Gromov-Witten invariant that "counts" the number of holomorphic maps $f : \Sigma_g \to X$ of degree d of a Riemann surface Σ_g of genus g into the Calabi-Yau manifold X.

To show that these invariants are very non-trivial and define some quantum geometry structure, it suffices to look at the simplest possible CY space $X = \mathbb{C}^3$. In that case only degree zero maps contribute. The corresponding Gromov-Witten invariants have been computed and can be expressed in terms of so-called Hodge integrals

$$GW_{g,0} = \int_{\overline{\mathcal{M}}_g} \lambda_{g-1}^3 = \frac{B_{2g} B_{2g-2}}{2g(2g-2)(2g-2)!}.$$

But in this case the full partition function Z_{top} simplifies considerably if it is expressed in terms of the strong coupling variable $q = e^{-g_s}$ instead of the weak coupling variable g_s:

$$Z_{top} = \exp \sum_g g_s^{2g-2} GW_{g,0} = \prod_{n>0} (1 - q^n)^{-n}.$$

In fact, this gives a beautiful reinterpretation in terms of a statistical mechanics model. The partition function can be written as a weighted sum over all planar partitions

$$Z_{top} = \sum_\pi q^{|\pi|}.$$

Here a planar partition π is a 3d version of the usual 2d partitions [25]

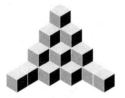

Clearly, this statistical model has a granular structure that is invisible in the perturbative limit $g_s \to 0$. In fact, these quantum crystals give a very nice model in which the stringy and quantum geometry regimes can be distinguished. Here one uses a toric description of $\mathbb{C}^3$ as a $\mathbb{T}^3$ bundle over the positive octant in $\mathbb{R}^3$. In terms of pictures we have (i) the statistical model, (ii) the so-called limit space, that captures the mirror manifold, (iii) and the classical geometry

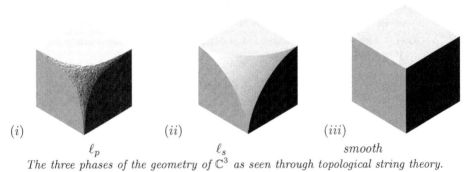

(i) $\qquad\qquad\qquad$ (ii) $\qquad\qquad\qquad$ (iii)

ℓ_p $\qquad\qquad\qquad\quad$ ℓ_s $\qquad\qquad\qquad$ *smooth*

The three phases of the geometry of $\mathbb{C}^3$ as seen through topological string theory.

4.1.4.3 *U-dualities*

Another way to probe non-perturbative effects in string theory is to investigate the symmetries (dualities). In the case of a compactification on a torus $\mathbb{T}$ the story becomes considerably more complicated then we saw in previous section. The lattice of quantum numbers of the various objects becomes larger and so do the symmetries. For small values of the dimension n of the torus $\mathbb{T}$ ($n \leq 4$) it turns out that the non-perturbative charge lattice M can be written as as the direct sum of the Narain lattice (the momenta and winding numbers of the strings) together with a lattice that keeps track of the homology classes of the branes

$$M = \Gamma^{n,n} \oplus H^{even/odd}(\mathbb{T})$$

Here we note that the lattice of branes (which are even or odd depending on the type of string theory that we consider)

$$H^{even/odd}(\mathbb{T}) \cong \Lambda^{even/odd} L^*$$

transform as half-spinor representations under the T-duality group $SO(n, n, \mathbb{Z})$. The full duality group turns out to be the exceptional group over the integers

$$E_{n+1}(\mathbb{Z}).$$

The lattice M will form an irreducible representation for this symmetry group. These so-called U-dualities will therefore permute strings with branes.

So we see that our hierarchy

$$\{Particles\} \subset \{Strings\} \subset \{Branes\}$$

is reflected in the corresponding sequence of symmetry (sub)groups

$$SL(n, \mathbb{Z}) \subset SO(n, n, \mathbb{Z}) \subset E_{n+1}(\mathbb{Z})$$

of rank $n - 1$, n, $n + 1$ respectively. Or, in terms of the Dynkin classification

$$A_{n-1} \subset D_n \subset E_{n+1}.$$

It is already a very deep (and generally unanswered) question what the 'right' mathematical structure is associated to a n-torus that gives rise to the exceptional group $E_{n+1}(\mathbb{Z})$.

4.1.5   *D-branes*

As we have mentioned, the crucial ingredient to extend string theory beyond perturbation theory are D-branes [22]. From a mathematical point of view D-branes can
be considered as a relative version of Gromov-Witten theory. The starting point
is now a pair of relative manifolds (X, Y) with X a d-dimensional manifold and
$Y \subset X$ closed. The string worldsheets are defined to be Riemann surfaces Σ with
boundary $\partial\Sigma$, and the class of maps $x : \Sigma \to X$ should satisfy

$$x(\partial\Sigma) \subset Y$$

That is, the boundary of the Riemann surfaces should be mapped to the subspace
Y.

Note that in a functorial description there are now two kinds of boundaries to
the surface. First there are the time-like boundaries that we just described. Here
we choose a definite boundary condition, namely that the string lies on the D-brane
Y. Second there are the space-like boundaries that we considered before. These
are an essential ingredient in any Hamiltonian description. On these boundaries we
choose initial value conditions that than propagate in time. In closed string theory
these boundaries are closed and therefore a sums of circles. With D-branes there is
a second kind of boundary: the open string with interval $I = [0, 1]$.

The occurrence of two kinds of space-like boundaries can be understood because
there are various ways to choose a 'time' coordinate on a Riemann surface with
boundary. Locally such a surface always looks like $S^1 \times \mathbb{R}$ or $I \times \mathbb{R}$. This ambiguity
how to slice up the surface is a powerful new ingredient in open string theory.

To the CFT described by the pair (X, Y) we will associate an extended modular
category. It has two kinds of objects or 1-manifolds: the circle S^1 (the closed
string) and the interval $I = [0, 1]$ (the open string). The morphisms between two 1-
manifolds are again bordisms or Riemann surfaces Σ now with a possible boundaries.
We now have to kinds of Hilbert spaces: closed strings $\mathcal{H}_{S^1}$ and open strings $\mathcal{H}_I$.

Semi-classically, the open string Hilbert space is given by

$$\mathcal{H}_I = L^2(Y, \mathcal{F})$$

with Fock space bundle

$$\mathcal{F} = \bigotimes_{n \geq 1} S_{q^n}(TX)$$

Note that we have only a single copy of the Fock space $\mathcal{F}$, the boundary conditions
at the end of the interval relate the left-movers and the right-movers. Also the
fields are sections of the Fock space bundle over the D-brane Y, not over the full
space-time manifold X. In this sense the open string states are localized on the
D-brane.

4.1.5.1 *Branes and matrices*

One of the most remarkable facts is that D-branes can be given a multiplicity N which naturally leads to a non-abelian structure [26].

Given a modular category as described above there is a simple way in which this can be tensored over the $N \times N$ hermitean matrices. We simply replace the Hilbert space $\mathcal{H}_I$ associated to the interval I by

$$\mathcal{H}_I \otimes Mat_{N \times N}$$

with the hermiticity condition

$$(\psi \otimes M_{IJ})^* = \psi^* \otimes M_{JI}$$

The maps Φ_Σ are generalized as follows. Consider for simplicity first a surface Σ with a single boundary C. Let C contain n 'incoming' open string Hilbert spaces with states $\psi_1 \otimes M_1, \ldots, \psi_n \otimes M_n$. These states are now matrix valued. Then the new morphism is defined as

$$\Phi_\Sigma(\psi_1 \otimes M_1, \ldots, \psi_n \otimes M_n) = \Phi_\Sigma(\psi_1, \ldots, \psi_n)\mathrm{Tr}\,(M_1 \cdots M_n).$$

In case of more than one boundary component, we simply have an additional trace for every component.

In particular we can consider the disk diagram with three open string insertions. By considering this as a map

$$\Phi_\Sigma : \mathcal{H}_I \otimes \mathcal{H}_I \to \mathcal{H}_I$$

we see that this open string interaction vertex is now given by

$$\Phi_\Sigma(\psi_1 \otimes M_1, \psi_2 \otimes M_2) = (\psi_1 * \psi_2) \otimes (M_1 M_2).$$

So we have tensored the associate string product with matrix multiplication.

If we consider the geometric limit where the CFT is thought of as the semi-classical sigma model on X, the string fields that correspond to the states in the open string Hilbert space $\mathcal{H}_I$ will become matrix valued fields on the D-brane Y, *i.e.* they can be considered as sections of $\mathrm{End}(E)$ with E a (trivial) vector bundle over Y.

This matrix structure naturally appears if we consider N different D-branes $Y_1, \ldots, Y_N$. In that case we have a matrix of open strings that stretch from brane Y_I to Y_J. In this case there is no obvious vector bundle description. But if all the D-branes coincide $Y_1 = \ldots = Y_N$ a $U(N)$ symmetry appears.

4.1.5.2 *D-branes and K-theory*

The relation with vector bundles has proven to be extremely powerful. The next step is to consider D-branes with *non-trivial* vector bundles. It turns out that these configurations can be considered as a composite of branes of various dimensions [27]. There is a precise formula that relates the topology of the vector bundle E to the brane charge $\mu(E)$ that can be considered as a class in $H^*(X)$. (For convenience we consider first maximal branes $Y = X$.) It reads [28]

$$\mu(E) = ch(E)\widehat{A}^{1/2} \in H^*(X). \tag{3}$$

Here $ch(E)$ is the (generalized) Chern character $ch(E) = \text{Tr}\exp(F/2\pi i)$ and $\widehat{A}$ is the genus that appears in the Atiyah-Singer index theorem. Note that the D-brane charge can be fractional.

Branes of lower dimension can be described by starting with two branes of top dimension, with vector bundles E_1 and E_2, of opposite charge. Physically two such branes will annihilate leaving behind a lower-dimensional collection of branes. Mathematically the resulting object should be considered as a virtual bundle $E_1 \ominus E_2$ that represents a class in the K-theory group $K^0(X)$ of X [29]. In fact the map μ in (3) is a well-known correspondence

$$\mu : K^0(X) \to H^{even}(X)$$

which is an isomorphism when tensored with the reals. In this sense there is a one-to-one map between D-branes and K-theory classes [29]. This relation with K-theory has proven to be very useful.

4.1.5.3 *Example: the index theorem*

A good example of the power of translating between open and closed strings is the natural emergence of the index theorem. Consider the cylinder $\Sigma = S^1 \times I$ between two D-branes described by (virtual) vector bundles E_1 and E_2. This can be seen as closed string diagram with in-state $|E_1\rangle$ and out-state $|E_2\rangle$

$$\Phi_\Sigma = \langle E_2, E_1 \rangle$$

Translating the D-brane boundary state into closed string ground states (given by cohomology classes) we have

$$|E\rangle = \mu(E) \in H^*(X)$$

so that

$$\Phi_\Sigma = \int_X ch(E_1)ch(E_2^*)\widehat{A}$$

On the other hand we can see the cylinder also as a trace over the open string states, with boundary conditions labeled by E_1 and E_2. The ground states in $\mathcal{H}_I$ are sections of the Dirac spinor bundle twisted by $E_1 \otimes E_2^*$ This gives

$$\Phi_\Sigma = \text{Tr}\,_{\mathcal{H}_I}(-1)^F = \text{index}(D_{E_1 \otimes E_2^*})$$

So the index theorem follows rather elementary.

4.1.5.4 *Non-perturbative dualities*

We indicated that in M-theory we do not want to include only strings but also D-branes (and even further objects that I will suppress in this discussion such as NS 5-branes and Kaluza-Klein monopoles). So in the limit of small string coupling g_s the full (second quantized) string Hilbert space would look something like

$$\mathcal{H} = S^*(\mathcal{H}_{string}) \otimes S^*(\mathcal{H}_{brane}).$$

Of course our discussion up to now has been very skew. In the full theory there will be U-dualities that will exchange strings and branes.

We will give a rather simple example of such a symmetry that appears when we compactify the (Type IIA) superstring on a four-torus $\mathbb{T}^4 = \mathbb{R}^4/L$. In this case the charge lattice has rank 16 and can be written as

$$\Gamma^{4,4} \oplus K^0(\mathbb{T}^4).$$

It forms an irreducible spinor representation under the U-duality group

$$SO(5,5,\mathbb{Z}).$$

Notice that the T-duality subgroup $SO(4,4,\mathbb{Z})$ has three inequivalent 8-dimensional representations (related by triality). The strings with Narain lattice $\Gamma^{4,4}$ transform in the vector representation while the even-dimensional branes labeled by the K-group $K^0(\mathbb{T}^4) \cong \Lambda^{even}L^*$ transform in the spinor representation. (The odd-dimensional D-branes that are labeled by $K^1(T)$ and that appear in the Type IIB theory transform in the conjugate spinor representation.)

To compute the spectrum of superstrings we have to introduce the corresponding Fock space. It is given by

$$\mathcal{F}_q = \bigotimes_{n=1}^{\infty} S_{q^n}(\mathbb{R}^8) \otimes \Lambda_{q^n}(\mathbb{R}^8) = \bigoplus_{N \geq 0} q^N \mathcal{F}(N).$$

The Hilbert space of BPS strings with momenta $p \in \Gamma^{4,4}$ is then given by

$$\mathcal{H}_{string}(p) = \mathcal{F}(p^2/2).$$

For the D-branes we take a completely different approach. Since we only understand the system for small string coupling we have to use semi-classical methods. Consider a D-brane that corresponds to a K-theory class E with charge vector $\mu = ch(E) \in H^*(\mathbb{T})$. To such a vector bundle we can associate a moduli space $\mathcal{M}_\mu$ of self-dual connections. (If we work in the holomorphic context we could equally well consider the moduli space of holomorphic sheaves of this topological class.) Now luckily a lot is know about these moduli spaces. They are hyper-Kähler and (for primitive μ) smooth. In fact, they are topologically Hilbert schemes which are deformations of symmetric products

$$\mathcal{M}_\mu \cong \mathrm{Hilb}^{\mu^2/2}(\mathbb{T}^4) \sim S^{\mu^2/2}\mathbb{T}^4.$$

Computing the BPS states through geometric quantization we find that

$$\mathcal{H}_{brane}(\mu) = H^*(\mathcal{M}_\mu).$$

The cohomology of these moduli spaces have been computed [30] with the result that

$$\bigoplus_{N \geq 0} q^N H^*(\mathrm{Hilb}^N(\mathbb{T}^4)) = \mathcal{F}_q.$$

This gives the final result

$$\mathcal{H}_{brane}(\mu) = \mathcal{F}(\mu^2/2) \cong \mathcal{H}_{string}(p),$$

where μ and p are related by an $SO(5,5,\mathbb{Z})$ transformation.

This is just a simple example to show that indeed the same mathematical structures (representation theory of affine lie algebras, Virasoro algebras, *etc.*) can appear both in the perturbative theory of strings and non-perturbative brane systems. Again this is a powerfull hint that a more unified mathematical structure underlies quantum gravity.

4.1.6 *The Role of Mathematics*

In this rapporteur talk we have surveyed some deep connections between physics and mathematics that have stimulated much intellectual activity. Let me finish to raise some questions about these interactions.

- First of all, it must be said that despite all these nice results, there does seem to operate a principle of complementarity (in the spirit of Niels Bohr) that makes it difficult to combine physical intuition with mathematical rigor. Quite often, deep conjectures have been proven rigorously, not be making the physical intuition more precise, but by taking completely alternative routes.
- It is not at all clear what kind of mathematical structure Nature prefers. Here there seem to be two schools of thoughts. One the one hand one can argue that it is the most universal structures that have proven to be most successful. Here one can think about the formalism of calculus, Riemannian manifolds, Hilbert spaces, *etc.* On the other hand, the philosophy behind a Grand Unified Theory or string theory, is that our world is very much described by a single unique(?) mathematical structure. This point of view seems to prefer exceptional mathematical objects, such as the Lie algebra E_8, Calabi-Yau manifolds, *etc.* Perhaps in the end a synthesis of these two points of view (that roughly correspond to the laws of Nature versus the solutions of these laws) will emerge.
- Continuing this thought it is interesting to speculate what other mathematical fields should be brought into theoretical physics. One could think of number theory and arithmetic geometry, or logic, or even subjects that have not been developed at all.

- One could also question whether we are looking for a single overarching mathematical structure or a combination of different complementary points of view. Does a fundamental theory of Nature have a global definition, or do we have to work with a series of local definitions, like the charts and maps of a manifold, that describe physics in various "duality frames." At present string theory is very much formulated in the last kind of way.

- As a whole, the study of quantum geometry takes on the form of a mathematical program, very much like the Langlands Program. There are many non-trivial examples, strange relations, dualities and automorphic forms, tying together diverse fields, with vast generalizations, all in an open ended project that seems to encompass more and more mathematics.

- Finally, there should be word of caution. To which extend should mathematics be a factor in deciding the future of theoretical physics? Is mathematical elegance a guiding light or a Siren, whose song draws the Ship of Physics onto the cliffs? Only the future will tell us.

Acknowledgement

Let me end by thanking the organizers of the XXIII Solvay Conference, in particular Marc Henneaux and Alexander Sevrin for their tremendous efforts to arrange such a unique and stimulating scientific meeting following a distinguished tradition. Let me also congratulate the Solvay family on their long-time commitment to fundamental science by which they are setting an example for the whole world.

Bibliography

[1] J. Polchinski, *String Theory* (Cambridge Monographs on Mathematical Physics), Cambridge University Press, 1998.

[2] D.A. Cox and S. Katz, *Mirror Symmetry and Algebraic Geometry* (Mathematical Surveys and Monographs, No. 68.), AMS, 1999.

[3] K. Hori, S. Katz, A. Klemm, R. Pandharipande, R. Thomas, C. Vafa, R. Vakil, E. Zaslow, *Mirror symmetry.* Clay Mathematics Monographs, AMS 2003.

[4] Yu.I. Manin, *Frobenius Manifolds, Quantum Cohomology, and Moduli Spaces* AMS, 1999.

[5] Ph. Di Francesco, P. Mathieu, and D. Senechal, *Conformal Field Theory* (Graduate Texts in Contemporary Physics), Spinger, 1996.

[6] E. Wigner, *The Unreasonable Effectiveness of Mathematics in the Natural Sciences,* Comm. Pure Appl. Math. **13** (1960), 1–14.

[7] *The Millennium Prize Problems,* Edited by: J. Carlson, A. Jaffe, and A. Wiles, AMS, 2006.

[8] A. Connes and D. Kreimer, *Renormalization in quantum field theory and the Riemann-Hilbert problem. I: The Hopf algebra structure of graphs and the main theorem,* Commun. Math. Phys. **210** (2000) 249.

[9] E. Witten, *Perturbative gauge theory as a string theory in twistor space,* Commun. Math. Phys. **252**, 189 (2004).

[10] Z. Bern, L. J. Dixon and D. A. Kosower, $N = 4$ *super-Yang-Mills theory, QCD and collider physics,* Comptes Rendus Physique **5**, 955 (2004) [arXiv:hep-th/0410021].

[11] C. Vafa and E. Witten, *A Strong coupling test of S duality,* Nucl. Phys. B **431**, 3 (1994) [arXiv:hep-th/9408074].

[12] N. Seiberg, *Electric - magnetic duality in supersymmetric non Abelian gauge theories,* Nucl. Phys. B **435**, 129 (1995) [arXiv:hep-th/9411149].

[13] N. Seiberg, in this volume.

[14] G. 't Hooft, *Dimensional reduction in quantum gravity,* arXiv:gr-qc/9310026.
     L. Susskind, *The world as a hologram,* J. Math. Phys. **36**, 6377 (1995) [arXiv:hep-th/9409089].

[15] E. Witten, *(2+1)-Dimensional gravity as an exactly soluble system,* Nucl. Phys. B **311**, 46 (1988).

[16] A. Ashtekar, *New variables for classical and quantum gravity,* Phys. Rev. Lett. **57** (1986) 2244.

[17] R. Dijkgraaf, S. Gukov, A. Neitzke and C. Vafa, *Topological M-theory as unification of form theories of gravity,* Adv. Theor. Math. Phys. **9**, 593 (2005) [arXiv:hep-th/0411073].

[18] B. Zwiebach, *Closed string field theory: quantum action and the B-V master equation,* Nucl. Phys. **B390** (1993) 33–152, hep-th/9206084.

[19] G. Segal, *The definition of conformal field theory,* preprint; *Two dimensional conformal field theories and modular functors,* in IXth International Conference on Mathematical Physics, . B. Simon, A. Truman and I.M. Davies Eds. (Adam Hilger, Bristol, 1989).

[20] R. Dijkgraaf, *Les Houches Lectures on Fields, Strings and Duality,* in *Quantum Symmetries,* les Houches Session LXIV, Eds. A. Connes, K. Gawedzki, and J. Zinn-Justin, North-Holland, 1998, hep-th/9703136.

[21] P. Candelas, P. Green, L. Parke, and X. de la Ossa, *A pair of Calabi-Yau manifolds as an exactly soluble superconformal field theory,* Nucl. Phys. **B359** (1991) 21, and in [2].

[22] J. Polchinski, *Dirichlet-branes and Ramond-Ramond charges,* Phys. Rev. Lett. **75** (1995) 4724–4727, hep-th/9510017.

[23] J. Maldacena, *The large N limit of superconformal field theories and supergravity,* Adv. Theor. Math. Phys. bf 2 (1998) 231–252, hep-th/9711200.

[24] E. Witten, *String theory in various dimensions,* Nucl. Phys. **B 443** (1995) 85, hep-th/9503124.

[25] A. Okounkov, N. Reshetikhin and C. Vafa, *Quantum Calabi-Yau and classical crystals,* arXiv:hep-th/0309208.
     A. Iqbal, N. Nekrasov, A. Okounkov and C. Vafa, *Quantum foam and topological strings,* arXiv:hep-th/0312022.

[26] E. Witten, *Bound states of strings and p-branes,* Nucl. Phys. **B460** (1996) 335, hep-th/9510135.

[27] M. Douglas, *Branes within branes,* hep-th/9512077.

[28] M. Green, J. Harvey, and G. Moore, *I-brane inflow and anomalous couplings on D-branes,* Class. Quant. Grav. **14** (1997) 47–52, hep-th/9605033.

[29] E. Witten, *D-branes and K-theory,* JHEP **9812** (1998) 019, hep-th/9810188.

[30] L. Göttsche, *The Betti numbers of the Hilbert scheme of points on a smooth projective surface,* Math. Ann. **286** (1990) 193–207; *Hilbert Schemes of Zero-dimensional Subschemes of Smooth Varieties,* Lecture Notes in Mathematics **1572**, Springer-Verlag, 1994.

4.2 Discussion

G. Horowitz At the very beginning you asked: what is quantum geometry? You said it should involve non-commutative geometry, non-commuting coordinates. It is not obvious to me that it has to involve that. Later on, you gave this example of a melting crystal and it was unclear to me where non-commutative geometry came in in that example.

R. Dijkgraaf Actually, I do not think the details have been worked out but it has a very nice interpretation: again there is a matrix model description of this melting crystal which is the reduction of supersymmetric Yang-Mills theory to zero dimension. So it is a three-matrix model where the action is $trX[Y, Z]$. If we look at the critical point of that action, it corresponds exactly to this kind of crystal configurations. So again in this example, there is a D-brane interpretation. I must say that I deliberately did not make that argument exact because usually people argue something like : "If you have to measure space you have to concentrate energy. A little black hole will form and it will have some uncertainty because it will evaporate". I never felt very comfortable with that argument.

H. Ooguri In that particular example, you are exhibiting already half of the space. In the total space, of six dimensions, you do, in fact, see the non-commutative structures?

R. Dijkgraaf Yes. I guess that is a good point.

H. Ooguri There are three torus directions and three directions he was exhibiting which are non-commuting. This quantum structure of the space, the state being represented by blocks in this three dimensions, is a reflection of this non-commutative structure. So I think the example that Dijkgraaf was describing exactly demonstrates the non-commutative feature of space time where the Planck constant is replaced by $e^{-1/g}$, with g the coupling constant.

G. Horowitz I certainly agree, any quantum description will involve non-commutativity, like some x and p do not commute. But were you suggesting that it is always some sort of X with X not commuting?

R. Dijkgraaf In this case it is indeed so. To follow what Ooguri was saying: I was talking about a six dimensional space and drawing a three dimensional picture. In some sense I was using kind of a symplectic space and I was only showing you the coordinates, not the momenta.

N. Seiberg You wrote that time does not commute with space? What did you have in mind?

R. Dijkgraaf What I just said. I am just repeating folklore here: making precise space-time measurements will create small black holes that will evaporate and give a time uncertainty. I do not have any example here. All the non-commutativity that I was discussing here was non-commutativity in the space-like directions. I did not have any examples.

N. Seiberg Do you have in mind time being an operator which does not commute with something?

R. Dijkgraaf I am completely ignorant about that. I think you know more than I.

H. Ooguri Talking about a new direction to go, I think stringy Lorentzian geometry is completely uncharted territory that we need to explore. We have gained lots of insights into quantum geometry but those are all mostly static geometries and the important question about how Lorentzian geometry can be quantized needs to be understood.

A. Ashtekar Partly going back to what Horowitz and Seiberg were saying: in loop quantum gravity we do have a quantum geometry. The coordinates are commuting. There is no problem with that, the manifold is as it is. It is the Riemannian structures which are not commuting. So for example, areas of surfaces which intersect with each other are not commuting and therefore you cannot measure the areas arbitrarily accurately, for example. So there is this other possibility also. Namely that observable quantities such as areas, etc, are not commuting, but there is commutativity for the manifold itself. The manifold itself does not go away.

R. Dijkgraaf I think that is important. One thing I did not mention, but also a good open question, is just to go to three dimensional gravity because there are many ways in which all these approaches connect. Of course, from the loop quantum gravity point of view, three dimensional gravity, written as a Chern-Simon theory, is very interesting. In fact many of these topological theories, when we reduce them down to three dimensions, you get also some Chern-Simon theories. But again, there are many open issues: "Are these Chern-Simons theories really well defined? Do they really correspond to semi-classical quantum gravity theories?" If you want to think about more precise areas, I feel that that point should be developed.

E. Rabinovici I would like to make two comments. One is that when we use strings as probes, it seems that all mathematical concepts we are used to somehow become ambiguous. When you describe T-duality, geometry becomes ambiguous or symmetric. You have two totally different representations of the same geometry. The same goes for topology and the number of dimensions. It also applies to the questions: "Is some manifold singular or not?" and "Is a manifold commutative or not commutative?" So I was wondering: "Is there anything which remains non-ambiguous when we study it with strings?" That was the first comment.

The second comment, which also relates to a discussion we had yesterday, relates to what you said about algebra and geometry. When we use the relation between affine Lie algebras and their semi-classical geometrical description, we can sometimes treat systems which have curvature singularities and large R^2 corrections. Even if we do not know what the Einstein equations are nor what the effective Lagrangian is to all orders in R^2, we still know what the answer is for g-string equals zero, that is on the sphere. For such systems, we sometimes even know the answer on the torus and for higher genus surfaces. In this way we can somehow circumvent the α'-correction problem.

R. Dijkgraaf I do not really know what to say in response. Concerning your first remark, we have indeed often asked ourselves "What is string theory?". But now, we have to ask "What is not string theory?" That is really a big question because, in some sense, finding the upper bound, finding structures which definitely are not connected in any way, is probably now more challenging than the other way round.

P. Ramond You talked, glibly, about exceptional structures. The next frontier in a sense is that, besides non-commutativity, non-associativity occurs with specific systems. Have you encountered any need for this? Does it come up geometrically?

R. Dijkgraaf That is again speculative. Roughly, look at the three levels. At the first level of particles, we have gauge fields. At the second level where we have strings, we have these kind of B-fields. That is a 2-form field that is not related to a gauge field but to something that is called gerbes. And then, in M-theory, there is the 3-form field. So the B-field is clearly related to non-commutativity. But the 3-form field, if with anything, it might be related to non-associativity. But again, there have been many isolated ideas, but I do not think there is something that ties everything together. In fact, it is also a question to the mathematicians. For instance, in string theory, the B-field is intimately connected to K-theory. It is related to some periodicity *mod* 2. The 2 is the same 2 of the B-field. In fact, in M-theory, there is almost something like a periodicity *mod* 3. So again, the question is what replaces this structure. I do not think we have any hint. There are some suggestions that E_8 plays an important rôle. So there is a direct connection between gauge theories of E_8 and 3-forms. Greg Moore has very much pushed this. But again, these are only questions.

A. Polyakov This fusion of physics and mathematics which occurred in the last few decades and which you discussed so beautifully is quite amazing of course. But it is actually bothersome to me because it seems that we start following the steps of mathematicians. I believe we are supposed to invent our own mathematics even though modern mathematics is very seductive. Mathematicians are clever enough to do their own job, I suppose. Anyway that is general philosophy. More concretely, I think there is an unexplored domain which is important for understanding gauge-string duality. It is differential geometry in loop space. In gauge theory we get some interesting differential operators in loop space, which are practically unexplored. Very few, primitive, things are known about these operators and the loop equations for gauge theories. There is obviously some deep mathematical structure. There could be some Lax pairs or integrability in the loop space. Understanding what is the right way to construct these operators is not standard mathematics, but I believe, it is important for physics.

H. Ooguri From that point of view, string field theory has been one of the directions which has been pursued in order to take these ideas seriously. Do you have any comment about that?

A. Polyakov Of course it is all very closely related to string theory. It is essentially the question how the Schrödinger equation for string theory looks like, how to effectively write it down correctly. For example, how should one write the Schrödinger equation when the boundary data are given at infinity, like in AdS space. How should one write the Wheeler-DeWitt equation? Normally, we have some boundary and the Wheeler-DeWitt equation is the equation for the partition function as a functional of this boundary. But if this boundary is at infinity then, instead of the usual Laplacian, there should be some first order operator like the loop Laplacian. How should one make this concrete? I think these are important questions, both for string theory and for gauge theory, of course.

L. Faddeev After the "insult" to mathematics from my friend, I must say that I think there are two intuitions. There is the physical intuition and the mathematical intuition. And now, in this field, they compete and they help each other. There was a try to make a kind of consensus in Princeton five years ago, not with real success. But I think we have still to work somehow together to try to have some consensus of these two different intuitions.

––––––––––––

H. Ooguri We will hear from three panelists now and then I will open the floor for discussions. I would like to suggest that the discussion today should focus on mathematical structures of string vacua, and in particular on what we have learnt about the space of vacua and on what we need to do to better understand it. We should leave the discussion on the physical implications, such as the multi-universes and the anthropic principle to the session on cosmology tomorrow.

4.3 Prepared Comments

4.3.1 *Renata Kallosh: Stabilization of moduli in string theory*

Stabilization of moduli is necessary for string theory to describe the effective 4-dimensional particle physics and cosmology. This is a long-standing problem. Recently a significant progress towards its solution was achieved: a combination of flux compactification with non-perturbative corrections leads to stabilization of moduli, with de Sitter vacua [1]. Such vacua with positive cosmological constant can be viewed as the simplest possibility for the string theory to explain the observable dark energy. This is a prerequisite for the Landscape of String Theory [2]. Main progress was achieved in type II string theories, the heterotic case still remains unclear, which is a serious problem for particle physics models related to string theory.

In type IIB string theory flux compactification with $W_{flux} = \int G_3 \wedge \Omega_3$ leads to stabilization of the dilaton and complex structure moduli [3], whereas gaugino condensation and/or instanton corrections $W_{non-pert} = Ae^{-(\text{Vol}+i\alpha)}$ stabilize the remaining Kahler moduli. The basic steps here are: i) Using the warped geometry of the compactified space and nonperturbative effects one can stabilize all moduli in anti-de-Sitter space. ii) One can uplift the AdS space to a metastable dS space by adding anti-D3 brane at the tip of the conifold (or D7 brane with fluxes [4]). More recently there was a dramatic progress in moduli stabilization in string theory and various successful possibilities were explored [5]-[6].

Examples of new tools include the recently discovered criteria, in presence of fluxes, for the instanton corrections due to branes, wrapping particular cycles [7]. This has allowed one of the simplest models with all moduli stabilized. We have found that M-theory compactified on K3xK3 is incredibly simple and elegant [6]. Without fluxes in the compactified 3d theory there are two 80-dimensional quaternionic Kähler spaces, one for each K3. With non-vanishing primitive (2,2) flux, (2,0) and (0,2), each K3 becomes an attractive K3: one-half of all moduli are fixed. 40 in each K3 still remain moduli and need to be fixed by instantons. There are 20 proper 4-cycles in each K3. They provide instanton corrections from M5-branes wrapped on these cycles: Moduli space is no more...

All cases of moduli stabilization in black holes and in flux vacua which are due to fluxes in string theory can be described by the relevant attractor equations, the so-called "new attractors" [8].

It is known for about 10 years that in extremal black holes the moduli of vector multiplets are stabilized near the horizon where they become fixed function of fluxes (p, q) independently of the values of these moduli far away from the black hole horizon. This is known as a black hole attractor mechanism [9].

$$t_{fix} = t(p, q) , \qquad \bar{t}_{fix} = \bar{t}(p, q) . \tag{1}$$

Stabilization of moduli is equivalent to minimization of the black hole potential

$$V_{BH} = |DZ|^2 + |Z|^2 \tag{2}$$

defined by the central charge Z.

In case of BPS black holes the attractor equation relating fluxes to fixed values of moduli is

$$F_3 = 2\mathrm{Im}\,(Z\,\bar{\Omega}_3)_{DZ=0} \ . \tag{3}$$

It has been studied extensively over the last 10 years and many interesting solutions have been found. One of the most curious solutions of the black hole attractor equation is the so-called STU black holes with three moduli [10]. It was discovered recently [11] that the entropy of such black holes is given by the Caley's hyper-determinant of the 2x2x2 matrix describing also the 3-qubit system in quantum information theory.

The non-BPS black holes under certain conditions also exhibit the attractor phenomenon: the moduli near the horizon tend to fixed values defined by fluxes [9, 12, 13]. The corresponding attractor equation is

$$F_3 = 2\mathrm{Im}\left[\, Z\,\bar{\Omega}_3 - D^I Z\,\bar{D}_I\bar{\Omega}_3\right]_{\partial V_{BH}=0} \tag{4}$$

This equation can be also used in the form

$$H_3 = 2\mathrm{Im}\left[Z\,\bar{\Omega}_3 - \frac{(\overline{\mathcal{D}}_{\bar{a}}\overline{\mathcal{D}}_{\bar{b}}\bar{Z})g^{\bar{a}a}g^{\bar{b}b}\mathcal{D}_b Z}{2Z}\mathcal{D}_a\Omega_3\right] \ . \tag{5}$$

Stabilization of moduli is equivalent to minimization of effective N=1 supergravity potential

$$V_{flux} = |DZ|^2 - 3|Z|^2 \tag{6}$$

defined by the effective central charge Z. All supersymmetric flux vacua in type IIB string theory compactified on a Calabi-Yau manifold are subject to the attractor equations defining the values of moduli in terms of fluxes.

$$F_4 = 2\mathrm{Re}\left[Z\,\bar{\Omega}_4 + D^{0I}Z\,\bar{D}_{0I}\bar{\Omega}_4\right]_{DZ=0} \tag{7}$$

We may rewrite these equations in a form in which it is easy to recognize them as generalized attractor equations. The dependence on the axion-dilaton τ is explicit, whereas the dependence on complex structure moduli is un-explicit in the section (L, M).

$$
\begin{pmatrix} p_h^a \\[4pt] q_{ha} \\[4pt] p_f^a \\[4pt] q_{fa} \end{pmatrix}
=
\begin{pmatrix} \bar{Z}L^a + Z\bar{L}^a \\[4pt] \bar{Z}M_a + Z\bar{M}_a \\[4pt] \tau\bar{Z}L^a + \bar{\tau}ZL^a \\[4pt] \tau\bar{Z}M_a + \bar{\tau}Z\bar{M}_a \end{pmatrix}_{DZ=0}
+
\begin{pmatrix} \bar{Z}^{0I}D_I L^a + Z^{0I}\bar{D}_I\bar{L}^a \\[4pt] \bar{Z}^{0I}D_I M_a + Z^{0I}\bar{D}_I\bar{M}_a \\[4pt] \bar{\tau}\bar{Z}^{0I}D_I L^a + \tau Z^{0I}\bar{D}_I L^a \\[4pt] \bar{\tau}\bar{Z}^{0I}D_I M_a + \tau Z^{0I}\bar{D}_I\bar{M}_a \end{pmatrix}_{DZ=0}
\tag{8}
$$

The second term in this equation is absent in BH case. In the black hole case $Z = 0$ and $DZ = 0$ conditions lead to null singularity and runaway moduli. In flux vacua, the presence of the term proportional to chiral fermion masses, $\bar{\tau}\bar{Z}^{0I}$, permits the stabilization of moduli in Minkowski flux vacua.

Finally, we can also describe all non-supersymmetric flux vacua which minimize the effective potential (6) by the corresponding attractor equation

$$F_4 = 2\text{Re}\left[M_{3/2}\bar{\Omega}_4 - F^A \bar{D}_A \Omega_4 + M^{0I}\,\bar{D}_{0I}\bar{\Omega}_4\right]_{\partial V_{\text{flux}}=0} \tag{9}$$

We have presented the common features and differences in stabilization of moduli near the black hole horizon and in flux vacua.

There is an apparent similarity between non-BPS extremal black holes with stabilized moduli and the O'Raifeartaigh model of spontaneous supersymmetry breaking. In models of this type the system cannot decay to a supersymmetric ground state since such a state does not exist, so the non-SUSY vacuum is stable. The same is true of the non-BPS black hole — there is a choice of fluxes which leads to an effective superpotential such that V_{BH} does not admit a supersymmetric minimum of the potential but does admit a non-supersymmetric one, see [12], [13]. It remains a challenge to construct the analog of the stable non-BPS extremal black holes in dS flux vacua.

Bibliography

[1] S. Kachru, R. Kallosh, A. Linde and S. P. Trivedi, *De Sitter vacua in string theory*, Phys. Rev. D **68** (2003) 046005 hep-th/0301240.

[2] L. Susskind, *The anthropic landscape of string theory*, hep-th/0302219.

[3] S. B. Giddings, S. Kachru and J. Polchinski, *Hierarchies from fluxes in string compactifications*, Phys. Rev. D **66** (2002) 106006 hep-th/0105097.

[4] C. P. Burgess, R. Kallosh and F. Quevedo, *de Sitter string vacua from supersymmetric D-terms*, JHEP **0310** (2003) 056 hep-th/0309187.

[5] F. Denef, M. R. Douglas and B. Florea, *Building a better racetrack*, JHEP **0406** (2004) 034 hep-th/0404257; A. Saltman and E. Silverstein, *A new handle on de Sitter compactifications*, JHEP **0601** (2006) 139 hep-th/0411271; V. Balasubramanian, P. Berglund, J. P. Conlon and F. Quevedo, *Systematics of moduli stabilisation in Calabi-Yau flux compactifications*, JHEP **0503** (2005) 007 hep-th/0502058; B. S. Acharya, F. Denef and R. Valandro, *Statistics of M theory vacua*, JHEP **0506** (2005) 056 hep-th/0502060; F. Denef, M. R. Douglas, B. Florea, A. Grassi and S. Kachru, *Fixing all moduli in a simple F-theory compactification*, hep-th/0503124; J. P. Derendinger, C. Kounnas, P. M. Petropoulos and F. Zwirner, *Fluxes and gaugings: N = 1 effective superpotentials*, Fortsch. Phys. **53** (2005) 926 hep-th/0503229; O. DeWolfe, A. Giryavets, S. Kachru and W. Taylor, *Type IIA moduli stabilization*, JHEP **0507** (2005) 066 hep-th/0505160; D. Lust, S. Reffert, W. Schulgin and S. Stieberger, *Moduli stabilization in type IIB orientifolds. I: Orbifold limits*, hep-th/0506090.

[6] P. S. Aspinwall and R. Kallosh, *Fixing all moduli for M-theory on K3 x K3*, JHEP **0510** (2005) 001 hep-th/0506014.

[7] E. Bergshoeff, R. Kallosh, A. K. Kashani-Poor, D. Sorokin and A. Tomasiello, *An index for the Dirac operator on D3 branes with background fluxes*, JHEP **0510** (2005) 102 hep-th/0507069. R. Kallosh, A. K. Kashani-Poor and A. Tomasiello, *Counting fermionic zero modes on M5 with fluxes*, JHEP **0506** (2005) 069 hep-th/0503138.

[8] R. Kallosh, *New attractors*, JHEP **0512** (2005) 022 hep-th/0510024; A. Giryavets, *New attractors and area codes*, hep-th/0511215.

[9] S. Ferrara, R. Kallosh and A. Strominger, *N=2 extremal black holes*, Phys. Rev. D **52** (1995) 5412 hep-th/9508072; A. Strominger, *Macroscopic Entropy of N = 2 Extremal Black Holes*, Phys. Lett. B **383** (1996) 39 hep-th/9602111; S. Ferrara and R. Kallosh, *Universality of Supersymmetric Attractors*, Phys. Rev. D **54** (1996) 1525 hep-th/9603090; S. Ferrara and R. Kallosh, *Supersymmetry and Attractors*, Phys. Rev. D **54** (1996) 1514 hep-th/9602136; S. Ferrara, G. W. Gibbons and R. Kallosh, *Black holes and critical points in moduli space*, Nucl. Phys. B **500** (1997) 75 hep-th/9702103; G. W. Moore, *Arithmetic and attractors*, hep-th/9807087; G. W. Moore, *Les Houches lectures on strings and arithmetic*, hep-th/0401049; F. Denef, B. R. Greene and M. Raugas, *Split attractor flows and the spectrum of BPS D-branes on the quintic*, hep-th/0101135.

[10] K. Behrndt, R. Kallosh, J. Rahmfeld, M. Shmakova and W. K. Wong, *STU black holes and string triality*, Phys. Rev. D **54** (1996) 6293 hep-th/9608059.

[11] M. J. Duff, *String triality, black hole entropy and Cayley's hyperdeterminant*, hep-th/0601134; R. Kallosh and A. Linde, *Strings, black holes, and quantum information*, hep-th/0602061.

[12] P. K. Tripathy and S. P. Trivedi, *Non-supersymmetric attractors in string theory*, hep-th/0511117; A. Sen, *Black hole entropy function and the attractor mechanism in higher derivative gravity*, JHEP **0509** (2005) 038 hep-th/0506177.

[13] R. Kallosh, N. Sivanandam and M. Soroush, *The non-BPS black hole attractor equation*, hep-th/0602005.

4.3.2 Dieter Lüst: A short remark on flux and D-brane vacua and their statistics

String compactifications provide a beautiful link between particle physics and the geometrical and topological structures of the corresponding background geometries. Already from the "old days" of heterotic string compactifications we know that they exist a large number of consistent string compactifications with or without (being tachyon free) space-time supersymmetry, non-Abelian gauge groups and chiral matter field representations, i.e. with more or less attractive phenomenological features. In particular, a concrete number of the order of $\mathcal{N}_{vac} \sim 10^{1500}$ of heterotic string models within the covariant lattice constructions was derived [1]. More recently, a detailed analysis of type II orientifold string compactifications with D-branes, their spectra, their effective actions and also of their statistical properties was performed, and also the study of heterotic string models and their landscape was pushed forward during the last years. In this note we will comment on type II orientifold compactifications with closed string background fluxes and with open strings ending on D-branes. Two questions will be central in our discussion: first we will briefly discuss the procedure of moduli stabilization due to background fluxes and non-perturbative superpotentials. Second, we will be interested in the question what is the fraction of all possible open string D-brane configurations within a given class of orientifold models (like the $Z_2 \times Z_2$ orientifold with background fluxes) that have realistic Standard Model like properties, such as gauge group $SU(3) \times SU(2) \times U(1)$, three generations of quarks and leptons, etc. More concretely, the following steps will be important:

- We begin with choosing a toroidal Z_N resp. $Z_N \times Z_M$ type II orbifold which preserves $\mathcal{N} = 2$ space-time supersymmtry in the closed string sector.
- A consistent orientifold projection has to be performed. This yields O-planes and in general changes the geometry. The bulk space-time supersymmetry is reduced to $\mathcal{N} = 1$ by the orientifold projection. The tadpoles due to the O-planes must be cancelled by adding D-branes and/or certain background fluxes. Then the resulting Ramond-Ramond tadpole equations as well as the NS-tadpoles, which ensure $\mathcal{N} = 1$ space-time supersymmetry on the D-branes, together with constraints from K-theory provide restrictions for the allowed D-brane configurations. For each of the allowed D-brane model one has to determine the corresponding open string spectrum, namely the gauge groups and the massless matter fields, where the chiral $\mathcal{N} = 1$ matter fields are located at the various brane intersections.
- In order to stabilize the moduli one is turning on certain background fluxes that generate a potential for the moduli. According to the KKLT proposal [2], 3-form fluxes in type IIB can fix all complex structure moduli and the dilaton. On the other hand, the Kähler moduli can be fixed by non-perturbative effects. In case the fluxes or the non-perturbative superpotential break $\mathcal{N} = 1$ supersymmetry,

one can compute the soft mass terms for the open string states on the D-branes.

- In the statistical search for D-brane models with Standard Model like properties one first has to count all possible solutions of the tadpole and K-theory constraints. Then one applies certain physical thresholds, i.e. counting those models with Standard Model gauge group or those models with a certain number of chiral matter fields. Of particular interest is the question whether certain physical observables are statistically correlated.

First let us give a few comments on the moduli stabilization process due to background fluxes and non-perturbative effects. To be specific consider type IIB Ramond and NS 3-form fluxes through 3-cycles of a Calabi-Yau space X. They give rise to the following effective flux superpotential in four dimensions [3–5]:

$$W_{\text{flux}}(\tau, U) \sim \int_X (H_3^R + \tau H_3^{NS}) \wedge \Omega. \tag{1}$$

It depends on the dilaton τ and also on the complex structure moduli U. However, since W_{flux} does not depend on the Kähler moduli, one needs additional non-perturbative contributions to the superpotential in order to stabilize them. These are provided by Euclidean D3-instantons [6], which are wrapped around 4-cycles (divisors) D inside X, and/or gaugino condensations in hidden gauge group sectors on the world volumes of D7-branes, which are also wrapped around certain divisors D. Both give rise to terms in the superpotential of the form

$$W_{\text{n.p.}} \sim g_i e^{-a_i V_i}, \tag{2}$$

where V_i is the volume of the divisor D_i, depending on the Kähler moduli T. Note that the prefactor g_i is in general not a constant, but rather depends on the complex structure moduli U. The generation of a non-perturbative superpotential crucially depends on the D-brane zero modes of the wrapping divisors, i.e. on the topology of the divisors together with their interplay with the O-planes and also with the background fluxes [7–10].

The moduli are stabilized to discrete values by solving the $\mathcal{N} = 1$ supersymmetry conditions

$$D_A W = 0 \quad (\text{vanishing F} - \text{term}). \tag{3}$$

Then typically the superpotentials of the form $W_{\text{flux}} + W_{\text{n.p.}}$ lead to stable supersymmetric AdS_4 minima. Additional restrictions on the form of the possible superpotential arise [11, 12] requiring that the mass matrix of all the fields (S, T, U) is already positive definite in the AdS vacuum (absence of tachyons), as it is necessary, if one wants to uplift the AdS vacua to a dS vacuum by a (constant) shift in the potential. These conditions cannot be satisfied in orientifold models without any complex structure moduli, i.e. for Calabi-Yau spaces with Hodge number $h^{2,1} = 0$. Alternatively one can also look for supersymmetric 4D Minkowski mimima which solve eq.(3) [13, 14]. They may exist if $W_{\text{n.p.}}$ is of the racetrack form. In this case the requirement that all flat directions are lifted in the Minkowski vacuum leads to similar constraints as the absence of tachyon condition in the AdS case.

In more concrete terms, the moduli stabilization procedure was studied in [15] for the $T^6/Z_2 \times Z_2$ orientifold, with the result that all moduli indeed can be fixed. Moreover in [12, 16, 17] all other Z_N and $Z_N \times Z_M$ orientfolds were studied in great detail, where it turns out that in order to have divisors, which contribute to the non-perturbative superpotential, one has to consider the blown-up orbifold geometries. Then the divisors originating from the blowing-ups give rise to D3-instantons and/or gaugino condensates, being rigid and hence satisfying the necessary topological conditions. As a result of this investigation of all possible orbifold models, it turns out that the $Z_2 \times Z_2$, $Z_2 \times Z_4$, Z_4, Z_{6-II} orientifolds are good candidates where all moduli can be completely stabilized.

The statistical approach to the flux vacua amounts to count all solutions of the $\mathcal{N} = 1$ supersymmetry condition eq.(3) refs. In fact, it was then shown that the number of flux vacua on a given background space is very huge [18, 19]: $\mathcal{N}_{vac} \sim 10^{500}$. In addition there is another method to assign a probability measure to flux compactifications via a black hole entropy functional $\mathcal{S}$. This method however does not apply to 3-form flux compactifications but rather to Ramond 5-form compactifications on $S^2 \times X$, hence leading to AdS_2 vacua. Specifically a connection between 4D black holes and flux compactifications is provided by type $\mathcal{N} = 2$ black hole solutions, for which the near horizon condition $DZ = 0$ can be viewed as the the extremization condition of a corresponding 5-form superpotential $W \sim \int_{S^2 \times X} (F_5 \wedge \Omega)$ [20]. In view of this connection, it was suggested in [20, 21] to interpret $\psi = e^{\mathcal{S}}$ as a probability distribution resp. wave function for flux compactifications, where ψ essentially counts the microscopic string degrees of freedom, which are associated to each flux vacuum. Maximization of the entropy $\mathcal{S}$ then shows that points in the moduli space, where a certain number of hypermultiplets become massless like the conifold point, are maxima of the entropy functional [22].

Now let us also include D-branes and discuss the statistics of D-brane models with open strings [23–25]. To be specific we discuss the toroidal type IIA orientifold $T^6/Z_2 \times Z_2$ at the orbifold point. We have to add D6-branes wrapping special Langrangian 3-cycles. They are characterized by integer-valued coefficients X^I, Y^I ($I = 0, \ldots, 3$). The supersymmetry conditions, being equivalent to the vanishing of the D-term scalar potential, have the form:

$$\sum_{I=0}^{3} \frac{Y^I}{U_I} = 0, \quad \sum_{I=0}^{3} X^I U_I > 0. \tag{4}$$

The Ramond tadpole cancellation conditions for k stacks of N_a D6-branes are given by

$$\sum_{a=1}^{k} N_a \vec{X}_a = \vec{L}, \tag{5}$$

where the L^I parameterize the orientifold charge. In addition there are some more constraints from K-theory. Chiral matter in bifundamental representations originate from open strings located at the intersection of two stacks of D6-branes with a multiplicity (generation) number given by the intersection number

$$I_{ab} = \sum_{I=0}^{3}(X_a^I Y_b^I - X_b^I Y_a^I).\tag{6}$$

Counting all possible solutions of the D-brane equations (4) and (5) leads to a total of $1.66 \cdot 10^8$ supersymmetric (4-stack) D-brane models on the $Z_2 \times Z_2$ orientifold. With this large sample of models we can ask the question which fraction of models satisfy several phenomenological constraints that gradually approach the spectrum of the supersymmetric MSSM. This is summarized in the following table: Therefore

Restriction	Factor
gauge factor $U(3)$	0.0816
gauge factor $U(2)/Sp(2)$	0.992
No symmetric representations	0.839
Massless $U(1)_Y$	0.423
Three generations of quarks	2.92×10^{-5}
Three generations of leptons	1.62×10^{-3}
Total	1.3×10^{-9}

only one in a billion models give rise to an MSSM like D-brane vacuum. Similar results can be obtained for models with $SU(5)$ GUT gauge group [26].

Finally we would like to pose the following question: is it possible to obtain an entropy resp. a probability wave function for D-brane vacua? To answer this question one might try to replace the D7-branes in IIB (D6-branes in IIA) by D5-branes (D4-branes). This will lead to cosmic strings in D=4. So reformulating this question would mean, can one associate an entropy to this type of cosmic string solutions? In this way one could derive, besides the statistical counting factor, a stringy probability measure for deriving the Standard Model from D-brane models.

Bibliography

[1] W. Lerche, D. Lüst and A. N. Schellekens, "Chiral Four-Dimensional Heterotic Strings From Selfdual Lattices," Nucl. Phys. B **287** (1987) 477.

[2] S. Kachru, R. Kallosh, A. Linde and S. P. Trivedi, "De Sitter vacua in string theory," Phys. Rev. D **68** (2003) 046005 [arXiv:hep-th/0301240].

[3] S. Gukov, C. Vafa and E. Witten, "CFT's from Calabi-Yau four-folds," Nucl. Phys. B **584** (2000) 69 [Erratum-ibid. B **608** (2001) 477] [arXiv:hep-th/9906070].

[4] T. R. Taylor and C. Vafa, "RR flux on Calabi-Yau and partial supersymmetry breaking," Phys. Lett. B **474** (2000) 130 [arXiv:hep-th/9912152].

[5] S. B. Giddings, S. Kachru and J. Polchinski, "Hierarchies from fluxes in string compactifications," Phys. Rev. D **66** (2002) 106006 [arXiv:hep-th/0105097].

[6] E. Witten, "Non-Perturbative Superpotentials In String Theory," Nucl. Phys. B **474** (1996) 343 [arXiv:hep-th/9604030].

[7] L. Görlich, S. Kachru, P. K. Tripathy and S. P. Trivedi, "Gaugino condensation and nonperturbative superpotentials in flux compactifications," JHEP **0412** (2004) 074 [arXiv:hep-th/0407130].

[8] P. K. Tripathy and S. P. Trivedi, "D3 brane action and fermion zero modes in presence of background flux," JHEP **0506** (2005) 066 [arXiv:hep-th/0503072].

[9] E. Bergshoeff, R. Kallosh, A. K. Kashani-Poor, D. Sorokin and A. Tomasiello, "An index for the Dirac operator on D3 branes with background fluxes," JHEP **0510** (2005) 102 [arXiv:hep-th/0507069].

[10] D. Lüst, S. Reffert, W. Schulgin and P. K. Tripathy, "Fermion zero modes in the presence of fluxes and a non-perturbative arXiv:hep-th/0509082.

[11] K. Choi, A. Falkowski, H. P. Nilles, M. Olechowski and S. Pokorski, "Stability of flux compactifications and the pattern of supersymmetry JHEP **0411** (2004) 076 [arXiv:hep-th/0411066].

[12] D. Lüst, S. Reffert, W. Schulgin and S. Stieberger, "Moduli stabilization in type IIB orientifolds. I: Orbifold limits," arXiv:hep-th/0506090.

[13] J. J. Blanco-Pillado, R. Kallosh and A. Linde, "Supersymmetry and stability of flux vacua," arXiv:hep-th/0511042.

[14] D. Krefl and D. Lüst, "On supersymmetric Minkowski vacua in IIB orientifolds," arXiv:hep-th/0603166.

[15] F. Denef, M. R. Douglas, B. Florea, A. Grassi and S. Kachru, "Fixing all moduli in a simple F-theory compactification," arXiv:hep-th/0503124.

[16] S. Reffert and E. Scheidegger, "Moduli stabilization in toroidal type IIB orientifolds," arXiv:hep-th/0512287.

[17] D. Lüst, S. Reffert, E. Scheidegger, W. Schulgin and S. Stieberger, paper to appear.

[18] R. Bousso and J. Polchinski, "Quantization of four-form fluxes and dynamical neutralization of the cosmological constant," JHEP **0006** (2000) 006 [arXiv:hep-th/0004134].

[19] S. Ashok and M. R. Douglas, "Counting flux vacua," JHEP **0401** (2004) 060 [arXiv:hep-th/0307049].

[20] H. Ooguri, C. Vafa and E. P. Verlinde, "Hartle-Hawking wave-function for flux compactifications," Lett. Math. Phys. **74** (2005) 311 [arXiv:hep-th/0502211].

[21] S. Gukov, K. Saraikin and C. Vafa, "The entropic principle and asymptotic freedom," arXiv:hep-th/0509109.

[22] G. L. Cardoso, D. Lüst and J. Perz, paper to appear

[23] R. Blumenhagen, F. Gmeiner, G. Honecker, D. Lüst and T. Weigand, "The statistics of supersymmetric D-brane models," Nucl. Phys. B **713** (2005) 83 [arXiv:hep-th/0411173].

[24] F. Gmeiner, R. Blumenhagen, G. Honecker, D. Lüst and T. Weigand, "One in a billion: MSSM-like D-brane statistics," JHEP **0601** (2006) 004 [arXiv:hep-th/0510170].

[25] F. Gmeiner, "Standard model statistics of a type II orientifold," arXiv:hep-th/0512190.

[26] F. Gmeiner and M. Stein, "Statistics of SU(5) D-brane models on a type II orientifold," arXiv:hep-th/0603019.

4.3.3 Michael Douglas: Mathematics and String Theory: Understanding the landscape

4.3.3.1 *Historical analogies*

At a conference with such a distinguished history, one cannot help but look for analogies between the present and the past. It is tempting to compare our present struggles to understand string theory, and to find clearer evidence for or against the claim that it describes our universe, with the deep issues discussed at past Solvay Conferences, particularly in 1911 and 1927.

As was beautifully described here by Peter Galison, the 1911 meeting focused on the theory of radiation, and the quantum hypotheses invented to explain black body radiation and the photoelectric effect. These were simple descriptions of simple phenomena, which suggested a new paradigm. This was to accept the basic structure of previous models, but modify the laws of classical mechanics by inventing new, somewhat *ad hoc* rules governing quantum phenomena. This paradigm soon scored a great success in Bohr's theory of the hydrogen atom. The discovery of the electron and Rutherford's scattering experiments had suggested modeling an atom as analogous to a planetary system. But while planetary configurations are described by continuous parameters, real atoms have a unique ground state, well-defined spectral lines associated with transitions from excited states, etc. From Bohr's postulate that the action of an allowed trajectory was quantized, he was able to deduce all of these features and make precise numerical predictions.

While very successful, it was soon found that this did not work for more complicated atoms like helium. A true quantum mechanics had to be developed. Most of its essential ideas had appeared by the 1927 meeting. Although the intuitions behind the Bohr atom turned out to be correct, making them precise required existing but unfamiliar mathematics, such as the theories of infinite dimensional matrices, and wave equations in configuration space.

Are there fruitful analogies between these long-ago problems and our own? What is the key issue we should discuss in 2005? What are our hydrogen atom(s)?

If we have them, they are clearly the maximally supersymmetric theories, whose basic physics was elucidated in the second superstring revolution of 1994–98. It's too bad we can't use them to describe real world physics. But they have precise and pretty formulations, and can be used to model one system we believe exists in our universe, the near-extremal black hole. We now have microscopic models of black holes, which explain their entropy.

Perhaps we can place our position as analogous to the period between 1913 and 1927.[2] Starting from our simple and attractive maximally supersymmetric theories, we are now combining their ingredients in a somewhat *ad hoc* way, to construct $N = 1$ and nonsupersymmetric theories, loose analogs of helium, molecules, and

---

[2]A similar analogy was made by David Gross in talks given around 2000. However, to judge from his talk here, he now has serious reservations about it.

more complicated systems. The Standard Model, with its 19 parameters, has a complexity perhaps comparable to a large atom or small molecule. The difficulty of our present struggles to reproduce its observed intricacies and the underlying infrastructure (moduli stabilization, supersymmetry breaking), discussed here by Kallosh, Lüst and others, are probably a sign that we have not yet found the best mathematical framework.

4.3.3.2 *The chemical analogy*

What might this "best mathematical framework" be? And would knowing it help with the central problems preventing us from making definite predictions and testing the theory?

In my opinion, the most serious obstacle to testing the theory is the problem of vacuum multiplicity. This has become acute with the recent study of the string/M theory landscape. We have a good reason to think the theory has more than 10^{122} vacua, the Weinberg-Banks-Abbott-Brown-Teitelboim-Bousso-Polchinski *et al* solution to the cosmological constant problem. Present computations give estimates more like 10^{500} vacua. We do not even know the number of candidate vacua is finite. Even granting that it is, the problem of searching through all of them is daunting. Perhaps *a priori* selection principles or measure factors will help, but there is little agreement on what these might be. We should furthermore admit that the explicit constructions of vacua and other arguments supporting this picture, while improving, are not yet incontrovertible.

We will shortly survey a few mathematical frameworks which may be useful in coming to grips with the landscape, either directly or by analogy. They are generally not familiar to physicists. I think the main reason for this is that analogous problems in the past were attacked in different, non-mathematical ways. Let us expand a bit on this point.

String theory is by no means the first example of an underlying simple and unique framework describing a huge, difficult to comprehend multiplicity of distinct solutions. There is another one, very well known, which we might consider as a source of analogies.

As condensed matter physicists never tire of reminding us, all of the physical properties of matter in the everyday world, and the diversity of chemistry, follow in principle from a well established "theory of everything," the Schrödinger equations governing a collection of electrons and nuclei. Learning even the rough outlines of the classification of its solutions takes years and forms the core of entire academic disciplines: chemistry, material science, and their various interdisciplinary and applied relatives.

Of course, most of this knowledge was first discovered empirically, by finding, creating and analyzing different substances, with the theoretical framework coming much later. But suppose we were given the Schrödinger equation and Coulomb potential without this body of empirical knowledge? Discovering the basics of chemistry would be a formidable project, and there are many more layers of structure to elucidate before one would reach the phenomena usually discussed in condensed matter physics: phase transitions, strong correlations, topological structures and defects, and so on.

As in my talk at String 2003, one can develop this analogy, by imagining beings who are embedded in an effectively infinite crystal, and can only do low energy experiments. Say they can observe the low-lying phonon spectrum, measure low frequency conductivity, and so on. Suppose among their experiments they can create electron-hole bound states, and based on phenomenological models of these they hypothesize the Schrödinger equation. They would have some empirical information, but not the ability to manipulate atoms and create new molecules. How long would it take them to come up with the idea of crystal lattices of molecules, and how much longer would it take them to identify the one which matched their data?

Now, consider the impressive body of knowledge string theorists developed in the late 1990's, assembling quasi-realistic compactifications out of local constituents such as branes, singularities, and so on. Individual constituents are simple, their basic properties largely determined by the representation theory of the maximal supersymmetry algebras in various dimensions. The rules for combining pairs of objects, such as intersecting branes or branes wrapping cycles – which combinations preserve supersymmetry, and what light states appear – are not complicated either. What is complicated is the combination of the whole required to duplicate the Standard Model, stabilize moduli, break supersymmetry and the rest. Perhaps all this is more analogous to chemistry than we would like to admit.

Other parallels can be drawn. For example, as Joe Polchinski pointed out in his talk, according to standard nuclear physics, the lowest energy state of a collection of electrons, protons and neutrons is a collection of Fe_{26} atoms, and thus almost all molecules in the real world are unstable under nuclear processes. Suppose this were the case for our crystal dwellers as well. After learning about these processes, they might come to a deep paradox: how can atoms other than iron exist at all? Of course, because of Coulomb barriers, the lifetime of matter is exceedingly long, but still finite, just as is claimed for the metastable de Sitter vacua of KKLT.

Perhaps all this is a nightmare from which we will awake, the history of Kekulé's dream being repeated as farce. If so, all our previous experience as physicists suggests that the key to the problem will be to identify some sort of **simplicity** which we have not seen in the problem so far. One might look for it in the physics of some dual or emergent formulation. But one might also look for it in mathematics. It is not crazy to suppose that the only consistent vacua are those which respect some principle or have some property which would only be apparent in an exact treatment. But what is that exact treatment going to look like? The ones we have now cannot be formulated without bringing in mathematics such as the geometry of Calabi-Yau manifolds, or the category theory underlying topological string theory. If we ever find exact descriptions of $N = 1$ or broken supersymmetry vacua, surely this will be by uncovering even more subtle mathematical structures.

But suppose the landscape in its present shape is real, and the key to the problem is to manage and abstract something useful out of its **complexity**. The tools we will need may not be those we traditionally associated with fundamental physics, but might be inspired by other parts of physics and even other disciplines. But such inspiration can not be too direct; the actual problems are too different. Again, we are probably better off looking to mathematical developments which capture the essence of the ideas and then generalize them, as more likely to be relevant.

On further developing these analogies, one realizes that we do not know even the most basic organizing principles of the stringy landscape. For the landscape of chemistry, these are the existence of atoms, the fact that each atom (independent of its type) takes up a roughly equal volume in three-dimensional space, and that binding interactions are local. This already determines the general features of matter, such as the fact that densities of solids range from 1–20 g/cm^3. Conjectures on the finite number of string vacua, on bounds on the number of massless fields or ranks of gauge groups, and so on, are suggestions for analogous general features of string vacua. But even knowing these, we would want organizing principles. The following brief overviews should be read with this question in mind.

4.3.3.3   *Two-dimensional CFT*

This is not everything, but a large swathe through the landscape. We do not understand it well enough. In particular, the often used concept of "the space of 2d CFT's," of obvious relevance for our questions, has never been given any precise meaning.

A prototype might be found in the mathematical theory of the space of all Riemannian manifolds. This exists and is useful for broad general statements. We recall Cheeger's theorem [5]:

A set of manifolds with metrics $\{X_i\}$, satisfying the following bounds,

(1) diameter(X_i) $< d_{max}$
(2) Volume (X_i) $> V_{min}$
(3) Curvature K satisfies $|K(X_i)| < K_{max}$ at every point,

contains a finite number of distinct homeomorphism types (and diffeomorphism types in $D \neq 4$).

Since (2) and (3) are conditions for validity of supergravity, while (1) with $d_{max} \sim 10\mu m$ follows from the validity of the gravitational inverse square law down to this distance, this theorem implies that there are finitely many manifolds which can be used for candidate supergravity compactifications [9, 2].

This and similar theorems are based on more general quasi-topological statements such as Cheeger-Gromov precompactness of the space of metrics – *i.e.*, infinite sequences have Cauchy subsequences, and cannot "run off to infinity." This is shown by constructions which break any manifold down into a finite number of coordinate patches, and showing that these patches and their gluing can be described by a finite amount of data.

Could we make any statement like this for the space of CFT's? (a question raised by Kontsevich). The diameter bound becomes a lower bound Δ_{min} on the operator dimensions (eigenvalues of $L_0 + \bar{L}_0$). We also need to fix c. Then, the question seems well posed, but we have no clear approach to it. Copying the approach in terms of coordinate patches does not seem right.

The key point in defining any "space" of anything is to put a topology on the set of objects. Something less abstract from which a topology can be derived is a distance between pairs of objects $d(X, Y)$ which satisfies the axioms of a metric, so that it can be used to define neighborhoods.

The usual operator approach to CFT, with a Hilbert space $\mathcal{H}$, the Virasoro algebras with $H = L_0 + \bar{L}_0$, and the operator product algebra, is very analogous to spectral geometry:

$$L_0 + \bar{L}_0 \text{ eigenvalues} \sim \text{spectrum of Laplacian } \Delta$$
$$\text{o.p.e. algebra} \sim \text{algebra of functions on a manifold}$$

Of course the o.p.e. algebra is not a standard commutative algebra and this is analogy, but a fairly close one.

A definition of a distance between a pair of manifolds with metric, based on spectral geometry, is given in Bérard, Besson, and Gallot [4]. The idea is to consider the entire list of eigenfunctions $\psi_i(x)$ of the Laplacian,

$$\Delta \psi_i = \lambda_i \psi_i,$$

as defining an embedding Ψ of the manifold into ℓ_2, the Hilbert space of semi-infinite sequences (indexed by i):

$$\Psi : x \to \{e^{-t\lambda_1}\psi_1(x), e^{-t\lambda_2}\psi_2(x), \ldots, e^{-t\lambda_n}\psi_n(x), \ldots\}.$$

We weigh by $e^{-t\lambda_i}$ for some fixed t to get convergence in ℓ_2.

Then, the distance between two manifolds M and M' is the Hausdorff distance d between their embeddings in ℓ_2. Roughly, this is the amount $\Psi(M)$ has to be "fuzzed out" to cover $\Psi(M')$.

In principle this definition might be directly adapted to CFT, where the x label boundary states $|x\rangle$ (which are the analog of points) and the $\psi_i(x)$ are their overlaps with closed string states $|\phi_i\rangle$,

$$x \to \langle \phi_i | e^{-t(L_0 + \bar{L}_0)} | x \rangle.$$

Another candidate definition would use the o.p.e. coefficients

$$\phi_i \phi_j \to \sum C_{ij}^k (z_i - z_j)^{\Delta_k - \Delta_i - \Delta_j} \phi_k$$

for all operators with dimensions between Δ_{min} and some Δ_{max} (one needs to show that this choice drops out), again weighted by $e^{-t(L_0 + \bar{L}_0)}$. The distance between a pair of CFT's would then be the ℓ_2 norm of the differences between these sets of numbers.

While abstract, this would make precise the idea of the "space of all 2D CFT's" and give a foundation for mapping it out.

4.3.3.4 *Topological open strings and derived categories*

This gives an example in which we actually know "the space of all X" in string theory. It is based on the discussion of boundary conditions and operators in CFT, which satisfy an operator product algebra with the usual non-commutativity of open strings. If we modify the theory to obtain a subset of dimension zero operators (by twisting to get a topological open string, taking the Seiberg-Witten limit in a B field, etc.), the o.p.e. becomes a standard associative but non-commutative algebra. This brings us into the realm of noncommutative geometry.

There are many types of noncommutative geometry. For the standard topological string obtained by twisting an $N = 2$ theory, the most appropriate is based on algebraic geometry. As described at the Van den Bergh 2004 Francqui prize colloquium, this is a highly developed subject, which forms the backdrop to quiver gauge theories, D-branes on Calabi-Yau manifolds, and so on.

One can summarize the theory of D-branes on a Calabi-Yau X in these terms as the "Pi-stable objects in the derived category $D(\text{Coh } X)$," as reviewed in [3]. Although abstract, the underlying idea is simple and physical. It is that all branes can be understood as bound states of a finite list of "generating branes," one for each generator of K theory, and their antibranes. The bound states are produced by tachyon condensation. Varying the Calabi-Yau moduli can vary masses of these condensing fields, and if one goes from tachyonic to massive, a bound state becomes unstable.

This leads to a description of all D-branes, and "geometric" pictures for all the processes of topology change which were considered "non-geometric" from the purely closed string point of view. For example, in a flop transition, an S^2 Σ is cut out and replaced with another S^2 Σ' in a topologically different embedding. In the derived category picture, what happens is that the brane wrapped on Σ, and all D0's (points) on Σ, go unstable at the flop transition, to be replaced by new branes on Σ'.

The general idea of combining classical geometric objects, using stringy rules of combination, and then extrapolating to get a more general type of geometry, should be widely useful.

4.3.3.5 *Computational complexity theory*

How hard is the problem of finding quasi-realistic string vacua? Computer scientists classify problems of varying degrees of difficulty:

- P can be solved in time polynomial in the size of the input.
- An NP problem has a solution which can be checked in polynomial time, but is far harder to find, typically requiring a search through all candidate solutions.
- An NP-complete problem is as hard as any NP problem – if any of these can be solved quickly, they all can.

It turns out that many of the problems arising in the search for string vacua are in NP or even NP-complete. [6] For example, to find the vacua in the Bousso-Polchinski model with cosmological constant $10^{-122} M_{Planck}^4$, one may need to search through 10^{122} candidates.

How did the universe do this? We usually say that the "multiverse" did it – many were tried, and we live in one that succeeded. But some problems are too difficult for the multiverse to solve in polynomial time. This is made precise by Aaronson's definition of an "anthropic computer." [1]

Using these ideas, Denef and I [7] have argued that the vacuum selected by the measure factor $\exp 1/\Lambda$ cannot be found by a quantum computer, working in polynomial time, even with anthropic postselection. Thus, if a cosmological model realizes this measure factor (and many other preselection principles which can be expressed as optimizing a function), it is doing something more powerful than such a computer.

Some cosmological models (e.g. eternal inflation) explicitly postulate exponentially long times, or other violations of our hypotheses. But for other possible theories, for example a field theory dual to eternal inflation, this might lead to a paradox.

4.3.3.6 *Conclusions*

We believe string theory has a set of solutions, some of which might describe our world. Even leaving aside the question of few vacua or many, and organizing principles, perhaps the most basic question about the landscape is whether it will turn out to be more like mathematics, or more like chemistry.

Mathematical analogy: like classification of Lie groups, finite simple groups, Calabi-Yau manifolds, etc. Characterized by simple axioms and huge symmetry groups. In this vision, the overall structure is simple, while the intricacies of our particular vacuum originate in symmetry breaking analogous to that of more familiar physical systems.

Chemical analogy: simple building blocks (atoms; here branes and extended susy gauge theory sectors) largely determined by symmetry. However, these are combined in intricate ways which defy simple characterization and require much study to master.

The current picture, as described here by Kallosh and Lüst, seems more like chemistry. Chemistry is a great science, after all the industrial chemistry of soda is what made these wonderful conferences possible. But it will surely be a long time (if ever) before we can manipulate the underlying constituents of our vacuum and produce new solutions, so this outcome would be less satisfying.

Still, our role as physicists is not to hope that one or the other picture turns out to be more correct, but to find the evidence from experiment and theory which will show us which if any of our present ideas are correct.

Acknowledgments I would like to thank the organizers and the Solvay Institute for a memorable meeting, and B. Acharya, F. Denef, M. Kontsevich, S. Zelditch and many others for collaboration and discussion of these ideas. This research was supported in part by DOE grant DE-FG02-96ER40959.

Bibliography

[1] S. Aaronson, "Quantum Computing, Postselection, and Probabilistic Polynomial-Time," arXiv:quant-ph/0412187.

[2] B. Acharya and M. R. Douglas, to appear.

[3] P. S. Aspinwall, "D-branes on Calabi-Yau manifolds," arXiv:hep-th/0403166.

[4] P. Bérard, G. Besson, and S. Gallot, "Embedding Riemannian manifolds by their heat kernel," Geom. Funct. Anal. 4 (1994), 373–398.

[5] J. Cheeger, "Finiteness theorems for Riemannian manifolds," Am. J. Math **92** (1970), 61–74.

[6] F. Denef and M. R. Douglas, "Computational Complexity of the Landscape I," hep-th/0602072.

[7] F. Denef and M. R. Douglas, "Computational Complexity of the Landscape II," to appear.

[8] M. R. Douglas, talk at Strings 2003.

[9] M. R. Douglas, talk at Strings 2005.

4.4 Discussion

A. Van Proeyen I want to make a remark or maybe a question about the structure of KKLT (Kachru, Kallosh, Linde and Trivedi). As far as I know, there is still no consistent supergravity framework for it. For the first step, there is no problem when one uses the superpotential. But then the uplifting needs D-terms which, as far as I heard, always needs the Fayet-Illiopoulos terms. But that is not consistent if one has already put a superpotential. So I do not know how this can be solved and brought into a consistent effective supergravity framework. Or is there another mechanism that someone sees? As far as I can see, the KKLT framework as a supergravity theory is just not yet consistent.

R. Kallosh In short, since 2003 when we attempted this uplifting, it was clear that the supergravity has Fayet-Illiopoulos terms, but to get them from string theory is rather difficult. So the best case we know today is when we have D7-branes at the tip of the conifold and fluxes on it. From the perspective of string theory this looks as close to a consistent D-term in supergravity as possible. At present we cannot do better, but I hope that somebody will. It needs to be done.

H. Ooguri In that spirit I would also like to note that the study of the landscape is still looking at a very limited range of the possible moduli space. There is a big territory that needs to be understood. In particular, in the context of developing mathematical tools to understand it, I would like to note that it is very important to understand the stringy corrections to this program. I hope to have some progress in that direction. I guess Kachru has a comment on that.

S. Kachru I completely agree with what you just said. This is more a comment about the status of constructions. The initial construction used some kind of configuration of branes to get a positive energy. But in fact if you look through the literature that has been generated in the last three years, there are now somehow an infinite number of proposals. The most mild one actually just uses the F-term potential coming from fluxes themselves to give the positive term. I actually do not see any possible inconsistency with embedding into supergravity. Explicit examples that give examples where the F-terms are non zero in the minima of the flux potential were actually constructed by Saltman and Silverstein. The statistics of Denef and Douglas that were quoted by Lüst in giving 10^{500} models, were actually more or less counting those vacua. So this may be relevant to Van Proeyen's question.

A. Van Proeyen As far as I can see, as long as one only has the F-terms, one still has vacua with a negative cosmological constant. You still need the uplifting.

S. Kachru No. My point was that the so-called uplifting can be done by non vanishing F-terms because of the fact that the F-terms contribute positively in the supergravity potential.

A. Strominger This sounded interesting but I did not catch the first statement: "There is no supergravity known with an F-I-term and a F-term in how many dimensions?" What was the problem you were referring to?

A. Van Proeyen $N = 1$ in four dimensions. If you want to add a Fayet-Illiopoulos term. As far as I can see, this is still necessary to uplift the potential to have it in a de Sitter vacuum. You cannot add just a Fayet-Illiopoulos term in supergravity if you have a non trivial superpotential, unless the superpotential is not invariant under the gauge symmetry that corresponds to the Fayet-Illiopoulos term. It is something which is not well known but it is a restriction in $N = 1$ supergravity.

A. Strominger And it is known to be impossible to do that or we just have not figured out how to do it yet?

A. Van Proeyen It is known to be impossible.

H. Ooguri It sounds like we need some response.

N. Seiberg Can you state very clearly what it is that is not possible?

A. Van Proeyen To add a Fayet-Illiopoulos term when you have a non trivial superpotential.

S. Kachru Can I make a comment that is relevant?

H. Ooguri That will be the last comment.

S. Kachru Of course what happens in the supergravities is that the Fayet-Illiopoulos terms become field dependent. Presumably in this model with the anti-brane, what happens is that the Kähler mode upon which the D-3-brane tension depends, which is included in the potential, has an axion partner. The coefficient of the superpotential that is used transforms by a shift under a gauge symmetry. I think this makes the structure that was used in the original model completely consistent with the field dependent F-I terms of supergravity. Do you agree with that possibility?

A. Van Proeyen Yes. I agree with the possibility. I have not seen a model, but I agree with the possibility.

J. Harvey The subject of the session is mathematical structures. I feel a certain tension between Dijkgraaf's beautiful talk about all the wonderful structures that come out and the comment that Douglas made about how our hydrogen atom is this maximally supersymmetric solutions. I have a feeling that, without actual hydrogen atoms and helium atoms and molecules, a string theorists faced with just the quantum mechanics of the hydrogen atom would discover you can use $SO(4, 2)$ as a spectrum generating algebra. He would then generalise it to $SO(p, q)$ rather than generalising to the helium atom. It is well recognised that one of the central problems facing string theory is how to narrow our research down to the investigation of the correct mathematical structures rather than this infinite sea of beautiful possibilities. The connections are wonderful and very inspiring. But how do we figure out which are the right directions to go without experiment, without data? This is not a new question, but I think it would be very welcome to have some discussion at this meeting of how we do this, how we try to connect string theory to data. Obviously these flux compactifications are an attempt, but so far they, and other things, have not really succeeded.

L. Randall I agree very much with the first part of the comments. I am surprised at the second part. Actually I thought what was surprising about this, is how narrowly defined this discussion is, in the sense that we are looking at the kind of things we know how to deal with mathematically at this point, namely things based on supersymmetric structures or even compactification. It seems to me that, before one starts restricting, one would want to make sure that we have actually covered the entire true landscape of what is possible.

———————————

H. Ooguri The second part of the program is about new geometrical structure that appears. Since string theory contains gravity, geometry has naturally played an important rôle. We expect that we learn more about geometrical structure from string theory. We first hear from Nekrasov about new applications of geometric ideas to physics.

4.5 Prepared Comments

4.5.1 *Nikita Nekrasov: On string theory applications in condensed matter physics*

Quantum field theorists have benefited from ideas originating in the condensed matter physics. In this note we present an interesting model of electrons living on a two dimensional lattice, interacting with random electric field, which can be solved using the knowledge accumulated in the studies of superstring compactifications.

4.5.1.1 *Electrons on a lattice, with noisy electric field*

Here is the model. Consider the hexagonal lattice with black and white vertices so that only the vertices of the different colors share a common edge. Let B, W denote the sets of black and white vertices, respectively. We can view the edges as the maps $e_i : B \to W$, $e_i^* : W \to B$, $i = 1, 2, 3$. The edge e_1 points northwise, e_2: southeast, and e_3 southwest. The set of edges, connecting black vertices with white ones will be denoted by E. We have two maps: $s : E \to B$ and $t : E \to W$, which send an edge to its source and target.

The free electrons on the lattice are described by the Lagrangian

$$L_0 = \sum_{b \in B} \sum_{i=1,2,3} \psi_b \psi_{e_i(b)}^* = \sum_{w \in W} \sum_{i=1,2,3} \psi_{e_i^*(w)} \psi_w^* \tag{1}$$

The variables ψ_b, ψ_w^* are fermionic variables. Our "electrons" will interact with the $U(1)$ gauge field A_e, where $e \in E$. Introduce three (complex) numbers $\varepsilon_1, \varepsilon_2, \varepsilon_3$, and their sum:$\varepsilon = \varepsilon_1 + \varepsilon_2 + \varepsilon_3$. We make the free Lagrangian (1) gauge invariant, by:

$$L_{\psi A} = \sum_{b \in B} \sum_{i=1}^{3} \psi_b e^{i \varepsilon A_{e_i}(b)} \psi_{e_i(b)}^* \tag{2}$$

The gauge transformations act as follows:

$$\psi_b \mapsto e^{i \varepsilon \theta_b}, \quad \psi_w^* \mapsto e^{-i \varepsilon \theta_w} \psi_w^*, \quad A_e \mapsto A_e + \theta_{t(e)} - \theta_{s(e)} \tag{3}$$

The Lagrangian (2) is invariant under (3) but the measure $\mathcal{D}\psi \mathcal{D}\psi^*$ is not, there is an "anomaly". It can be cancelled by adding the following Chern-Simons - like term to the Lagrangian (2)

$$L_{CS} = -i \sum_{b \in B} \sum_{i=1}^{3} \varepsilon_i A_{e_i}(b) \tag{4}$$

In continuous theory in two dimensions one can write the gauge invariant Lagrangian for the gauge field using the first order formalism:

$$\mathcal{L}_{2dYM} = \int_\Sigma \mathrm{tr} E F_A + \sum_k t_k \mathrm{tr} E^k \tag{5}$$

where E is the adjoint-valued scalar, the electric field. In the conventional Yang-Mills theory only the quadratic Casimir is kept in (5), t_2 playing the role of the (square) of the gauge coupling constant. In our case, the analogue of the Lagrangian (5) would be $\mathcal{L}_{\text{latticeYM}} = \sum_f \left(h_f \sum_{e \in \partial f} \pm A_e \right) + \sum_f \mathcal{U}(h_f)$. Note that in the continuous theory one could have added more general gauge invariant expression in E, i.e. involving the derivatives. The simplest non-trivial term would be: $\mathcal{L} = \mathcal{L}_{\text{YM}} + \int \text{tr}g(E)\Delta_A E$ where g is, say, polynomial. Such terms can be generated by integrating out some charged fields. Our lattice model has the kinetic term for the electric field, as well as the linear potential (it is possible in the abelian theory):

$$L_{Ah} = i \sum_f \left(h_f \sum_{e \in \partial f} \pm A_e \right) - \sum_f \mathcal{U}(h_f)(\Delta h)_f - t \sum_f h_f \qquad (6)$$

where Δ is the lattice Laplacian, and the "metric" $\mathcal{U}(x)$ is a random field, a gaussian noise with the dispersion law[3]:

$$\langle \mathcal{U}(x)\mathcal{U}(y) \rangle = D(x-y) \equiv \int_0^\infty dt \frac{e^{-t(x-y)}}{t(1 - e^{t\varepsilon_1})(1 - e^{t\varepsilon_2})(1 - e^{t\varepsilon_3})} \qquad (7)$$

The partition function of our model is (we should fix some boundary conditions, see below)

$$Z(t, \varepsilon_1, \varepsilon_2, \varepsilon_3) = \int \mathcal{D}\mathcal{U}e^{-\int \mathcal{U}(x)(D^{-1}\circ\mathcal{U})(x)} \int \mathcal{D}\psi \mathcal{D}\psi^* \mathcal{D}A\mathcal{D}h \, e^{L_{\psi A} + L_{CS} + L_{Ah}} \qquad (8)$$

4.5.1.2 *Dimers and three dimensional partitions*

We now proceed with the solution of the complicated model above. The idea is to expand in the kinetic term for the $\psi\psi^*$. The non-vanishing integral comes from the terms where every vertex, both black and white, is represented by the corresponding fermions, and exactly once. Thus the integral over ψ, ψ^* is the sum over dimer configurations [5],[6], weighted with the weight

$$\sum_{\text{dimers}} \prod_{e \in \text{dimer}} e^{i\varepsilon A_e} \qquad (9)$$

The gauge fields A_e enter now linearly in the exponential, integrating them out we get an equation $dh = \star\omega_{\text{dimer}}$ where ω_{dimer} is the one-form on the hexagonal lattice, whose value on the edge is equal to $\pm\varepsilon_{1,2,3}$ depending on its orientation $\pm\varepsilon$ depending on whether it belongs to the dimer configuration or not. Everything is arranged so the that at each vertex v the sum of the values of ω on the three incoming edges is equal to zero. The solution of the equation on h gives what is called *height function* in the theory of dimers. In our case it is the electric field. If we plot the graph of h_f and make it to a piecewise-linear function of two variables in an obvious way, we get a two dimensional surface – the boundary of a generalized three

---

[3]the integral is regularized via $\int \frac{dt}{t} \rightarrow \frac{d}{ds}\Big|_{s=0} \frac{1}{\Gamma(s)} \int \frac{dt}{t} t^s$.

dimensional partition. In order to make it a boundary of actual three dimensional (or plane) partition, we have to impose certain boundary conditions: asymptotically the graph of h_f looks like the boundary of the positive octant $\mathbf{R}_+^3$ [4]. Under these conditions, the final sum over dimers is equivalent to the sum over three dimensional partitions of the so-called *equivariant measure* [3]. The three dimensional partition is a (finite) set $\pi \subset \mathbf{Z}_+^3$ whose complement in $\bar{\pi} = \mathbf{Z}_+^3 \setminus \pi$ is invariant under the action of $\mathbf{Z}_+^3$. In other words, the space I_π of polynomials in three variables, generated by monomials $z_1^i z_2^j z_3^k$ where $(i,j,k) \in \bar{\pi}$ is an ideal, invariant under the action of the three dimensional torus $\mathbf{T}^3$. Let $ch_\pi = \sum_{(i,j,k)\in\pi} q_1^{i-1} q_2^{j-1} q_3^{k-1}$, $ch_{\bar\pi}(q) = \frac{1}{P(q)} - ch_\pi$, $|\pi| = ch_\pi(1)$, $P(q) = (1-q_1)(1-q_2)(1-q_3)$, $q_i = e^{\varepsilon_i}$. Define the "weights" x_α, y_α from $1/P(q) - P(q^{-1})ch_{\bar\pi}(q)ch_{\bar\pi}(q^{-1}) = \sum_\alpha e^{x_\alpha} - \sum_\alpha e^{y_\alpha}$. Then,

$$\mu_\pi(\varepsilon_1, \varepsilon_2, \varepsilon_3) = \prod_\alpha \frac{y_\alpha}{x_\alpha} \tag{10}$$

The partition function of our model reduces to:

$$Z(t, \varepsilon_1, \varepsilon_2, \varepsilon_3) = \sum_\pi \mu_\pi(\varepsilon_1, \varepsilon_2, \varepsilon_3) e^{-t|\pi|} \tag{11}$$

4.5.1.3 *Topological strings and S-duality*

The last partition function arises in the string theory context. The ideals I_π are the fixed points of the action of the torus $\mathbf{T}^3$ on the moduli space of zero dimensional D-branes in the topological string of B type on $\mathbf{C}^3$, bound to a single D5-brane, wrapping the whole space. The equivariant measure μ_π is the ratio of determinants of bosonic and fermionic fluctuations around the solution I_π in the corresponding gauge theory. The parameter t is the (complexified) theta angle, which couples to $\mathrm{tr}F^3$ instanton charge. This model is an infinite volume limit of a topological string on compact Calabi-Yau threefold. The topological string on Calabi-Yau threefold is the subsector of the physical type II superstring on Calabi-Yau $\times \mathbf{R}^4$. It inherits dualities of the physical string, like mirror symmetry and S-duality [4]. It maps the type B partition function (11) to the type A partition function. The latter counts holomorphic curves on the Calabi-Yau manifold. In the infinite volume limit it reduces to the two dimensional topological gravity contribution of the constant maps, which can be evaluated to be [3]:

$$Z(t, \varepsilon_1, \varepsilon_2, \varepsilon_3) = \exp\left(\frac{(\varepsilon_1+\varepsilon_2)(\varepsilon_3+\varepsilon_2)(\varepsilon_1+\varepsilon_3)}{\varepsilon_1\varepsilon_2\varepsilon_3}\right) \sum_{g=0}^\infty t^{2g-2} \frac{B_{2g-2}B_{2g}}{2g(2g-2)(2g-2)!} \tag{12}$$

$$= M(-e^{-it})^{-\frac{(\varepsilon_1+\varepsilon_2)(\varepsilon_3+\varepsilon_2)(\varepsilon_1+\varepsilon_3)}{\varepsilon_1\varepsilon_2\varepsilon_3}} \tag{13}$$

$$\tag{14}$$

where $M(q) = \prod_{n=1}^\infty (1-q^n)^{-n}$ is the so-called MacMahon function.

---

[4]i.e. as the function: $h(x,y) = \varepsilon_1 i + \varepsilon_2 j + \varepsilon_3 k$, $x = i - (j+k)/2$, $y = (j-k)/2$, $i,j,k \geq 0$, $ijk = 0$

4.5.1.4 *Discussion*

We have illustrated in the simple example that the string dualities can be used to solve for partition functions of interesting statistical physics problems. The obvious hope would be that the dualities are powerful enough to provide information on the correlation functions as well. One can consider more general lattices or boundary conditions (they correspond to different toric Calabi-Yau's), more sophisticated noise functions $D(x)$ (e.g. the one coming from Z-theory [7]) . Also, it is tempting to speculate that compact CYs correspond to more interesting condensed matter problems.

I am grateful to A.Okounkov for numerous fruitful discussions.

Bibliography

[1] A. Okounkov, N. Reshetikhin, C. Vafa, *Quantum Calabi-Yau and Classical Crystals*, hep-th/0309208.
[2] A. Iqbal, N. Nekrasov, A. Okounkov, C. Vafa, *Quantum Foam and Topological Strings*, hep-th/0312022.
[3] D. Maulik, N. Nekrasov, A. Okounkov, R. Pandharipande, *Gromov-Witten theory and Donaldson-Thomas theory, I,II*, math.AG/0312059, math.AG/0406092.
[4] N. Nekrasov, H. Ooguri, C. Vafa, *S-duality and topological strings*, JHEP 0410 (2004) 009, hep-th/0403167.
[5] R. Kenyon, A. Okounkov, S. Sheffield *Dimers and amoebas*, math-ph/0311005.
[6] R. Kenyon, A. Okounkov, *Planar dimers and Harnack curves*, math.AG/0311062.
[7] N. Nekrasov, *Z-Theory*, hep-th/0412021.

4.5.2   Shing-Tung Yau: Mathematical   Structures: Geometry of Six-Dimensional String

There has been a great deal of anxiety to find a suitable string vacuum solution or to perform statistics over the space of all such vacua. However, despite great successes in the twenty years since the first string revolution, our understanding of string vacua is far from complete. For starters, we have not achieved a satisfactory theory of computing supersymmetric cycles nor a good understanding of Hermitian-Yang-Mills fields and their instantons. Such issues pertain to deep problems in mathematics and ideas inspired from physical considerations have been essential for progress.

In the compactificaton of Candelas-Horowitz-Strominger-Witten [3], preserving supersymmetry with zero H-flux requires the compact six-manifold to be Kähler Calabi-Yau. While this class of manifold is quite large, it is believed to have a finite number of components with finite dimensions for its moduli space.

For the class of three-dimensional Kähler Calabi-Yau manifolds, there is a construction due to works of Clemens [4] and Friedman [6] where one takes a finite number of rational curves with negative normal bundle and pinch them to conifold points. Under suitable conditions for the homology class, one can deform the resulting (singular) manifold to a smooth manifold. The resulting manifold is in general non-Kähler. By repeating such procedures several times, one can obtain a smooth complex manifold with vanishing second Betti number (and hence clearly non-Kähler). If the homology of the original Calabi-Yau manifold has no torsion, a theorem of Wall [16] shows that the resulting manifold must be diffeomorphic to a connected sums of $S^3 \times S^3$. This type of manifold can be considered as a natural generalization of Riemann surfaces which are connected sums of handle bodies. These three-dimensional complex manifold also have a holomorphic three-form that is naturally inherited from the original Calabi-Yau.

There is a proposal of Reid [14] that the moduli space of all Calabi-Yau structures can be connected through such complex structures over handle bodies. Such a proposal may indeed be true. However, an immediate problem is that we are then required to analyze non-Kähler complex manifolds but we have virtually no theory for them. Non-Kähler manifolds have appeared naturally in string compactifications with fluxes. So perhaps a useful way to think about the construction of Clemens and Friedman is that the collapsing of the rational curves together with the deformation of the complex structure correspond to turning on a flux. Hence, just from mathematical considerations of the Calabi-Yau moduli space, we are led to study structures which contain fluxes and preserve supersymmetries.

A natural supersymmetric geometry with flux to consider is the one in heterotic string theory. The geometry is constrained by a system of differential equations worked out by Strominger [15] and takes the following form

$$d(\| \Omega \|_\omega \omega^2) = 0 \, ,$$
$$F^{2,0} = F^{0,2} = 0 \, , \quad F \wedge \omega^2 = 0 \, ,$$

$$dH = \sqrt{-1}\,\partial\bar\partial\omega = \alpha'(\mathrm{tr}R \wedge R - \mathrm{tr}F \wedge F)\,,$$

where ω is the hermitian metric, Ω the holomorphic three-form, and F the Hermitian-Yang-Mills field strength. In above, the H-flux is given by $H = \frac{\sqrt{-1}}{2}(\bar\partial - \partial)\omega$.

It would be nice to understand geometrically how the flux can be turned on from a thorough analysis of the Strominger system for the non-Kähler Calabi-Yau handle bodies. As a first step, Li-Yau [12] have shown in a rather general setting, that one can always obtain a solution to the above equations by perturbing the Calabi-Yau vacuum with the gauge bundle being a sum of the tangent bundle together with copies of the trivial bundle. The deformation to non-zero H-flux will mix together the tangent and trivial bundle parts of the gauge bundle. This allows Li-Yau to construct non-zero H-flux solutions with $SU(4)$ and $SU(5)$ gauge group. In the analysis of the deformations of such gauge bundles, the deformation space of the Kähler and complex structure of the Calabi-Yau naturally arised. Therefore, studying such deformation to non-zero H-flux systems can give insights into the moduli space of Calabi-Yau.

The first equation of the Strominger system calls for the existence of a balanced metric on such manifolds. These are n-dimensional complex manifolds which admit a hermitian metric ω that satisfies $d(\omega^{n-1}) = 0$ [13]. Balanced metrics satisfy many interesting properties such as being invariant under birational transformations as was observed by Alessandrini and Bassanelli [1]. Using parallel spinors, it is possible to decompose the space of differential forms similar to that of Hodge decomposition. This has been carried out by my student C. C. Wu in her thesis.

Presently, we do not know how large is the class of balanced manifolds. Michelsohn has shown that for the twistor space of anti-self-dual four manifolds, the natural complex structure is balanced [13]. It may be useful to identify such manifolds whose anti-canonical line bundle admits a holomorphic three-form. Another well-known class of non-Kähler manifolds that is balanced consists of $K3$ surfaces fibered with a twisted torus bundle. In this special case, there is a metric ansatz [5, 8] which enabled Fu-Yau [7] to demonstrate the existence of a solution to the Strominger system that is not connected to a Calabi-Yau manifold. The existence of such a solution is consistent with duality chasing arguments from M-theory that were first discussed in detail by Becker-Dasgupta [2].

As mentioned, the theory of complex non-Kähler manifolds has not been developed much. Similar to Calabi-Yau compactification, it will be important to rephrase the four-dimensional physical quantities like the types and number of massless fields or the Yukawa coupling in terms of the properties of the non-Kähler manifold. For example, can the number massless modes or geometric moduli be expressed purely in terms of certain geometrical quantities perhaps analogous to the Hodge numbers for the Kähler case. Here, trying to answer such physical questions will compel us to seek a deeper understanding of the differential structures of non-Kähler manifold than that known currently. It is likely that fluxes and in particular the H-flux (which is the torsion in the heterotic theory) will play a central role in non-Kähler stuctures.

More importantly, the study of complex non-Kähler manifolds is another step in understanding the whole space of string solutions or vacua. The space of string vacua contains both geometrical and non-geometrical regions. But even within the geometrical region, the compactification manifold need not be Kähler nor even complex (for type IIA theory [9]) when α' corrections and branes are allowed. This seems to give many possibilities for the geometry of the internal six-manifold for different types of string theories. However, since the six different string theories are related to each other through various dualities, the geometries and structures of six-dimensional compact manifolds associated with string vacua are most likely also subtlely related. This gives hope that the space of string vacua can indeed be understood well-enough such that we can confirm or rule out that there exists at least one string vacuum that can reproduce the four-dimensional standard model of our world. Given the recent successes of compactification with fluxes - from moduli fixing [10] to addressing the cosmological constant issue [11] - we can expect that the physical real world vacuum will involve fluxes and understanding the structures of non-Kähler manifolds may prove indispensable.

Bibliography

[1] L. Alessandrini and G. Bassanelli, *Modifications of compact balanced manifolds*, C. R. Acad. Sci. Paris Sér. I Math. **320** (1995) no. 12, 1517.

[2] K. Becker and K. Dasgupta, *Heterotic strings with torsion*, JHEP **0211** (2002) 006, hep-th/0209077.

[3] P. Candelas, G. T. Horowitz, A. Strominger and E. Witten, *Vacuum configurations for superstrings*, Nucl. Phys. B **258** (1985) 46.

[4] C. H. Clemens, *Double solids*, Adv. in Math. **47** (1983) no. 2, 107;
    C. H. Clemens, *Homological equivalence, modulo algebraic equivalence, is not finitely generated*, IHES Publ. Math. **58** (1983) 19.

[5] K. Dasgupta, G. Rajesh and S. Sethi, *M theory, orientifolds and G-flux*, JHEP **9908** (1999) 023, hep-th/9908088.

[6] R. Friedman, *Simultaneous resolution of threefold double points*, Math. Ann. **274** (1986) no. 4, 671.

[7] J.-X. Fu and S.-T. Yau, *Existence of supersymmetric Hermitian metrics with torsion on non-Kähler manifolds*, hep-th/0509028.

[8] E. Goldstein and S. Prokushkin, *Geometric model for complex non-Kähler manifolds with SU(3) structure*, Commun. Math. Phys. **251** (2004) 65, hep-th/0212307.

[9] M. Grana, R. Minasian, M. Petrini and A. Tomasiello, *Generalized structures of N = 1 vacua*, JHEP **0511** (2005) 020, hep-th/0505212.

[10] S. Gukov, C. Vafa and E. Witten, *CFT's from Calabi-Yau four-folds*, Nucl. Phys. B **584** (2000) 69; erratum-ibid. B **608** (2001) 477, hep-th/9906070.

[11] S. Kachru, R. Kallosh, A. Linde and S. P. Trivedi, *De Sitter vacua in string theory*, Phys. Rev. D **68** (2003) 046005, hep-th/0301240.

[12] J. Li and S.-T. Yau, *The existence of supersymmetric string theorey with torsion*, J. Differential Geom. **70** (2005) 143, hep-th/0411136.

[13] M. L. Michelsohn, *On the existence of special metrics in complex geometry*, Acta

Math. **149** (1982) no. 3-4, 261.

[14] M. Reid, *The moduli space of 3-folds with K = 0 may nevertheless be irreducible,* Math. Ann. **278** (1987) 329.

[15] A. Strominger, *Superstrings with torsion,* Nucl. Phys. B **274** (1986) 253.

[16] C. T. C. Wall, *Classification problems in differential topology. V. On certain 6-manifolds,* Invent. Math. **1** (1966) 355; corrigendum, ibid **2** (1966) 306.

4.6 Discussion

B. Greene Just in the spirit of the Mathematics-Physics interface which is the theme of the session, I might note that Dine and Seiberg some time ago have showed that the existence of certain R-symmetries allows you to prove that there can be exact flat directions. These directions can give rise to the kinds of deformations that Yau was talking about, and in particular, the one example on the three generation Calabi-Yau, the deformation $T \oplus O \oplus O$. You can in fact realise an example of that sort using the R-symmetries and prove that such a solution would exist. So you can have a physics proof, if you will, of that particular example that Yau was discussing.

A. Strominger Perhaps the most interesting new things are the ones that are not obtained by deformations like this last example and really cannot be understood by any such arguments. I have a question for Yau. Twenty years ago you made your famous estimate that there were ten thousand Calabi-Yau spaces. How many of these things do you think there are?

S.T. Yau More, I think, that is all I can say at this moment.

N. Seiberg I have two questions and I am glad that most of the relevant experts are in the audience. The first question is: What is the status of the non-perturbative existence of the topological string? The second question is: We have learned about many new Calabi-Yau spaces. How many of them look like the real world?

H. Ooguri I would like to personally respond to the first question. I can see at least two independent non-pertubative completions of the topological string in certain situations. In the case when you have an open string dual you can often use a matrix model to give a non-pertubative completion in the sense that you have a convergent matrix integral whose perturbative expansion gives rise to topological string amplitudes in the close string dual. On the other hand you can also propose to define topological string amplitudes in terms of black hole entropy, where the counting of number of states of black holes is well defined and the perturbative expansion of this counting, in particular the generating function, gives rise again to topological string partition function via the OSV conjecture. You can see in particular examples that these two give rise to different non-pertubative completions. One possible view is that the topological string is a tool to address various interesting geometric programs in physics. Depending on situations, there can be different non-perturbative completions. But there might be people with other views on that.

N. Seiberg I am not aware of an example where you have two systems which are the same to all orders in perturbation theory and their D-branes are the same, in the sense that you can probe the system with large classical field excursions, and yet they have more than one non-pertubative completion.

H. Ooguri Yes, so this might be a counterexample to that.

I. Klebanov I was also intrigued by Nekrasov's promise to repay some debts to condensed matter theorists. A question I had is: Is there any sign in the topological string of the resonating valence bond wave function, and if so, will there be a time when Phil Anderson will be learning topological string theory?

N. Nekrasov I am sure there is, but I do not know when it is.

I. Klebanov Is it a classical model or do you see the quantisation of these dimers?

N. Nekrasov Well, the model which I got is more like a statistical mechanical model.

H. Ooguri Statistical mechanical in the sense of a classical statistical mechanical model?

N. Nekrasov I am doing the functional integral over the fermions and the gauge fields and the rest. How do you call it? I call it quantum but some people may call it classical. When you reduce this problem to dimers, it is just a summation over dimer configurations. That is probably classical.

H. Ooguri At the end of the next part, I would like to also have some general discussion, and in particular I hope to identify some important physics program for which we would like to develop mathematical tools. So I would like to take some kind of informal call on this kind of questions. I have entitled the final part of this discussion session "What is M-theory?" Of course, finding a better formulation of M-theory is a very important project. We will hear two different points of view about this.

4.7 Prepared Comments

4.7.1 Hermann Nicolai: E_{10} and $K(E_{10})$: prospects and challenges

Definition of E_{10}: The maximal rank hyperbolic Kac-Moody algebra $\mathfrak{e}_{10} \equiv \mathrm{Lie}(E_{10})$ (in split real form) is defined via the so-called Chevalley Serre presentation in terms of generators h_i, e_i, f_i $(i = 1, \ldots, 10)$ with relations [1]

$$[h_i, h_j] = 0, \quad [e_i, f_j] = \delta_{ij} h_i, \quad [h_i, e_j] = A_{ij} e_j, \quad [h_i, f_j] = -A_{ij} f_j,$$

$$(\mathrm{ad}\, e_i)^{1-A_{ij}} e_j = 0, \quad (\mathrm{ad}\, f_i)^{1-A_{ij}} f_j = 0 \ .$$

where $\{h_i\}$ span the Cartan subalgebra $\mathfrak{h}$. The entries of the Cartan matrix A_{ij} can be read off from the Dynkin diagram displayed in Figure 1 below. With all other Kac-Moody algebras, $\mathfrak{e}_{10}$ shares the following key properties:

- *Root space decomposition:* for any root $\alpha \in Q(E_{10}) = \mathrm{II}_{1,9}$ (= the unique even self-dual Lorentzian lattice in ten dimensions), we have

$$(\mathfrak{e}_{10})_\alpha = \big\{ x \in \mathfrak{e}_{10} : [h, x] = \alpha(h)x \ \text{for all}\, h \in \mathfrak{h} \big\}$$

  One distinguishes *real* roots $(\alpha^2 = 2)$ and *imaginary* roots $(\alpha^2 \leq 0)$; the latter can be further subdivided into lightlike (null) roots $(\alpha^2 = 0)$ and timelike roots $(\alpha^2 < 0)$.

- *Triangular decomposition:* this is a generalization of the well known decomposition of finite dimensional matrices into (strictly) upper and lower triangular $(\mathfrak{n}^\pm)$, and diagonal matrices $(\mathfrak{h})$, respectively.

$$\mathfrak{e}_{10} = \mathfrak{n}_- \oplus \mathfrak{h} \oplus \mathfrak{n}_+ \ , \quad \text{with}\ \mathfrak{n}_\pm := \bigoplus_{\alpha \gtrless 0} (\mathfrak{e}_{10})_\alpha$$

  This is the feature that ensures *computability* in the present context, via choice of a *triangular gauge* for the 'vielbein' $\mathcal{V}(t) \in E_{10}/K(E_{10})$.

- Existence of an *invariant bilinear form:*

$$\langle h_i | h_j \rangle = A_{ij} \quad , \quad \langle e_i | f_j \rangle = \delta_{ij} \quad , \quad \langle [x, y] | z \rangle = \langle x | [y, z] \rangle.$$

  This is the feature which, in the present context, allows for the formulation of an *action principle*. Because $\dim \mathfrak{e}_{10} = \infty$, this quadratic form is, in fact, the only polynomial Casimir invariant, ensuring the (essential) *uniqueness* of the σ-model action below [2, 3].

Compact subalgebra $\mathfrak{k}\mathfrak{e}_{10}$: The Chevalley involution is defined by

$$\omega(e_i) = -f_i, \quad \omega(f_i) = -e_i, \quad \omega(h_i) = -h_i$$

and extends to all of $\mathfrak{e}_{10}$ by $\omega([x, y]) = [\omega(x), \omega(y)]$. The fixed point set $\mathfrak{k}\mathfrak{e}_{10} = \big\{ x \in \mathfrak{e}_{10} : \omega(x) = x \big\}$ is a subalgebra of $\mathfrak{e}_{10}$, which is called the *compact subalgebra*. Note that $\mathfrak{k}\mathfrak{e}_{10}$ is *not* a Kac–Moody algebra [4].

Level decomposition: No closed formulas exist for the dimensions of the root spaces, although the root multiplicities are in principle computable recursively [1].

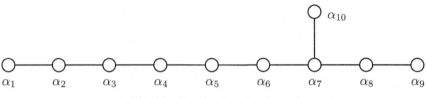

Fig. 4.1   Dynkin diagram of $\mathfrak{e}_{10}$.

However, it is known that, generically, they *grow exponentially* as $\alpha^2 \to -\infty$, like the number of massive string states. In order to get a handle at least on low-lying generators, one analyzes $\mathfrak{e}_{10}$ w.r.t. certain finite-dimensional regular subalgebras by means of a *level decomposition*: pick one special node α_0, and write a given root as $\alpha = \sum_j m^j \alpha_j + \ell \alpha_0$ with $\alpha_j \in$ subalgebra, and where ℓ is the 'level'. For instance, decomposing w.r.t. the subalgebra $A_9 \equiv \mathfrak{sl}_{10}$ (*i.e.* $\alpha_0 = \alpha_{10}$) we obtain the following table for $\ell \leq 3$:

ℓ	A_9 module	Tensor
0	$[100000001] \oplus [000000000]$	$K^a{}_b$
1	$[000000100]$	E^{abc}
2	$[000100000]$	$E^{a_1 \dots a_6}$
3	$[010000001]$	$E^{a_1 \dots a_8 \mid a_9}$

These tensors correspond to the bosonic fields of $D = 11$ supergravity and their 'magnetic' duals. Similar low level decompositions of $\mathfrak{e}_{10}$ w.r.t. its other distinguished rank-9 subalgebras $D_9 \equiv \mathfrak{so}(9,9)$ and $A_8 \oplus A_1 \equiv \mathfrak{sl}_9 \oplus \mathfrak{sl}_2$ yield the correct bosonic multiplets, again with 'magnetic' duals, of (massive) type IIA and type IIB supergravity, respectively. Furthermore, for the D_9 decomposition, one finds that the (Neveu-Schwarz)^2 states (at even levels) and the (Ramond)^2 states (at odd levels), respectively, belong to tensorial and spinorial representations of the T-duality group $SO(9,9)$, and that the truncation to even levels contains the rank-10 hyperbolic Kac–Moody algebra DE_{10}, corresponding to type-I supergravity, as a subalgebra.

Dynamics (cf. [2, 3, 5]): The equations of motion are derived from the following (essentially unique) 'geodesic' σ-model over $E_{10}/K(E_{10})$:

$$\int dt\, \mathcal{L}(t) = \int \frac{dt}{n(t)} \langle \mathcal{P}(t) | \mathcal{P}(t) \rangle \,,$$

where $\mathcal{P} := \dot{\mathcal{V}}\mathcal{V}^{-1} - \omega(\dot{\mathcal{V}}\mathcal{V}^{-1})$ is the 'velocity', $\langle . | . \rangle$ is the standard invariant bilinear form, and $n(t)$ a one-dimensional 'lapse' needed to ensure (time) reparametrisation invariance. When truncated to levels ≤ 3, the corresponding equations of motion coincide with the appropriately truncated bosonic supergravity equations of motion, where only first order spatial gradients are retained [2]. Analogous results hold for the D_9 and $A_8 \times A_1$ decompositions. Remarkably, for the bosonic sector of these theories, E_{10} yields exactly the same information as local supersymmetry, namely

- unique (bosonic) actions with the correct (Chern–Simons) couplings;
- incompatibility of $D = 11$ supergravity with a cosmological constant;
- self-duality of the 5-form field strength in IIB supergravity;
- information about the higher order $R^4, R^7, \dots$ corrections to M theory.

This may indicate, perhaps surprisingly, that (maximal local) supersymmetry may not play such a prominent role at the most fundamental level. Indeed, in a theory, where space and time are 'emergent', the distinction between bosons and fermions must also be regarded an 'emergent' phenomenon, and not as a feature of the theory at its most basic level [5].

Towards higher levels: Understanding the (exponentially growing) spectrum of higher level representations and their correct physical interpretation is the key problem; on the mathematical side, this is the (still unsolved) problem of finding a manageable realization for indefinite Kac–Moody algebras. An analysis of the higher level representations w.r.t. A_9 has revealed the existence of a distinguished series of representations, the so-called *gradient representations*, which we tentatively associate with (non-local functionals of) the higher order spatial gradients, corresponding to the differential operators $\partial_{a_1} \cdots \partial_{a_k}$ acting on the $D = 11$ supergravity fields.

ℓ	A_9 module	Tensor
$3k + 1$	$[k00000100]$	$E_{a_1 \dots a_k}{}^{b_1 b_2 b_3}$
$3k + 2$	$[k00100000]$	$E_{a_1 \dots a_k}{}^{b_1 \dots b_6}$
$3k + 3$	$[k10000001]$	$E_{a_1 \dots a_k}{}^{b_1 \dots b_8 \vert b_9}$

(All tensors in the above table are *symmetric* in the lower indices $a_1, \dots, a_k$.) We thus face the following questions:

- Can the complete evolution of $D = 11$ supergravity (or some M theoretic extension thereof) be mapped to a *null geodesic motion* on the infinite-dimensional coset space $E_{10}/K(E_{10})$?
- Can the expansion in spatial gradients be understood in terms of a 'zero tension limit', where space-time is regarded as some kind of 'elastic medium'? And is the level expansion (or an expansion in the height of the roots) the correct mathematical framework for studying this limit?
- Does E_{10} thereby provide a new *Lie algebraic mechanism* for the emergence of space [6], and can the initial singularity be described in these terms as the place

---

[5] This point of view might receive further support if, contrary to widespread expectations, no evidence for supersymmetry is found at the Large Hadron Collider.

[6] Whereas time could emerge 'operationally', as it is expected to in canonical approaches based on a Wheeler-DeWitt equation.

where space 'de-emerges'?

Even if some variant of the gradient hypothesis turns out to be correct, there remains the question how to interpret the remaining (M theoretic?) degrees of freedom. E.g., up to level $\ell = 28$, there are already $4\,400\,752\,653$ representations of A_9, out of which only 28 qualify as gradient representations!

Fermions: The above considerations can be extended to the fermionic sector, and it can be shown that $K(E_{10})$ indeed plays the role of a generalised 'R symmetry' [6]. Because $K(E_{10})$ is not a Kac–Moody group, many of the standard tools are not available. This applies in particular to *fermionic (i.e.* double-valued) representations which cannot be obtained from (or lifted to) representations of E_{10}. Remarkably, it has been shown very recently [6, 7] that the gravitino field of $D = 11$ supergravity (at a fixed spatial point) can be promoted to a *bona fide*, albeit unfaithful, spinorial representation of $K(E_{10})$. This result strengthens the evidence for the correspondence proposed in [2], and for the existence of a map between the time evolution of the bosonic and fermionic fields of $D = 11$ supergravity and the dynamics of a *massless spinning particle* on $E_{10}/K(E_{10})$. However, the existence (and explicit construction) of a *faithful* spinorial representation, which might also accommodate spatially dependent fermionic degrees of freedom, remains an open problem.

For further references and details on the results reported in this comment see [2, 3, 6]. The potential relevance of E_{10} was first recognized in [8]; an alternative proposal based on E_{11} has been developed in [9].

Acknowledgments: It is a great pleasure to thank T. Damour, T. Fischbacher, M. Henneaux and A. Kleinschmidt for enjoyable collaborations and innumerable discussions, which have shaped my understanding of the results reported here.

Bibliography

[1] V. Kac, *Infinite Dimensional Lie Algebras*, 3rd edition, C.U.P. (1990).
[2] T. Damour, M. Henneaux and H. Nicolai, Phys. Rev. Lett. **89** (2002) 221601, hep-th/0207267.
[3] T. Damour and H. Nicolai, hep-th/0410245.
[4] A. Kleinschmidt and H. Nicolai, hep-th/0407101.
[5] T. Damour, prepared comment in this volume.
[6] T. Damour, A. Kleinschmidt and H. Nicolai, hep-th/0512163.
[7] S. de Buyl, M. Henneaux and L. Paulot, hep-th/0512292.
[8] B. Julia, in: Lectures in Applied Mathematics, Vol. 21 (1985), AMS-SIAM, p. 335; preprint LPTENS 80/16.
[9] P. C. West, Class. Quant. Grav. **18** (2001) 4443, hep-th/0104081; hep-th/0407088.

4.7.2   Michael Atiyah: Beyond string theory?

String theory (and M-theory) is a remarkably sophisticated mathematical structure, as yet incomplete. While it has led to exciting new results in mathematics it is not yet clear what shape the final physical theory will take. Perhaps it will only be a modest extension of what we now have, but perhaps it will have to undergo some radical reformulation.

As we know the great aim is to combine quantum theory with gravity. String theory maintains the formal apparatus of quantum theory unchanged but modifies Einstein's theory of gravitation. Perhaps both sides need modification?

I confess to being a believer in Occam's razor, in which the simple solution is always preferred. Although string theory is an impressive structure it still lacks the overall simplicity that we should aim at.

We may need radically new ideas and I think it is worth investigating whether **retarded differential equations** should be seriously considered at a fundamental level. This idea has been put forward by Raju [1], although my ideas are somewhat different.

Consider as a simple example the linear equation for the function $x(t)$.

$$\dot{x}(t) + kx(t - r) = 0 \qquad\qquad r > 0 \qquad\qquad (1)$$

where r, k are fixed constants. Such an equation can be solved for $t > 0$ with **initial data** being a function on the interval $[-r, 0]$. This is very different from the usual differential equations of mathematical physics (for which $r = 0$), which have been our paradigm since Isaac Newton.

While this equation makes sense for $x(t)$ in Minkowski space, with t being proper time along the trajectory, it is not clear how to extend it to wave-propagation in a relativistic manner. For this purpose note that $t \longrightarrow t - r$ has infinitesimal generator $-r\dfrac{\partial}{\partial t}$, so that translation by $-r$ is just $\exp\left(-r\dfrac{\partial}{\partial t}\right)$. Guided by this we can consider formally the relativistically invariant retarded Dirac operator (as suggested to me by G. Moore).

$$i\ D - mc + k\exp\left(-rD\right) \qquad\qquad (2)$$

where D is the Dirac operator in Minkowski space. In the non-relativistic limit D reduces to $\dfrac{\gamma^\circ}{c}\dfrac{\partial}{\partial t}$ where γ° is the 4×4 matrix $\begin{pmatrix} 1 & 0 \\ 0 & -1 \end{pmatrix}$, showing that we have a retarded equation for positive energy and an advanced equation for negative energy.

There are problems in interpreting $\exp\left(-rD\right)$ which I will pass over, but applied to the plane wave solution

$$\exp\left(\frac{-iEt}{}\right)\begin{pmatrix} \phi \\ 0 \end{pmatrix}$$

(2) leads to the dispersion relation

$$\frac{E}{c} - mc + k \exp\left(\frac{irE}{c}\right) = 0 \tag{3}$$

giving the quantization condition

$$r = \frac{nc}{E}\pi \qquad (n \text{ an integer}) \tag{4}$$

which shows that r is an integer multiple of the Compton wavelength of the electron, so is of order 10^{-12} m.

This elementary argument giving the scale of the retardation parameter r/c is striking. Note that the real part of the equation gives

$$E = mc^2 + (-1)^{n-1}kc \tag{5}$$

so that the constant k will have to be extremely small.

Analysis of $\exp(-rD)$ raises many challenging problems which are related to the mixing of positive and negative energy states. On the other hand the equation makes sense on a curved background. Coupling the Dirac operator to other gauge fields can be treated in a similar way, though a gravitational counterpart presents further problems.

I finish by simply asking whether these ideas have any relevance to string theory or M-theory. In particular is the initial data problem for retarded equations related in some way to quantum theory?

My purpose is not to make any claims but to stimulate thought[†].

Bibliography

[1] C.K. Raju, *Time: Towards a Consistent Theory*, Kluwer Academic, Dordrecht, (1994) Fundamental Theories of Physics, Vol. 65.

[†] I am grateful to the participants of the Solvay Conference for a number of cogent and constructive criticisms.

4.8 Discussion

N. Seiberg I have a question to the current speaker. Putting derivatives in exponentials is very common when you consider fields in non-commutative spaces. Is there any connection to your work?

M. Atiyah I do not know. Non-commutative geometry/analysis is a very interesting part of mathematics and physics has a link to all these things. It could well be. I approach this from a very different point of view. I just naively ask certain philosophical questions and I am led to this by the nature of the formalism. I have not had time to search the literature, but I would be delighted if it links up with anything else you or anybody else knows in physics, or in non-commutative geometry. The hope is, of course, that all the ideas we have been talking about, string theory, non-commutative geometry, and so on, are obviously related in some way. We want to clear the ground and find out what the real relations are. If this plays any role, I would be delighted.

H. Ooguri I would like to reserve some time for general discussions. I recently read the history of this conference and there is a preface that was written by Werner Heisenberg who commented that this conference has been held for the purpose of attacking problems of unusual difficulty rather than exchanging the results of recent scientific work. In that spirit I would like to raise the question: What would be the important physics programs that are still waiting for some new mathematical tool? Or maybe are there some hidden tools that we are not aware of, that we should try to make use of?

M. Douglas I am coming back actually to answer the question of Harvey and also Seiberg's second question, which I hope are the kind of general questions you were asking. There is all this wealth of mathematics and structures, but we are physicists and we have to address some physical question to make progress. The basic physical questions are the combination of the ones that we started to work on twenty years ago: trying to get the standard model out of string compactifications. This has made twenty years of progress and inspired a lot of the mathematics that Dijkgraaf talked about. Also, there is the recent discovery of the dark energy, which in the simplest models is a positive cosmological constant. Those are the experimental facts that seem the most salient to the type of work that we were discussing.

Let me turn then to Seiberg's second question. We have this big number of Calabi-Yaus, and this potentially vaster number of non Calabi-Yaus. We have an even vaster number of flux vacua. What number of them looks like the real world? That is obviously a very hard question. I think this mathematics is relevant because it gives us tools for addressing problems that we, as physicists, have had very little experience with. Namely, exploring this vast mathematical space of possibilities in string theory. Experimental input is essential, such as the standard model and what will be discovered at LHC, but also this mathematical space to some extent has to be explored. Then, just to give a glib answer to Seiberg's question: if you make the zeroth order sorts of estimates that I made in my work of two years ago and if you would incorporate both the issues that Lüst talked about, then the difficulty of getting the standard model is one in a billion. It is one in 10^{200} taking into account the cosmological constant and other factors. You then decide that out of this number of Calabi-Yaus and fluxes and the rest, something like one in 10^{200} to 10^{300} should work. Although you may say: "How can you say such a thing?", the arguments have the virtue that they are very simple. They are using very little input and so one can take them as a zeroth order starting point. I certainly hope that is not the right answer and that there are far more features that have not been exploited yet in terms of the structure of the string vacua. To the extent that there is structural peaking, then there might be far fewer that match the standard model, or far more. Both of those possibilities would be interesting. It could be that there is information from early cosmology. Just the difficulty of getting a viable cosmology that will fit the data, or these more speculative considerations about the wave function measure factors and so forth, all this could drastically affect this calculation. All I am saying by throwing out a number like this is the following: here is a framework in which to think about the problem of combining these many disparate ingredients and talk about them together.

D. Gross It seems to be clear that the one thing that was missing from this session, in which we focused partly on mathematical structures that might reveal the nature of space and time, was time, except in the last talk which could have been delivered back in 1911. It is clear that elliptic equations are easier than hyperbolic, and Euclidean metrics easier than Lorentzian, and ignoring time easier than taking it into account. But for example the discussion of what some people mistakenly call vacua, which are really metastable states, should illustrate that they are discussing things which are of course time dependent. Yet, nothing is known about the time evolution. This indicates that, surely, the big open issue in string theory is time. What does it mean for time to be emergent? Non-locality in time, how do we deal with that? How do we make or have a causal structure? How do we discuss metastable states whose beginning we know nothing about, and so on. It seems that a lot of tools we are focusing on are avoiding the tough questions because the mathematics is simpler. As I said, we prefer to study elliptic equations rather than the hard case of hyperbolic equations.

N. Arkani-Hamed I just wanted to ask something about the status of large numbers of vacua in the heterotic context. It seems that a lot of the studies of the statistics for type IIB vacua for example, which are definitely interesting, are less likely to apply to the real world if we take the hint from gauge coupling unification seriously. Everything seemed to be going swimmingly in the perturbative heterotic string, except for not being able to find a mechanism to find a small cosmological constant. So, I would like to know what the status of statistics is in the heterotic context.

S. Kachru I can say one thing. One of the huge advances in the mid-nineties was this duality revolution. After the duality revolution, Freedman, Morgan, Witten and many others developed very powerful techniques to take heterotic strings on certain Calabi-Yau three-folds, elliptic Calabi-Yaus, and Fourier transform them over to type II strings, F-theory or type IIB string theory. Actually the groups that have been making the most progress in constructing realistic GUT models in the heterotic string, the Penn group for instance, worked precisely in this elliptic Calabi-Yaus. Now you can then ask: "What happens if you dualise these over to the type II context where people have started counting vacua with fluxes?" It seems quite clear that the same kind of structure will emerge in the heterotic theory. The reason that it is much easier to study in the type II context is that what these fluxes in the type II theory map back to in the heterotic theory correspond to deformations of the Calabi-Yau into a non-Kähler geometry. As should have been clear from Yau's talk, although that is a very interesting subject, we know almost nothing about it. That makes it clear that you can import the best features of GUTs into the type II context and the best features of the type II context back into the heterotic theory. But different sides are definitely better suited to describing different phenomena.

A. Strominger I think that is very misleading. There is no reason to believe that there are not also non-Kähler types of geometries on the type II side. We just are sticking with those because we are looking under the lamp post.

S. Kachru I completely agree with what you have said.

A. Polyakov Two brief comments. First of all it seems to me that one should not be overfixated on Calabi-Yau compactifications, or on compactifications at all. We do not really know how string theory applies to the real world. It is quite possible that we have some non-critical string theory working directly in four dimensions. I think that the important thing to do is to realise what mechanisms, what possibilities, we have in string theory, and it might be premature to try to get too much, to directly derive the standard model, etc. That is the first comment. The second comment is about a physical problem. It is just to inform you about things that are not widely known. There is an interesting application of the methods which we developed in string theory and conformal field theory in the theory of turbulence. There are recent numerical results which indicate that, in two dimensional turbulence, in some cases there are very conclusive signs of conformal theories. So that might be repaying debts which Nekrasov mentioned.

B. Julia I would like just to make a general remark in the same spirit of being useful. I think the most interesting thing I have learned, and everybody has learned, here is that the classical world does not exist. There is no unique classical limit. I think we should really develop more powerful tools to decide how good any classical limit is and which classical limit applies in any given computation. That would be useful for many people.

P. Ramond I do not know about being useful. First a remark about landscape. I just saw in a museum of ancient art Hieronymous Bosch's temptation of Saint Anthony, let me not say more.

The second thing is about some no-go theorems that we should be aware of. There are no-go theorems about massless particles of spin higher than two. If they are taken one at a time, there is no doubt that the no go-theorems apply. If there are an infinite amount, it is an open question. Moreover, there are mathematical structures based on the coset $F_4/SO(9)$, which is of great mathematical interest actually, that basically contain the fields of $N = 1$ supergravity in eleven dimension as its lowest level. One should not think this is completely about things that are known. This presents great difficulties, but one should keep an open mind and try to look at these things without giving up.

J. Harvey I think of the areas of mathematics that have been discussed here, the one that to me seems to have some of the strongest hints, but the least understood, is the general question of E_{10}, Borcherds algebras, generalized Kac-Moody algebras. In Nicolai's talk, to the degree I understood it, there definitely seems to be some structure going on, where E_{10} is reflected in eleven dimensional supergravity. But when he was talking about higher level in the landscape of these representations, I think it was about one in a billion that actually fit into the structure that we know so far, and we don't know what the rest of these are doing. In calculations I did with Greg Moore number of years ago, we found denominator formulas for Borcherds algebras coming out of definite one-loop string integrals. We tried to give an algebraic explanation of this but we failed. I do not think that these things can be coincidences. I think there is a very general algebraic structure which will allow us to have much greater control over some supersymmetric theories. I do not think it will solve the real world problem of what we do without supersymmetry or time evolution. But it seems to me to be an example of not drilling where the board is thinnest, but where it is thin enough that we might actually get through it.

Session 5

Emergent Spacetime

Chair: *Jeffrey Harvey*, University of Chicago, USA
Rapporteur: *Nathan Seiberg*, IAS Princeton, USA
Scientific secretaries: *Ben Craps* (Vrije Universiteit Brussel) and *Frank Ferarri* (Université Libre de Bruxelles)

5.1 Rapporteur talk: Emergent Spacetime, by Nathan Seiberg

5.1.1 *Introduction*

The purpose of this talk is to review the case for the idea that space and time will end up being emergent concepts; i.e. they will not be present in the fundamental formulation of the theory and will appear as approximate semiclassical notions in the macroscopic world. This point of view is widely held in the string community and many of the points which we will stress are well known.

Before we motivate the idea that spacetime should be emergent, we should discuss the nature of space in string theory. We do that in section 2, where we review some of the ambiguities in the underlying geometry and topology. These follow from the dualities of string theory. T-duality leads to ambiguities at the string length l_s and the quantum dualities lead to ambiguities at the Planck length $l_p \ll l_s$. All these ambiguities in the geometry are associated with the fact that as we try to probe the space with increasing resolution, the probes we use become big and prevent us from achieving the desired accuracy.

The discussion about ambiguities in space will lead us to make some comments about locality. In particular, we will ask whether to expect locality in a space or in one of its duals.

In section 3 we will briefly mention some of the peculiar non-gravitational theories which are found as certain limits of string theory. Some of them are expected to be standard field theories, albeit without a Lagrangian. Others, like theories on a noncommutative space or little string theory, are not local quantum field theory. They exhibit interesting nonlocal behavior.

In section 4 we will make the case that general covariance is likely to be a derived concept.

Section 5 will present several examples of emergent space. First we will discuss the simplest examples which do not involve gravity. Then we will turn to four classes of examples of emergent space: the emergent two-dimensional (worldsheet) gravity from the matrix model, the celebrated gauge/gravity duality, linear dilaton backgrounds, and the BFSS matrix model. We will discuss some of their properties and will stress the similarities and the differences between them. In particular, we will discuss their finite temperature behavior as a diagnostic of the system in extreme conditions.

Section 6 will be devoted to emergent time. Here we do not have concrete examples. Instead, we will present some of the challenges and confusions that this idea poses. We will also mention that understanding how time emerges will undoubtedly shed new light on some of the most important questions in theoretical physics including the origin of the Universe.

We will summarize the talk in section 7 where we will also present some general speculations.

Before we start we should mention some important disclaimers. As we said, most of the points which will be discussed here are elementary and are well known in the string community. We apologize for boring you with them. Other points will be inconclusive because they reflect our confusions. Also, not all issues and all points of view will be presented. Instead, the presentation will be biased by my prejudice and my own work. For example, the discussion will focus on string theory (for textbooks, see [1], [2]), and other approaches to quantum gravity will not be reviewed. Since this talk is expected to lead to a discussion, we will present certain provocative and perhaps outrageous ideas. Finally, there will be very few references, mostly to reviews of the subject, rather than to original papers.

5.1.2 *Ambiguous space*

5.1.2.1 *Ambiguous space in classical string theory*

We start this section by discussing the ambiguities in the geometry and the topology which exist already at string tree level. These are usually referred to as *T-duality* (for reviews, see e.g. [3], [4]).

Consider strings propagating in some background fields (e.g. metric). Clearly, these background fields should satisfy the equations of motion. Then, it turns out that different backgrounds can lead to the same physics without any observable difference between them. Therefore, there is no unique answer to the question: *"What is the background metric?"* and the background geometry is ambiguous.

Intuitively, these ambiguities arise from the extended nature of the string. Features in the geometry which are smaller than the string length $l_s = \sqrt{\alpha'}$ cannot be detected using a string probe whose characteristic size is l_s.[1]

---

[1] D-branes [2] which are smaller than l_s can sometime lead to a more precise metric, but different kinds of D-branes lead to different answers and therefore the ambiguity is not resolved.

The simplest and most widely known example of this ambiguity is the equivalence between a circle with radius R and a circle with radius α'/R. A slightly more peculiar example is the equivalence between a circle with radius $R = 2\sqrt{\alpha'}$ and a $\mathcal{Z}_2$ quotient of a circle (a line segment) with $R = \sqrt{\alpha'}$. This example demonstrates that even the topology is ambiguous. Furthermore, we can start with a circle of radius R, smoothly change it to $R = 2\sqrt{\alpha'}$, then use the duality with the line segment and then change the length of the line segment. This way we start with a circle which is not dual to a line segment and we continuously change its topology to a line segment which is not dual to a circle.

A characteristic feature of these dualities is the role played by momentum and winding symmetries. In the example of the two circles with radii R and α'/R momentum conservation in one system is mapped to winding conservation in the other. Momentum conservation arises from a geometric symmetry (an isometry) of the circle. It is mapped to winding conservation which is a *stringy symmetry*. This is a manifestation of the stringy nature of T-duality and it makes it clear that it is associated with the extended nature of the string.

In some situations there exists a description of the system in terms of a *macroscopic background*; i.e. the space and all its features are larger than l_s. This is the most natural description among all possible dual descriptions. However, two points should be stressed about this case. First, even though this description is the most natural one, there is nothing wrong with all other T-dual descriptions and they are equally valid. Second, it is never the case that there is more than one such macroscopic and natural description.

More elaborate and richer examples of this fundamental phenomenon arise in the study of Calabi-Yau spaces. Here two different Calabi-Yau spaces which are a "mirror pair" (for a review, see e.g. [5]) lead to the same physics. Furthermore, it is often the case that one can continuously interpolate between different Calabi-Yau spaces with different topology. These developments had dramatic impact on mathematics (see e.g. [5], [6]).

Another kind of T-duality is the cigar/Sine-Liouville duality [7]. One side of the duality involves the cigar geometry: a semi-infinite cylinder which is capped at one side. It has a varying dilaton, such that the string coupling at the open end of the cigar vanishes. This description makes it clear that the shift symmetry around the cigar leads to conserved momentum. However, the string winding number is not conserved, because wound strings can slip through the capped end of the cigar. The other side of this duality involves an infinite cylinder. Here the winding conservation is broken by a condensate of wound strings. The cigar geometry is described by a two-dimensional field theory with a nontrivial metric but no potential, while its dual, the Sine-Liouville theory, is a theory with a flat metric but a nontrivial potential. This example again highlights the importance of the winding modes. It also demonstrates that the T-duality ambiguity is not limited to compact dimensions.

Here the ambiguity is between two different non-compact systems (an infinite and a half infinite cylinder).

From the worldsheet point of view T-duality represents an exact equivalence between different two-dimensional conformal field theories. Therefore, the phenomenon of T-duality persists beyond classical string theory, and extends to all orders in perturbation theory. Furthermore, in some situations one can argue that T-duality is a gauge symmetry. This observation means that T-duality is exact and it cannot be violated non-perturbatively.

The phenomenon of T-duality leads us to ask two interesting questions. First, is l_s a minimum length; i.e. is the notion of distance ill defined below l_s? Second, is the theory local in one space, or in its T-dual space, or in neither? We will return to these questions below.

Before we leave the topic of ambiguities in classical string theory we would like to mention another important stringy phenomenon which is associated with the extended nature of the string. The high energy density of string states is such that the canonical ensemble of free strings does not exist above a certain temperature $T_H \sim \frac{1}{l_s}$, which is known as the *Hagedorn temperature* [1], [2]. The relevant modes which lead to this phenomenon are long strings. They have large entropy and hence the partition function diverges at T_H. Equivalently, when Euclidean time is compactified on a circle of radius $R = \frac{1}{2\pi T}$ (with thermal boundary conditions) an instability appears when $R \leq \frac{1}{2\pi T_H}$. This instability is associated with strings which are wound around the Euclidean time circle. T_H could be a limiting temperature, beyond which the theory does not exist. Alternatively, this phenomenon could mean that the system undergoes a first order phase transition to another phase. That phase could exhibit the fundamental degrees of freedom more clearly. Again we see that the theory tries to hide its short distance behavior.

5.1.2.2   *Ambiguous space in quantum string theory*

Quantum mechanics introduces new ambiguities in space which are related to new dualities (for reviews, see e.g. [2], [4]). These ambiguities go beyond the obvious ambiguities due to the quantum fluctuations. Here the characteristic length scale is the Planck length $l_p \ll l_s$.

An intuitive argument explaining the origin of these ambiguities is the following. If we want to explore space with resolution of order r, the uncertainly principle tells us that we need to use energy $E > \frac{1}{r}$. This energy has to be concentrated in a region of size r. But in the presence of gravitational interactions, this concentration of energy creates a black hole unless $r > l_p$. Therefore, *we cannot explore distances smaller than the Planck length.*

It is important to stress that although the ambiguities in the quantum theory are often described as of different nature than the ambiguities in the classical theory, fundamentally they are quite similar. Both of them are associated with the breakdown of the standard small distance/high energy connection – as we try to increase the energy of a probe it becomes bigger and does not allow us to explore short distances.

The *quantum dualities*, which are also known as S-duality or U-duality, extend the classical T-duality and lead to a beautiful and coherent picture of stringy dualities. These exchange highly quantum situations with semiclassical backgrounds, exchange different branes, etc. As in the classical dualities, among all dual descriptions there is at most one description which is natural because it is semiclassical. All other dual descriptions are very quantum mechanical.

5.1.2.3 *Comments about locality*

We now turn to some comments about locality in string theory.

Quantum field theory is local. This locality guarantees that the theory is causal. We would like string theory also to be causal or at least macroscopically causal. Furthermore, we know that at long distances string theory behaves like quantum field theory and therefore it is macroscopically local. But is string theory local also over short distances?

One piece of evidence in favor of locality is the analyticity of the perturbative string S-matrix. Normally, causality and locality lead to analyticity. Since the string S-matrix is analytic, it is likely that string theory is local. However, it is logically possible that a slightly weaker condition than locality and therefore of causality can also guarantee the analyticity of the S-matrix.

One reason string theory might not be local in a standard way is the extended nature of the interacting objects, the strings. At the most naive and intuitive level locality of string interactions is not obvious. Even though two strings interact at a point to form a third string, this interaction is nonlocal when viewed from the point of view of the center of masses of the interacting strings. It is known that this nonlocality is harmless and is consistent with the analyticity of the S-matrix.[2]

We would like to comment about locality and the cosmological constant. The old fashioned point of view of the cosmological constant problem suggested that its value is related to some kind of a UV/IR mixing and to violation of naive locality – the short distance theory somehow reacts to long distance fluctuations and thus sets the value of the cosmological constant. A more modern point of view on the subject is that the cosmological constant is set anthropically (see, e.g. [8]). It remains to be seen whether the cosmological constant is a hint about some intrinsic nonlocality in the theory.

The ambiguities we discussed above might hint at some form of nonlocality. We have stressed that increasing the energy of a probe does not lead to increased resolution. Instead, the probe becomes bigger and the resolution is reduced. This point is at the heart of the various dualities and ambiguities in the background. We have already asked whether we expect locality in a space, or in its dual space.

---

[2]In open string field theory a basis based on the string midpoint replaces the basis based on the center of mass and then the interaction appears to be local.

It is hard to imagine that the theory can be simultaneously local in both of them. Then, perhaps it is local in neither. Of course, when a macroscopic weakly coupled natural description exists, we expect the theory to be at least approximately local in that description.

It is important to stress that although intuitively the notion of locality is obvious, this is not the case in string theory or in any generally covariant theory. The theory has no local observables. Most of the observables are related to the S-matrix or other objects at infinity. These do not probe the detailed structure of the theory in the interior. Therefore, without local observables it is not clear how to precisely define locality.

We will argue below that space and time should be emergent concepts. So if they are not fundamental, the concept of locality cannot be fundamental as well. It is possible that locality will end up being ill defined, and there will be only an approximate notion of locality when there is an approximate notion of spacetime.

5.1.3 *Non-standard theories without gravity*

Next, let us digress slightly to review some of the non-standard theories without gravity that were found by studying various limits of string theory. These theories exhibit interesting and surprising new phenomena. We expect that these theories and their peculiar phenomena will be clues to the structure of the underlying string theory. Since they are significantly simpler than string theory, they could be used as efficient laboratories or toy models.

The first kind of surprising theories are new local field theories which cannot be given a standard Lagrangian description. These are superconformal field theories in five or six dimensions with various amount of supersymmetry. The most symmetric examples are the six-dimensional (2,0) theories (for a review, see e.g. [9]). They are found by taking an appropriate scaling limit of string theory in various singularities or on coincident 5-branes. The existence of these theories calls for a new formulation of local quantum field theory without basing it on a Lagrangian.

Another class of interesting non-gravitational theories are *field theories on non-commutative spaces* (for a review, see e.g. [10]). These theories do not satisfy the standard rules of local quantum field theory. For example, they exhibit a UV/IR mixing which is similar to the UV/IR mixing in string theory – as the energy of an object is increased its size becomes bigger.

The most enigmatic theories which are derived from string theory are the *little string theories* (for a review, see e.g. [11]). These non-gravitational theories exhibit puzzling stringy behavior. The stringy nature of these theories arises from the fact that they appear by taking a certain scaling limit of string theory (in the presence of NS5-branes or some singularities) while keeping α' fixed. One stringy phenomenon they exhibit is T-duality. This suggests that despite the lack of gravity, these theories do not have a local energy momentum tensor. Otherwise, there should have been several different energy momentum tensors which are related by T-duality. It was also argued that because of their high energy behavior these theories cannot have local observables. Finally, these theories exhibit Hagedorn spectrum with a Hagedorn temperature which is below T_H of the underlying string theory. It was suggested that this Hagedorn temperature is a limiting temperature; i.e. the canonical ensemble does not exist beyond that temperature.

5.1.4 *Derived general covariance*

The purpose of this section is to argue that general covariance which is the starting point of General Relativity might not be fundamental. It could emerge as a useful concept at long distances without being present in the underlying formulation of the theory.

General covariance is a *gauge symmetry*. As with other gauge symmetries, the term "symmetry" is a misnomer. Gauge symmetries are not symmetries of the Hilbert space; the Hilbert space is invariant under the entire gauge group. Instead, gauge symmetries represent a redundancy in our description of the theory. (It is important to stress, though, that this is an extremely useful redundancy which allows us to describe the theory in simple local and Lorentz invariant terms.)

Indeed, experience from duality in field theory shows that gauge symmetries are not fundamental. It is often the case that a theory with a gauge symmetry is dual to a theory with a different gauge symmetry, or no gauge symmetry at all. A very simple example is Maxwell theory in 2+1 dimensions. This theory has a $U(1)$ gauge symmetry, and it has a dual description in terms of a free massless scalar without a local gauge symmetry. More subtle examples in higher dimensions were found in supersymmetric theories (for reviews, see e.g. [12], [13]).

If ordinary gauge symmetries are not fundamental, it is reasonable that general covariance is also not fundamental. This suggests that the basic formulation of the theory will not have general covariance. General covariance will appear as a derived (and useful) concept at long distances.

An important constraint on the emergence of gauge symmetries follows from the Weinberg-Witten theorem [14]. It states that if the theory has massless spin one or spin two particles, these particles are gauge particles. Therefore, the currents that they couple to are not observable operators. If these gauge symmetries are not present in some formulation of the theory, these currents should not exist there. In particular, it means that if an ordinary gauge symmetry emerges, the fundamental theory should not have this symmetry as a global symmetry. In the context of emergent general covariance, this means that the fundamental theory cannot have an energy momentum tensor.

If we are looking for a fundamental theory without general covariance, it is likely that this theory should not have an underlying spacetime. This point is further motivated by the fact that General Relativity has no local observables and perhaps no local gauge invariant degrees of freedom. Therefore, there is really no need for an underlying spacetime. Spacetime and general covariance should appear as approximate concepts which are valid only macroscopically.

5.1.5   *Examples of emergent space*

5.1.5.1   *Emergent space without gravity*

The simplest examples of emergent space are those which do not involve gravity. Here the starting point is a theory without a fundamental space, but the resulting answers look approximately like a theory on some space. The first examples of this kind were the *Eguchi-Kawai* model and its various variants (for a review, see e.g. [15]). Here a d dimensional $SU(N)$ gauge theory is formulated at one point. The *large N* answers look like a gauge theory on a macroscopic space.

Certain extensions of the (twisted) Eguchi-Kawai model are theories on a *noncommutative space* (for a review, see e.g. [10]). Here the coordinates of the space do not commute and are well defined only when they are macroscopic.

A physical realization of these ideas is the *Myers effect* [16]. Here we start with a collection of N branes in some background flux. These branes expand and become a single brane of higher dimension. The new dimensions of this brane are not standard dimensions. They form a so-called "fuzzy space." In the *large N limit* the resulting space becomes macroscopic and its fuzzyness disappears.

5.1.5.2   *Emergent space with gravity: matrix model of 2d gravity*

The first examples of emergent space with gravity and general covariance arose from the *matrix model of random surfaces* (for a review, see e.g. [17]). Here we start with a certain matrix integral or matrix quantum mechanics and study it in perturbation theory. Large Feynman diagrams of this perturbation expansion can be viewed as discretized two-dimensional surfaces.

This system is particularly interesting when the size of the matrices N is taken to infinity together with a certain limit of the parameters of the matrix integral. In this double scaling limit the two-dimensional surfaces become large and smooth and the system has an effective description in terms of random surfaces. The degrees of freedom on these surfaces are local quantum fields including a dynamical metric and therefore this description is generally covariant.

The formulation of these theories as matrix models does not have a two-dimensional space nor does it have general covariance. These concepts emerge in the effective description.

In addition to being interesting and calculable models of two-dimensional gravity, these are concrete examples of how space and its general covariance can be emergent concepts.

5.1.5.3 *Emergent space with gravity: Gauge/Gravity duality*

The most widely studied examples of emergent space with gravity are based on the AdS/CFT correspondence [18], [19], [20], [21]. This celebrated correspondence is the duality between string theory in AdS space and a conformal field theory at its boundary. Since other speakers in this conference will also talk about it, we will only review it briefly and will make a few general comments about it.

The bulk theory is a theory of gravity and as such it does not have an energy momentum tensor. The dual field theory on the boundary has an energy momentum tensor. This is consistent with the discussion above about emergent gravity (section 4), because the energy momentum tensor of the field theory is in lower dimensions than the bulk theory and reflects only its boundary behavior.

The operators of the boundary theory are mapped to string states in the bulk. A particularly important example is the energy momentum tensor of the boundary theory which is mapped to the bulk graviton. The correlation functions of the conformal field theory are related through the correspondence to string amplitudes in the AdS space. (Because of the asymptotic structure of AdS, these are not S-matrix elements.) When the field theory is deformed by relevant operators, the background geometry is slightly deformed near the boundary but the deformation in the interior becomes large. This way massive field theories are mapped to nearly AdS spaces.

The radial direction in AdS emerges without being a space dimension in the field theory. It can be interpreted as the renormalization group scale, or the energy scale used to probe the theory. The asymptotic region corresponds to the UV region of the field theory. This is where the theory is formulated, and this is where the operators are defined. The interior of the space corresponds to the IR region of the field theory. It is determined from the definition of the theory in the UV.

A crucial fact which underlies the correspondence, is the infinite warp factor at the boundary of the AdS space. Because of this warp factor, finite distances in the field theory correspond to infinite distances in the bulk. Therefore, a field theory correlation function of finitely separated operators is mapped to a gravity problem which infinitely separated sources.

An important consequence of this infinite warp factor is the effect of finite temperature. The boundary field theory can be put at finite temperature T by compactifying its Euclidean time direction on a finite circle of radius $R = \frac{1}{2\pi T}$. At low temperature, the only change in the dual asymptotically AdS background it to compactify its Euclidean time. Because of the infinite warp factor, the radius of the Euclidean time circle in the AdS space is large near the boundary, and it is small only in a region of the size of the AdS radius R_{AdS}. Therefore, most of the bulk of the space is cold. Only a finite region in the interior is hot. As the system is heated up, the boundary theory undergoes a thermal deconfinement phase transition. In the bulk it is mapped to the appearance of a Schwarzschild horizon at small radius and the topology is such that the Euclidean time circle becomes contractible. For a CFT on a 3-sphere, this phase transition is the Hawking-Page transition, and the dual high temperature background is AdS-Schwarzschild. Both above and below the transition the bulk asymptotes to (nearly) AdS. Most of it remains cold and it is not sensitive to the short distance behavior of string theory.

While the boundary field theory is manifestly local, locality in the bulk is subtle. Because of the infinite warp factor, possible violation of locality in the bulk over distances of order l_s could be consistent with locality at the boundary. In fact, it is quite difficult to find operators in the field theory which represent events in the bulk which are localized on scales of order R_{AdS} or smaller. This underscores the fact that it is not clear what we mean by locality, if all we can measure are observables at infinity.

These developments have led to many new insights about the two sides of the duality and the relation between them (for a review, see [21]). In particular, many new results about gauge theories, including their strong coupling phenomena like thermal phase transitions, confinement and chiral symmetry breaking were elucidated. The main new insight about gravity is its *holographic nature* – the boundary theory contains all the information about the bulk gravity theory which is higher dimensional. Therefore, the number of degrees of freedom of a gravity theory is not extensive. This is consistent with the lack of local observables in gravity.

5.1.5.4 *Emergent space with gravity: linear dilaton backgrounds*

Generalities   Another class of examples of an emergent space dimension involves backgrounds with a linear dilaton direction. The string coupling constant depends on the position in the emergent direction, parameterized by the spatial coordinate ϕ, through $g_s(\phi) = e^{\frac{Q\phi}{2}}$ with an appropriate constant Q. Therefore, the string coupling constant vanishes at the boundary $\phi \to -\infty$. The other end of the space at $\phi \to +\infty$ is effectively compact.

Like the AdS examples, here the bulk string theory is also dual to a theory without gravity at the boundary. In that sense, this is another example of holography. However, there are a few important differences between this duality and the AdS/CFT duality.

In most of the linear dilaton examples the holographic theory is not a standard local quantum field theory. For example, the near horizon geometry of a stack of NS5-branes is a linear dilaton background which is holographic to the little string theory (for a review, see e.g. [11]). The stringy, non-field theoretic nature of the holographic theory follows from the fact that it has nonzero α', and therefore it exhibits T-duality.

Because of the vanishing interactions at the boundary of the space, the interactions take place in an effectively compact region (the strong coupling end). Therefore, we can study the S-matrix elements of the bulk theory. These are the observables of the boundary theory.

Unlike the AdS examples, the string metric does not have an infinite warp factor.

Here finite distances in the boundary theory correspond to finite distances (in string units) in the bulk. Therefore, it is difficult to define *local* observables in the boundary theory and as a result, the holographic theory is not a local quantum field theory.

This lack of the infinite warp factor affects also the finite temperature behavior of the system. Finite temperature in the boundary theory is dual to finite temperature in the entire bulk. Hence, the holographic theory can exhibit Hagedorn behavior and have maximal temperature.

Matrix model duals of linear dilaton backgrounds Even though the generic linear dilaton theory is dual to a complicated boundary theory, there are a few simple cases where the holographic theories are very simple and are given by the large N limit of certain matrix models.

The simplest cases involve strings in one dimension ϕ with a linear dilaton. The string worldsheet theory includes a Liouville field ϕ and a $c < 1$ minimal model (or in the type 0 theory a $\hat{c} < 1$ superminimal model). The holographic description of these *minimal string theories* is in terms of the large N limit of matrix integrals (for a review, see e.g. [22]).

Richer theories involve strings in two dimensions: a linear dilaton direction ϕ and time x (for a review, see e.g. [23]). Here the holographic theory is the large N limit of matrix quantum mechanics.

These two-dimensional string theories have a finite number of particle species. The bosonic string and the supersymmetric 0A theory have one massless boson, and the 0B theory has two massless bosons. Therefore, these theories do not have the familiar Hagedorn density of states of higher dimensional string theories, and correspondingly, their finite temperature behavior is smooth.

One can view the finite temperature system as a system with compact Euclidean time x. Then, the system has $R \to \alpha'/R$ T-duality which relates high and low temperature. As a check, the smooth answers for the thermodynamical quantities respect this T-duality.

It is important to distinguish the two different ways matrix models lead to emergent space. Above (section 5.2) we discussed the emergence of the two-dimensional string worldsheet with its worldsheet general covariance. Here, we discuss the target space of this string theory with the emergent holographic dimension ϕ.

Since the emergence of the holographic direction in these systems is very explicit, we can use them to address various questions about this direction. In particular, it seems that there are a number of inequivalent ways to describe this dimension. The most obvious description is in terms of the Liouville field ϕ. A second possibility is to use a free worldsheet field which is related to ϕ through a nonlocal transformation (similar to T-duality transformation). This is the Backlund field of Liouville theory. A third possibility, which is also related to ϕ in a nonlocal way arises more naturally out of the matrices as their eigenvalue direction. These different descriptions of the emergent direction demonstrate again that the ambiguity in the description of space which we reviewed above (section 2) is not limited to compact dimensions. It also highlights the question of locality in the space. In which of these descriptions do we expect the theory to be local? Do we expect locality in one of them, or in all of them, or perhaps in none of them?

2d heterotic strings   We would like to end this subsection with a short discussion of the heterotic two-dimensional linear dilaton system. Even though there is no known holographic matrix model dual of this system, some of its peculiar properties can be analyzed.

As with the two-dimensional linear dilaton bosonic and type 0 theories, this theory also has a finite number of massless particles. But here the thermodynamics is more subtle. We again compactify Euclidean time on a circle of radius R. The worldsheet analysis shows that the system has $R \to \alpha'/2R$ T-duality. Indeed, the string amplitudes respect this symmetry. However, unlike the simpler bosonic system, here the answers are not smooth at the selfdual point $R = \sqrt{\alpha'/2}$. This lack of smoothness is related to long macroscopic strings excitations [24].

What is puzzling about these results is that they cannot be interpreted as standard thermodynamics. If we try to interpret the Euclidean time circle as a thermal ensemble with temperature $T = \frac{1}{2\pi R}$, then the transition at $R = \sqrt{\alpha'/2}$ has negative latent heat. This violates standard thermodynamical inequalities which follow from the fact that the partition function can be written as a trace over a Hilbert space $\text{Tr}\, e^{-H/T}$ for some Hamiltonian H. Therefore, we seem to have a contradiction between compactified Euclidean time and finite temperature. The familiar relation between them follows from the existence of a Hamiltonian which generates *local time evolution*. Perhaps this contradiction means that we cannot simultaneously have locality in the circle and in its T-dual circle. For large R the Euclidean circle answers agree with the thermal answers with low temperature. But while these large R answers can be extended to smaller R, the finite temperature interpretation ceases to make sense at the selfdual point. Instead, for smaller R we can use the T-dual circle, which is large, and describe the T-dual system as having low temperature.

5.1.5.5   *Emergent space in the BFSS matrix model*

As a final example of emergent space we consider the BFSS matrix model (for a review, see e.g. [25]). Its starting point is a large collection of D0-branes in the lightcone frame. The lightcone coordinate x^+ is fundamental and the theory is an ordinary quantum mechanical system with x^+ being the time.

The transverse coordinates of the branes x^i are the variables in the quantum mechanical system. They are not numbers. They are N dimensional matrices. The standard interpretation as positions of the branes arises only when the branes are far apart. Then the matrices are approximately diagonal and their eigenvalues are the positions of the branes. In that sense the transverse dimensions emerge from the simple quantum mechanical system.

The remaining spacetime direction, x^-, emerges holographically. It is related to the size of the matrices $N \sim p_-$ where p_- is the momentum conjugate to x^-.

5.1.6 *Emergent time*

After motivating the emergence of space it is natural to ask whether time can also emerge. One reason to expect it is that this will put space and time on equal footing – if space emerges, so should time. This suggests that time is also not fundamental. The theory will be formulated without reference to time and an approximate (classical) notion of macroscopic time, which is our familiar "time", will emerge. Microscopically, the notion of time will be ill defined and time will be fuzzy.

There are several obvious arguments that time should not be emergent:

(1) Even though we have several examples of emergent space, we do not have a single example of emergent time.
(2) We have mentioned some of the issues associated with locality in emergent space. If time is also emergent we are in danger of violating locality in time and that might lead to violation of causality.
(3) It is particularly confusing what it means to have a theory without fundamental time. Physics is about predicting the outcome of an experiment *before* the experiment is performed. How can this happen without fundamental time and without notions of "before and after"? Equivalently, physics is about describing the evolution of a system. How can systems evolve without an underlying time? Perhaps these questions can be avoided, if some order of events is well defined without an underlying time.
(4) More technically, we can ask how much of the standard setup of quantum mechanics should be preserved. In particular, is there a wave function? What is its probabilistic interpretation? Is there a Hilbert space of all possible wave functions, or is the wave function unique? What do we mean by unitarity (we cannot have unitary evolution, because without time there is no evolution)? Some of these questions are discussed in [26].

My personal prejudice is that these objections and questions are not obstacles to emergent time. Instead, they should be viewed as challenges and perhaps even clues to the answers.

Such an understanding of time (or lack thereof) will have, among other things, immediate implications for the physics of space-like and null singularities (for a review, see e.g. [27]) like the black hole singularity and the cosmological singularity. We can speculate that understanding how time emerges and what one means by a wave function will explain the meaning of *the wave-function of the Universe*. Understanding this wave function, or equivalently understanding the proper initial conditions for the Universe, might help resolving some of the perplexing questions of vacuum selection in string theory. For a review of some aspects of these questions see [8].

5.1.7 *Conclusions and speculations*

We have argued that spacetime is likely to be an emergent concept. The fundamental formulation of the theory will not have spacetime and it will emerge as an approximate, classical concept which is valid only macroscopically.

One challenge is to have emergent spacetime, while preserving some locality – at least macroscopic locality, causality, analyticity, etc. Particularly challenging are the obstacles to formulating physics without time. It is clear that in order to resolve them many of our standard ideas about physics will have to be revolutionized. This will undoubtedly shed new light on the fundamental structure of the theory.

Understanding how time emerges will also have other implications. It will address deep issues like the cosmological singularity and the origin of the Universe.

We would like to end this talk with two general speculative comments.

Examining the known examples of a complete formulation of string theory, like the various matrix models, AdS/CFT, etc., a disturbing fact becomes clear. It seems that many different definitions lead to a consistent string theory in some background. In particular, perhaps every local quantum field theory can be used as a boundary theory to define string theory in (nearly) AdS space. Perhaps every quantum mechanical system can be the holographic description of string theory in 1+1 dimensions. And perhaps even every ordinary integral defines string theory in one Euclidean dimension. With so many different definitions we are tempted to conclude that we should not ask the question: *"What is string theory?"* Instead, we should ask: *"Which string theories have macroscopic dimensions?"* Although we do not have an answer to this question, it seems that *large N* will play an important role in the answer.

Our second general comment is about reductionism – the idea that science at one length scale is derived (at least in principle) from science at shorter scales. This idea has always been a theme in all branches of science. However, if there is a basic length scale, below which the notion of space (and time) does not make sense, we cannot derive the principles there from deeper principles at shorter distances. Therefore, once we understand how spacetime emerges, we could still look for more basic fundamental laws, but these laws will not operate at shorter distances. This follows from the simple fact that the notion of "shorter distances" will no longer make sense. This might mean the *end of standard reductionism*.

Acknowledgments: We would like to thank the organizers of the 23rd Solvay Conference in Physics for arranging such an interesting and stimulating meeting and for inviting me to give this talk. We also thank T. Banks, I. Klebanov, J. Maldacena, and D. Shih for useful comments and suggestions about this rapporteur talk. This research is supported in part by DOE grant DE-FG02-90ER40542.

Bibliography

[1] M. B. Green, J. H. Schwarz and E. Witten, "Superstring Theory. Vol. 1: Introduction," Cambridge, Uk: Univ. Pr. (1987), "Superstring Theory. Vol. 2: Loop Amplitudes, Anomalies And Phenomenology," Cambridge, Uk: Univ. Pr. (1987).

[2] J. Polchinski, "String theory. Vol. 1: An introduction to the bosonic string," Cambridge, UK: Univ. Pr. (1998), "String theory. Vol. 2: Superstring theory and beyond," Cambridge, UK: Univ. Pr. (1998).

[3] A. Giveon, M. Porrati and E. Rabinovici, "Target space duality in string theory," Phys. Rept. **244** (1994) 77 [arXiv:hep-th/9401139].

[4] A. Sen, "An introduction to duality symmetries in string theory," *Prepared for Les Houches Summer School: Session 76: Euro Summer School on Unity of Fundamental Physics: Gravity, Gauge Theory and Strings, Les Houches, France, 30 Jul - 31 Aug 2001.*

[5] K. Hori, S. Katz, A. Klemm, R. Pandharipande, R. Thomas, C. Vafa, R. Vakil and E. Zaslow, "Mirror Symmetry" Clay Mathematics Monographs Vol 1 (AMS, 2003).

[6] R. Dijkgraaf, Rapporteur talk at this conference.

[7] V. Fateev, A.B. Zamolodchikov and Al.B. Zamolodchikov, unpublished.

[8] J. Polchinski, Rapporteur talk at this conference.

[9] E. Witten, "Conformal field theory in four and six dimensions," *Prepared for Symposium on Topology, Geometry and Quantum Field Theory (Segalfest), Oxford, England, United Kingdom, 24-29 Jun 2002.*

[10] M. R. Douglas and N. A. Nekrasov, "Noncommutative field theory," Rev. Mod. Phys. **73** (2001) 977 [arXiv:hep-th/0106048].

[11] O. Aharony, "A brief review of 'little string theories'," Class. Quant. Grav. **17** (2000) 929 [arXiv:hep-th/9911147].

[12] N. Seiberg, "The power of duality: Exact results in 4D SUSY field theory," Int. J. Mod. Phys. A **16** (2001) 4365 [arXiv:hep-th/9506077].

[13] K. A. Intriligator and N. Seiberg, "Lectures on supersymmetric gauge theories and electric-magnetic duality," Nucl. Phys. Proc. Suppl. **45BC** (1996) 1 [arXiv:hep-th/9509066].

[14] S. Weinberg and E. Witten, "Limits On Massless Particles," Phys. Lett. B **96** (1980) 59.

[15] Y. Makeenko, "Methods of contemporary gauge theory," Cambridge, UK: Univ. Pr. (2002).

[16] R. C. Myers, "Dielectric-branes," JHEP **9912** (1999) 022 [arXiv:hep-th/9910053].

[17] P. H. Ginsparg and G. W. Moore, "Lectures on 2-D gravity and 2-D string theory," arXiv:hep-th/9304011.

[18] J. M. Maldacena, "The large N limit of superconformal field theories and supergravity," Adv. Theor. Math. Phys. **2** (1998) 231 [Int. J. Theor. Phys. **38** (1999) 1113] [arXiv:hep-th/9711200].

[19] S. S. Gubser, I. R. Klebanov and A. M. Polyakov, "Gauge theory correlators from non-critical string theory," Phys. Lett. B **428** (1998) 105 [arXiv:hep-th/9802109].

[20] E. Witten, "Anti-de Sitter space and holography," Adv. Theor. Math. Phys. **2** (1998) 253 [arXiv:hep-th/9802150].

[21] O. Aharony, S. S. Gubser, J. M. Maldacena, H. Ooguri and Y. Oz, "Large N field theories, string theory and gravity," Phys. Rept. **323** (2000) 183 [arXiv:hep-th/9905111].

[22] N. Seiberg and D. Shih, "Minimal string theory," Comptes Rendus Physique **6** (2005) 165 [arXiv:hep-th/0409306].

[23] I. R. Klebanov, "String theory in two-dimensions," arXiv:hep-th/9108019.

[24] N. Seiberg, "Long strings, anomaly cancellation, phase transitions, T-duality and locality in the 2d heterotic string," arXiv:hep-th/0511220.

[25] T. Banks, "TASI lectures on matrix theory," arXiv:hep-th/9911068.

[26] J. Hartle, Rapporteur talk at this conference.

[27] G. Gibbons, Rapporteur talk at this conference.

5.2 Discussion

S. Shenker I would like to speak in defense of redundancy. It is certainly true as you say that gauge symmetry really does reflect just a redundancy in the description. But sometimes, as we know, this redundancy is very useful. Formulating quantum electrodynamics with four gauge fields is good, because you can choose different ways of resolving the redundancy to make things like unitarity, or Lorentz covariance, clear in different gauges. And so, well, some of us have been worrying for a long time, and Gerard 't Hooft made these ideas quite explicit, perhaps what we should be thinking about is a description where we vastly enlarge the number of degrees of freedom in our description of whatever the thing we're trying to describe is, quantum gravity. And so we would then be able to resolve this redundancy in a way in which locality, to the extent it exists, is manifest, or holography is manifest in another gauge, or some other property is manifest. And then it would be too much to ask for a description in which all the desirable properties of the theory are clear, in one presentation.

N. Seiberg I would like to respond to that. I sympathize with your point of view, but recall that you wrote a paper about the matrix model, where the local reparametrization on the world sheet was absent. You had a description of two dimensional gravity, in the sense of the world sheet, without reparametrization freedom. So this is an example where gravity and general covariance emerge.

S. Shenker What is this statement: to do as I say, not as I do.

J. Harvey There is also saying that consistency is the hobgoblin of small minds.

E. Silverstein I just have a small comment on non-locality appearing in string theories. So another example occurs in AdS/CFT. You said we don't have any argument against non-locality at the string scale, but there is evidence for non-locality at a much larger scale, if you consider multi-trace deformations of the field theory. So if you include the internal space, the simplest versions being spheres or other Einstein spaces, the boundary conditions are grossly non-local on those dimensions, at a scale of order the AdS radius scale. So this is another indication that we might need to incorporate non-locality in a serious way.

S. Weinberg I have a comment and a question. The comment is that for a long time, I thought that general covariance was a red herring, because any particle with mass zero and spin two would have the properties that we derive from general covariance, as shown also by Feynman, and the great thing done by string theory, in this area, is to show that there has to be a particle of mass zero and spin two, while before string theory we didn't know why there had to be one. My question had to do with your remark that the S matrix in string theory is more analytic than in quantum field theory. I thought that in quantum field theory the S matrix was as analytic as it possibly could be, given the constraints of unitarity – I do not see how anything...

N. Seiberg What I had in mind is the good high energy behaviour. And I think the person who should really answer this sits next to you [G. Veneziano].

S. Weinberg I see what you mean. So you include the point at infinity in the analyticity.

N. Seiberg Yes.

S. Weinberg Oh well OK, thank you.

N. Seiberg The sign of my comment was not in this direction, it was actually in the opposite. This usually signals locality and causality. But I am not aware, maybe you can enlighten me on that, of an argument that this must mean that there is underlying locality. There might be some weaker statement than locality and causality which leads to the same kind of analyticity.

S. Weinberg Well, I have never been able to elevate it into a theorem; in my courses, I taught quantum field theory in such a way that locality emerges out of the requirements of Lorentz invariance plus the cluster decomposition principle, and then the cluster decomposition principle is more fundamental than locality. In fact, I think it would be an interesting challenge, to understand how cluster decomposition emerges from something like string theory.

N. Seiberg That is a very interesting point, because in particular it means that there could be something fuzzy and non-local at short distance, as long as when you separate things, things are well-behaved.

S. Weinberg That is the real test. The locality, we can live without, but I do not see how science is possible without the cluster decomposition principle.

M. Douglas As a candidate to a model with emergent time, how about 2D quantum gravity coupled to c greater than twenty-five matter, where the Liouville becomes time-like.

J. Harvey How about it?...

W. Fischler What is the requirement of analyticity, if the emergent space-time and its asymptopia do not allow for an S matrix?

N. Seiberg Well, if you have this asymptotic region, you can scatter particles...

W. Fischler No, I am saying, what if, there is no asymptopia that allows for an S matrix. Or do we say that maybe the requirement of analyticity forbids such space-times to be solutions of quantum gravity?

N. Seiberg I do not see anything wrong with compact space. It is very confusing, but I do not see anything wrong with that.

W. Fischler Let me just give you an example: de Sitter space-time, there is no S matrix. Asymptopia does not allow you to define such an object. So what requirement do we have about analyticity in this case? Or do we say de Sitter space-time is not a viable quantum mechanical solution to whatever theory of quantum gravity?

N. Seiberg I am not an expert on the subject. Some people believe that this class of questions and confusions perharps are trying to tell us that de Sitter space is not a stable solution. Other people, including some people in this room, strongly disagree with that.

W. Fischler It was a question, I just do not know the answer.

G. Horowitz I wanted to mention a couple of other examples of emergent time. I think one is $2+1$ gravity, which as you know can be written as a Chern-Simons theory, there is a state of zero metric, no space, no time, anything, and that's perfectly, you know, allowed and understood, and yet we see how we can get classical space-times out of that theory. And in string theory there is sort of at least a formal generalization of all of that in the purely cubic action in string field theory.

N. Seiberg Well, Chern-Simons theory has time, right? You write the integral $\int \mathrm{d}^2 x \, \mathrm{d}t$ of something. So t appears there. I think we would like something which is a little bit more dramatic than this.

J. Harvey Yes, I would have thought space-time was there, it was just rewriting the action in a different form.

G. Horowitz There is a state of zero metric, so I am not sure what time or space would mean if the metric is zero.

A. Ashtekar In terms of this emergent time, there is a lot of literature, in the relativity circles, quantum gravity circles, about emergent time. Basically there is a relation to dynamics, and how one of the variables, under circumstances when there is actually a semiclassical space-time, can be taken to be time. But in general there would be no such variable, and then we would not have a good notion of time. In a particular example that I sketched, which had to do with scalar fields coupled with gravity in cosmology, in that example a scalar field is a good notion of time, in quantum theory I am talking about. But near the Big-Bang singularity, it becomes very very fuzzy, we do not have the standard notion, and again, it reemerges as the standard notion on the other side. So it seems to me that there are definite examples, certainly not a completely general theorem or anything like that, but lots and lots of such examples, that exist in the literature.

J. Harvey I think that this might be a good point to stop, have a coffee break, and I would like to continue this discussion, so if you have things to say, please make a note of that and we will continue in half an hour and have time for discussion after the next four talks.

5.3 Prepared Comments

5.3.1 *Tom Banks: The Holographic Approach to Quantum Gravity*

String Theory provides us with many consistent models of quantum gravity in space-times which are asymptotically flat or AdS. These models are explicitly holographic: the observables are gauge invariant boundary correlation functions. Typical cosmological situations do not have well understood asymptotic boundaries. They begin with a Big Bang, and can end with *e.g.* asymptotic de Sitter (dS) space. In order to formulate a string theory of cosmology, we have to find a more general formulation of the theory.

Einstein gravity has the flexibility to deal with a wide variety of asymptotic behaviors for space-time. It describes space-time in terms of a local, gauge variant, variable, the metric tensor, $g_{\mu\nu}(x)$. The corresponding object in the quantum theory is a preferred algebra of operators for a causal diamond in space-time. An *observer* is a large quantum system with a wealth of semi-classical observables. Our mathematical model of observers is a cut-off quantum field theory, with volume large in cutoff units. The semi-classical observables are averages of local fields over large volumes. Tunneling transitions between different values of these observables are suppressed by exponentials of the volume.

Experiment teaches us that there are many such observers in the real world, and that they travel on time-like trajectories. A theory of quantum gravity should reproduce this fact as a mathematical theorem, but it is permissible to use the idea of an observer in the basic formulation of the theory. A pair of points $P > Q$ on the trajectory of an observer defines a causal diamond: the intersection of the interior of the backward light cone of P with that of the forward light cone of Q. Conversely, a dense sequence of nested causal diamonds completely defines the trajectory.

The covariant entropy bound[5][6][7] associates an entropy with each causal diamond. For sufficiently small proper time between P and Q the entropy is always finite. Fischler and the present author have argued that the only general ansatz one can make about the density matrix corresponding to this entropy is that it is proportional to the unit matrix. This hypothesis provides us with a dictionary for translating concepts of Lorentzian geometry into quantum mechanics. A nested sequence of causal diamonds, describing the trajectory of an observer, is replaced by a sequence of finite dimensional Hilbert spaces, $\mathcal{H}_N$, with $\mathcal{H}_{n-1}$ a tensor factor in $\mathcal{H}_n$. The precise mapping of this sequence into space-time is partly a gauge choice. We will concentrate on Big Bang space-times, where it is convenient to choose the initial point of every causal diamond to lie on the Big Bang hypersurface. Each Hilbert space $\mathcal{H}_n$ is equipped with a sequence of time evolution transformations $U(k,n)$ with $1 \le k \le n$. A basic consistency condition is that $U(k,n) = U(k,m) \otimes V(k,m)$ if $k \le m < n$. The unitary $V(k,m)$ operates on the tensor complement of $\mathcal{H}_m$ in $\mathcal{H}_n$. This condition guarantees that the notion of *particle horizon* usually derived from local field theory, is incorporated into holographic cosmology.

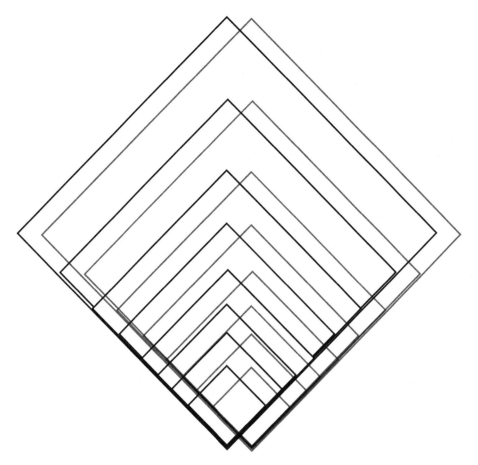

Fig. 5.1 A Sequence of Causal Diamonds in a Big Bang Space-time Defines an Observer. A Lattice of Overlapping Diamonds Defines Space-Time.

The particle horizon condition guarantees that a certain set of degrees of freedom will interact only among themselves before a given fixed time. It makes the apparent increase of spatial volume with cosmological time, compatible with quantum unitarity and a Planck scale cutoff. In fact, the discretization of time implicit in this formalism is not a simple cut-off. Subsequent Hilbert spaces in the sequence have dimension increasing by a fixed factor, which we will specify below. This implies a fixed area cutoff in space-time, which generically corresponds to a time cut-off which goes to zero with the size of causal diamonds.

To model an entire space-time we need a collection of time-like observers, with overlapping causal diamonds. We introduce a spatial lattice, which defines the topology of the non-compact dimensions of space on the initial time-slice[3]. This topology is conserved in time, and for the moment we will take it to be that of flat

---

[3]The compact dimensions will be dealt with in a completely different manner below.

$d-1$ dimensional space. The geometry of the lattice should be thought of as another gauge choice. The formalism must be built in such a way that true semi-classical measurements do not depend on it. It is not clear whether any real measurement will depend on the geometric structure of the spatial lattice. Asymptotically infinite space-times are defined as limits, and the boundary observables will be independent of the structure of the lattice. Finite space-times will have a built in restriction on the accuracy of measurements, stemming from the inability of a finite quantum system to make arbitrarily precise measurements on itself. It may be that this *a priori* lack of precision will make the micro-structure of the spatial lattice truly unmeasurable.

To model a family of time-like observers, we introduce a sequence of Hilbert spaces $\mathcal{H}_n(\mathbf{x})$ (an observer) for each point $\mathbf{x}$ on the lattice. It is convenient to choose an *equal area time slicing* in which the dimension of $\mathcal{H}_n(\mathbf{x})$ depends only on n. For each pair of space-time points we introduce an overlap Hilbert space $\mathcal{O}(m, \mathbf{x}, n, \mathbf{y})$ which encodes the physics of the maximal causal diamond in the overlap of the causal diamonds described by $\mathcal{H}_m(\mathbf{x})$ and $\mathcal{H}_n(\mathbf{y})$. Each Hilbert space is equipped with a sequence of unitaries and there are an infinite number of complicated consistency conditions relating the time evolution in different Hilbert spaces. The basic claim is that every solution of these consistency conditions is a quantum space-time[4] In the limit where all causal diamonds are large it should determine a solution of Einstein's equations, coupled to sensible matter. *Note that there is no sense in which the quantum system itself can be thought of as a sum over space-time histories. Metrics appear only as a semi-classical artifact, and are truly emergent.*

5.3.1.1 *The variables of quantum gravity*

The holographic principle suggests that the fundamental geometrical object in Lorentzian space-time is the holographic screen of a causal diamond. This is a spatial $d-2$ surface on which all the information in the diamond is projected. Consider a little area or *pixel* on the holographic screen. It determines a null direction, which penetrates the pixel, and the screen element transverse to this null ray. The Cartan-Penrose equation

$$\bar{\psi}\gamma_\mu \psi \gamma^\mu \psi = 0,$$

encodes all of this information in a pure spinor ψ, determined up to a real or complex constant, depending on the dimension.

It is natural to quantize the real independent components, S_a of the pure spinor associated with a pixel by writing

$$[S_a, S_b]_+ = \delta_{ab}$$

---

[4]A quantum Big Bang cosmology for the choice we have made of the relation between Hilbert spaces for a single observer.

which is the most general rule giving a finite number of states per pixel and co-variant under the transverse $SO(d-2)$ little group of the null vector $\bar{\psi}\gamma^\mu\psi$. It breaks the projective invariance of the CP equation to a Z_2 (for each pixel), which we treat as a gauge symmetry and identify with fermion parity $(-1)^F$, enforcing the usual connection between spin and statistics. This quantization rule implements the Bekenstein Hawking relation between quantum entropy, and area. The logarithm of the dimension of the Hilbert space of the irreducible representation of this algebra is the area of a fundamental pixel, in d dimensional Planck units. Compact dimensions lead to an enlarged pixel algebra, incorporating charges for Kaluza-Klein Killing symmetries, their magnetic duals, and wrapped brane configurations. Note that it is precisely these quantum numbers which remain unchanged under the topology changing duality transformations of string/M-theory. We will ignore the complications of compactification in the brief description which follows.

After performing a Klein transformation using the $(-1)^F$ gauge symmetry, the operator algebra of an entire causal diamond takes the form,

$$[S_a(n), S_b(m)]_+ = \delta_{ab}\delta_{mn}.$$

The middle alphabet labels stand for individual pixels. More generally we say that the geometry of the holographic screen is pixelated by replacing its algebra of functions by a finite dimensional algebra, and these labels stand for a general basis in that algebra. If we use finite dimensional non-abelian function algebras, we can have finite causal diamonds with exact rotational invariance, which would be appropriate for describing the local physics in asymptotically symmetric space-times.

The $S_a(n)$ operators should be thought of as transforming in the spinor bundle over the holographic screen. Informally, we can say that the algebra of operators of a pixel on the holoscreen, is described by the degrees of freedom of a massless super-particle which exits the holoscreen via that pixel. In a forthcoming paper[1], I will describe how an infinite dimensional limit of such a construction can reproduce the Fock space of eleven dimensional supergravity. The basic idea is to find a sequence of algebras which converges to

$$\mathcal{A}_{11} \equiv R_{[0,1]} \otimes M(S^9),$$

where $R_{[0,1]}$ is the unique, hyperfinite Type II_∞ von Neumann factor, and $M(S^9)$ the algebra of measurable functions on the sphere. We take our quantum operator algebra to be operator valued linear functionals $S(q)$, $q \in \mathcal{S}[\mathcal{A}_{11}]$, which are invariant under inner automorphisms of $\mathcal{A}_{11}$. $\mathcal{S}$ is the spinor bundle over the algebra. The projectors in $R_{[0,1]}$ are characterized, up to inner automorphism, by their trace, which is a real number between 0 and ∞. A general invariant linear functional is determined by its value on a finite sum of projectors. Thus, the quantum algebra consists of finite collections of operators of the form $S_a(p_i)q^a(\mathbf{\Omega}_i)$ where p_i is a positive number and $\mathbf{\Omega}_i$ a direction on S^9. These parametrize a null momentum

$p_i(1, \mathbf{\Omega}_i)$. It can be shown that the resulting Hilbert space on which the quantum algebra is represented is the SUGRA Fock space. Massless particles and the overall scale of null momenta appear in a manner reminiscent of Matrix Theory[4], while S^9 parametrizes the direction of null momenta. The holographic formalism can thus reproduce the kinematics of M-theory. One would like to show that the dynamical consistency conditions lead to equations which determine the scattering matrix uniquely.

5.3.1.2 *Holographic cosmology*

Only one solution of the dynamical consistency conditions of holographic cosmology has been found[9]. In this model, the dynamics of a given observer is described, at each time t by a Hamiltonian which is a random *irrelevant* perturbation of a random bilinear Hamiltonian

$$H = \sum S_a(m) h_{mn} S_a(n).$$

The term irrelevant is used because, for large $t \times t$ matrices h, the bilinear dynamics approaches that of free massless $1 + 1$ dimensional fermions.

A simple prescription for the overlap Hilbert spaces, combined with this ansatz for the Hamiltonian, satisfies all the consistency conditions. One then observes the emergence, at large times, of a flat FRW geometry, with equation of state $p = \rho$. That is, one can define a distance function on the lattice in terms of which lattice points are causally disconnected at a given time. This geometry satisfies the scaling laws of the FRW universe. Moreover, in the large time limit, the exact quantum dynamics of the system is invariant under the conformal Killing symmetry of this geometry.

The $p = \rho$ cosmology was previous introduced by Fischler and the author[8] in terms of a heuristic picture of a *dense black hole fluid*. This is a system in which black holes continuously coalesce to form a single horizon filling black hole. The random fermion model above is a precise mathematical realization of this idea. Based on the heuristic picture, one can develop an ansatz for a more normal universe, with normal regions originally arising as small defects in the black hole fluid. This leads to a cosmology which can solve all of the standard cosmological puzzles, with only a small amount of inflation just before nucleosynthesis. Inflation is necessary only to stretch the scale of fluctuations generated during the $p = \rho$ era, to the size of our current horizon. The resulting cosmological model depends on only a few parameters, and in one parameter range the fluctuations are entirely generated in the $p = \rho$ era. They are *exactly* scale invariant, between sharp infrared and ultraviolet cutoffs. This fluctuation spectrum is, in principle, distinguishable from that of inflationary models. It might explain the apparent disagreement between inflationary models and the data at low L. We do not yet have predictions for the gravitational wave spectrum in this parameter range.

In other parameter regimes, inflation must generate the observed fluctuations, and this implies some kind of hybrid model, since the scale of inflation must be quite low. In this regime there are no observable tensor fluctuations. Even in this regime, holographic cosmology is an advance over standard inflation, because the primordial $p = \rho$ regime sets up the right initial conditions for inflation, starting from a fairly generic primordial state. Indeed, Penrose[10] and others have argued that conventional inflation models *do not* resolve the question of why the universe began in a low entropy state. In holographic cosmology, this might be resolved by the following line of argument: the most generic initial condition is the uniform $p = \rho$ fluid. The more normal universe, initially consists of defects in this fluid: regions where not all of the degrees of freedom in a horizon volume are excited. This has lower entropy, but can evolve into a stable normal universe if the following two conditions are met:

- The initial matter density in the normal regions is a dilute fluid of black holes. This fluid must be sufficiently homogeneous that black hole collisions do not result in a recollapse to the $p = \rho$ phase.
- The initial normal region either is a finite volume fraction of the infinite $p = \rho$ system, or contains only a finite number of degrees of freedom in total. The latter case, which is entropically favored, evolves to a de Sitter universe. Thus, holographic cosmology predicts a de Sitter universe with the largest cosmological constant compatible with the existence of observers. If the gross features of the theory of a small Λ universe is uniquely determined by Λ, then this may predict a universe with physics like our own.

5.3.1.3 *de Sitter space*

Holographic cosmology predicts that the asymptotic future is a de Sitter space, so it behooves us to construct a quantum theory for that symmetric space-time. We will restrict attention to 4 dimensions, which may be the only case where the quantum theory of de Sitter space is defined. In four dimensions, the holographic screen of the maximal causal diamond of any observer following a time-like geodesic, is a two sphere of radius R. Our general formalism tells us that we must pixelate the surface of this sphere in order to have a finite number of quantum states[3][11]. The most elegant pixelation is given by the fuzzy sphere.

The spinor bundle over the fuzzy sphere consists of complex $N \times N + 1$ matrices ψ_i^A, transforming in the $[N] \otimes [N+1]$ dimensional representation of $SU(2)$. If these are quantized as fermions:

$$[\psi_i^A, (\psi^\dagger)_B^j]_+ = \delta_i^j \delta_B^A,$$

then the Hilbert space of the system has entropy $N(N+1)\ln 2$, which will agree with the area formula for large N if $N \propto R$.

The natural Hamiltonian, H, for a geodesic observer in dS space would seem to be the one which generates motion along the observer's time-like Killing vector.

However, once quantum mechanics is taken into account, there are no stable local-ized states. Everything decays back into the dS vacuum. Classically, the vacuum has zero eigenvalue of H, but the semiclassical results of Gibbons and Hawking can be explained if we assume instead that the spectrum of H is spread more or less randomly between 0 and something of order the dS temperature, $T_{dS} = \frac{1}{2\pi R}$, with level density $e^{-\pi(RM_P)^2}$. The vacuum state corresponding to empty dS space is the thermal density matrix $\rho = e^{-\frac{H}{T_{dS}}}$. This ansatz explains the thermal nature of dS space, as well as its entropy, which for large R is very close to the log of the dimension of the Hilbert space.

However, the spectrum of H has no relation to our familiar notions of energy. In [12] the latter concept was argued to emerge only in the large R limit. It is an operator P_0 which converges to the Poincaré Hamiltonian of the limiting asymptoti-cally flat space-time, in the reference frame of the static observer. The conventional argument that $H \to P_0$ is wrong. This is true for the mathematical action of Killing vectors on finite points in dS space. However, physical generators in GR are defined on boundaries of space-time. The cosmological horizon of dS space converges to null infinity in asymptotically flat space, and the two generators act differently on the boundary. The boundary action motivates the approximate commutation relation:

$$[H, P_0] \approx \frac{1}{R} P_0,$$

which says that eigenvalues of P_0 much smaller than the maximal black hole mass, are approximately conserved quantum numbers, which resolve the degeneracy of the spectrum of H. The corresponding eigenstates of P_0 are localized in the observer's horizon volume. Semiclassical physics indicates that

$$\mathrm{tr}\; e^{-H/T_{dS}} \delta(P_0 - E) \approx e^{-\frac{E}{T_{dS}}} \mathrm{tr}\; e^{-H/T_{dS}}.$$

This relation can be explained if, for small P_0 eigenvalue, E, the entropy deficit of the corresponding eigenspace, relative to the dS vacuum is equal to $(2\pi R)E$. This formula can be explicitly verified for black hole states, if we identify the mass parameter in the Kerr-Newman- de Sitter black hole with the Poincare eigenvalue.

Based on this picture it is relatively easy to identify black hole states in terms of the fermionic pixel operators. We work in the approximation in which all states of the vacuum ensemble are exactly degenerate, as well as all black hole eigenstates corresponding to the same classical solution. The vacuum is, in this approximation, just the unit density matrix. A black hole of radius K is the density matrix of all states satisfying

$$\psi_i^A | BH >= 0,$$

for $0 \le i \le K$ and $0 \le A \le K + 1$. This satisfies the geometrical relation between radius and entropy. For $K \ll N$ we can define the Hamiltonian P_0 (in T_{dS} units) to be the entropy deficit, as explained above. The choice of which K and $K + 1$

indices are used in the above equation is a gauge choice, equivalent to the choice of a particular horizon volume.

A similar analysis leads to a guess about the description of particle states in dS space. If, using ideas from quantum field theory, we ask for the maximal entropy states in dS space which contain no black holes whose radius goes to infinity with R, then we find that they are made from massless particles (or other conformal field theory degrees of freedom) with a typical momentum of order $(\frac{M_P}{R})^{-\frac{1}{2}}$. The entropy of such states scales like $(RM_P)^{3/2}$. These are the only states of the dS theory which are kept in the Poincare invariant $R \to \infty$ limit. We can view the full entropy of dS space as built up from $(RM_P)^{1/2}$ independent horizon volumes[2], each filled with such maximal entropy particle states.

In terms of the matrix ψ_i^A, we model these degrees of freedom as follows. Write the block decomposition

$$\psi = \begin{pmatrix} 1 & 2 & 3 & \dots & M \\ M & 1 & 2 & \dots & M-1 \\ \dots & \dots & \dots & & \dots \\ 2 & 3 & 4 & \dots & 1 \end{pmatrix},$$

where each block is an independent $M \times M$ matrix, with $M \sim N^{1/2}$. The integer labels on the blocks refer to a given horizon volume, of which there are of order M. In a future publication [13] I hope to show, using Matrix Theory ideas, similar to those described in [1], that the degrees of freedom in a single block correspond, in the large M limit to those of a single 4 dimensional supergraviton. The integer M will be the longitudinal momentum of the supergraviton in units of the cutoff.

If this idea works, it is clear that corrections to the commutator $[P_0, Q_\alpha]$ for the super-Poincare algebra will be of order $N^{-\frac{1}{2}}$, which implies the scaling law for the gravitino mass postulated in [3].

The holographic approach to quantum gravity is thus a promising generalization of string theory which is applicable to cosmological backgrounds. Much work remains to be done to refine its principles, make more explicit contact with string theory, find non-perturbative equations for S-matrices in asymptotically flat space, and solve the quantum theory of de Sitter space.

Bibliography

[1] T. Banks, *von Neumann algebras and the holographic description of 11D SUGRA*, manuscript in preparation.
[2] T. Banks, W. Fischler and S. Paban, JHEP **0212** (2002) 062 [arXiv:hep-th/0210160].
[3] T. Banks, arXiv:hep-th/0007146. ; T. Banks, Int. J. Mod. Phys. A **16** (2001) 910.
[4] T. Banks, W. Fischler, S. H. Shenker and L. Susskind, Phys. Rev. D **55** (1997) 5112 [arXiv:hep-th/9610043].
[5] W. Fischler and L. Susskind, arXiv:hep-th/9806039.
[6] R. Bousso, JHEP **9907** (1999) 004 [arXiv:hep-th/9905177].
[7] R. Bousso, Class. Quant. Grav. **17** (2000) 997 [arXiv:hep-th/9911002].

[8] T. Banks and W. Fischler, arXiv:hep-th/0111142. ;T. Banks and W. Fischler, Phys. Scripta **T117** (2005) 56 [arXiv:hep-th/0310288].

[9] T. Banks, W. Fischler and L. Mannelli, Phys. Rev. D **71** (2005) 123514 [arXiv:hep-th/0408076].

[10] R. Penrose, *The Road to Reality*, Knopf, February 2005.

[11] W. Fischler, *Taking de Sitter Seriously*, talk at the conference for G. West's 60th Birthday, Scaling Laws in Biology and Physics, Santa Fe, July 2000.

[12] T. Banks, arXiv:hep-th/0503066.

[13] T. Banks, *Particles and blackholes from matrices in a quantum theory of de Sitter space*, work in progress.

5.3.2 Igor Klebanov: Confinement, Chiral Symmetry Breaking and String Theory

The AdS/CFT duality [1–3] provides well-tested examples of emergent spacetimes. The best studied example is the emergence of an $AdS_5 \times \mathbf{S}^5$ background of type IIB string theory, supported by N units of quantized Ramond-Ramond flux, from the $\mathcal{N} = 4$ supersymmetric $SU(N)$ gauge theory. The emergent spacetime has radii of curvature $L = (g_{\mathrm{YM}}^2 N)^{1/4}\sqrt{\alpha'}$. In the 't Hooft large N limit, $g_{\mathrm{YM}}^2 N$ is held fixed [4]. This corresponds to the classical limit of the string theory on $AdS_5 \times \mathbf{S}^5$ (the string loop corrections proceed in powers of $1/N^2$). The traditional Feynman graph perturbative expansion is in powers of the 't Hooft coupling $g_{\mathrm{YM}}^2 N$. The AdS/CFT duality allows us to develop a completely different perturbation theory that works for large 't Hooft coupling where the emergent spacetime is weakly curved. The string scale corrections to the supergravity limit proceed in powers of $\frac{\alpha'}{L^2} = (g_{\mathrm{YM}}^2 N)^{-1/2}$.

The metric of the Poincaré wedge of AdS_d is

$$ds^2 = \frac{L^2}{z^2}\left(dz^2 - (dx^0)^2 + \sum_{i=1}^{d-2}(dx^i)^2 \right) . \tag{1}$$

Here $z \in [0, \infty)$ is an emergent dimension related to the energy scale in the gauge theory. In the AdS/CFT duality, fields in AdS space are dual to the gauge invariant local operators [2, 3]. The fundamental strings are dual to the chromo-electric flux lines in the gauge theory, providing a string theoretic set-up for calculating the quark anti-quark potential [5]. The quark and anti-quark are placed near the boundary of Anti-de Sitter space ($z = 0$), and the fundamental string connecting them is required to obey the equations of motion following from the Nambu action. The string bends into the interior ($z > 0$), and the maximum value of the z-coordinate is proportional to the separation r between quarks. An explicit calculation of the string action gives an attractive Coulombic $q\bar{q}$ potential [5]. Historically, a dual string description was expected mainly for confining gauge theories where long confining flux tubes have string-like properties. In a pleasant surprise, we have seen that a string description can apply to non-confining theories as well, due to the presence of extra dimensions in the string theory.

It is also possible to generalize the AdS/CFT correspondence in such a way that the $q\bar{q}$ potential is linear at large distances. In an effective 5-dimensional approach [6] the necessary metric is

$$ds^2 = \frac{dz^2}{z^2} + a^2(z)\left(- (dx^0)^2 + (dx^i)^2 \right) \tag{2}$$

and the space must end at a maximum value of z where the "warp factor" $a^2(z_{\mathrm{max}})$ is finite. Placing widely separated probe quark and anti-quark near $z = 0$, we find that the string connecting them bends toward larger z until it stabilizes at z_{max} where its tension is minimized at the value $\frac{a^2(z_{\mathrm{max}})}{2\pi\alpha'}$. Thus, the confining flux tube is described by a fundamental string placed at $z = z_{\max}$ parallel to one of the x^i-directions. This establishes a duality between "emergent" chromo-electric flux tubes and fundamental strings in certain curved string theory backgrounds.

Several 10-dimensional supergravity backgrounds dual to confining gauge theories are now known, but they are somewhat more complicated than (2) in that the compact directions are "mixed" with the 5-d (x^μ, z) space. Witten [7] constructed a background in the universality class of non-supersymmetric pure glue gauge theory. While in this background there is no asymptotic freedom in the UV, hence no dimensional transmutation, the background has served as a simple model of confinement where many infrared observables have been calculated using the classical supergravity. For example, the lightest glueballs correspond to normalizable fluctuations around the supergravity solution.

Introduction of $\mathcal{N} = 1$ supersymmetry facilitates construction of confining gauge/string dualities. A useful method to generate $\mathcal{N} = 1$ dualities (for reviews, see [8, 9]) is to place a stack of N D3-branes at the tip of a Calabi-Yau cone, whose base is Y_5. In the near-horizon limit, one finds the background $AdS_5 \times Y_5$, which is conjectured to be dual to the superconformal gauge theory on the D3-branes. Furthermore, for spaces Y_5 whose topology is $\mathbf{S}^2 \times \mathbf{S}^3$, the conformal invariance may be broken by adding M D5-branes wrapped over the $\mathbf{S}^2$ at the tip of the cone. The gauge theory on such a combined stack is no longer conformal; it exhibits a novel pattern of quasi-periodic renormalization group flow, called a duality cascade [10, 9].

To date, the most extensive study of a cascading gauge theory has been carried out for a 6-d cone called the conifold. Here one finds a $\mathcal{N} = 1$ supersymmetric $SU(N) \times SU(N+M)$ theory coupled to chiral superfields A_1, A_2 in the $(\mathbf{N}, \overline{\mathbf{N+M}})$ representation, B_1, B_2 in $(\overline{\mathbf{N}}, \mathbf{N+M})$, with a quartic superpotential [11]. The M wrapped D5-branes create M units of R-R flux through the 3-cycle in the conifold. This flux creates a "geometric transition" to the deformed conifold $\sum_{a=1}^{4} w_a^2 = \epsilon^2$, where the 3-cycle is blown up. An exact non-singular supergravity solution dual to the cascading gauge theory, incorporating the 3-form and the 5-form R-R field strengths and their back-reaction on the geometry, is the *warped deformed conifold* [10]

$$ds^2 = h^{-1/2}(\tau)\left(-(dx^0)^2 + (dx^i)^2\right) + h^{1/2}(\tau)d\tilde{s}_6^2 , \qquad (3)$$

where $d\tilde{s}_6^2$ is the Calabi-Yau metric of the deformed conifold with radial coordinate τ. The 5-from R-R field, which is dual to N, decreases as the theory flows to the infrared (towards smaller τ).

What is the field theoretic interpretation of this effect? After a finite amount of RG flow, the $SU(N + M)$ group undergoes a Seiberg duality transformation [12]. After this transformation, and an interchange of the two gauge groups, the new gauge theory is $SU(\tilde{N}) \times SU(\tilde{N} + M)$ with the same matter and superpotential, and with $\tilde{N} = N - M$. The self-similar structure of the gauge theory under the Seiberg duality is the crucial fact that allows this pattern to repeat many times.

If $N = (k + 1)M$, where k is an integer, then the duality cascade stops after k steps, and we find a $SU(M) \times SU(2M)$ gauge theory. This IR gauge theory exhibits a multitude of interesting effects visible in the dual supergravity background. One of them is confinement, which follows from the fact that the warp factor h is finite and non-vanishing at the smallest radial coordinate, $\tau = 0$, which roughly corresponds to $z = z_{\max}$ in an effective 5-d approach (2). This implies that the $q\bar{q}$ potential grows linearly at large distances. The confinement scale is proportional to $\epsilon^{2/3}$. The geometric transition that generates ϵ is dual in the gauge theory to a non-perturbative quantum deformation of the moduli space of vacua, which originates from *dimensional transmutation*. It breaks the Z_{2M} chiral R-symmetry, which rotates the complex conifold coordinates w_a, down to Z_2.

The string dual also incorporates the Goldstone mechanism due to a spontaneous breaking of the $U(1)$ baryon number symmetry [13]. Because of the $\mathcal{N} = 1$ SUSY, this produces a moduli space of confining vacua. In the $SU(M) \times SU(2M)$ gauge theory there exist baryonic operators $\mathcal{A} = A_1^M A_2^M$, $\mathcal{B} = B_1^M B_2^M$, which satisfy the "baryonic branch" relation $\mathcal{A}\mathcal{B} = \text{const}$. If the gauge theory were treated classically, this constant would vanish and the baryon symmetry would be unbroken. In the full quantum theory the constant arises non-perturbatively and deforms the moduli space [14]. The warped deformed conifold of [10] is dual to the locus $|\mathcal{A}| = |\mathcal{B}|$ in the gauge theory. Remarkably, the more general "throat" backgrounds, the *resolved warped deformed conifolds* corresponding to the entire baryonic branch, were constructed in [15]. These backgrounds were further studied in [16] where various observables were calculated along the baryonic branch. It was shown that a D3-brane moving on a resolved warped deformed conifold has a monotonically rising potential that asymptotes to a constant value at large radius. Therefore, when such a construction is embedded into a string compactification, it may serve as a model of inflationary universe, with the position of the 3-brane on the throat playing the role of the inflaton field, as in [17].

Throughout its history, string theory has been intertwined with the theory of strong interactions. The AdS/CFT correspondence [1–3] succeeded in making precise connections between conformal 4-dimensional gauge theories and superstring theories in 10 dimensions. This duality leads to many dynamical predictions about strongly coupled gauge theories. Extensions of the AdS/CFT correspondence to confining gauge theories provide new geometrical viewpoints on such important phenomena as chiral symmetry breaking, dimensional transmutation, and quantum deformations of moduli spaces of supersymmetric vacua. They allow for studying glueball spectra, string tensions, and other observables. The throat backgrounds that arise in this context may have applications also to physics beyond the standard model, and to cosmological modeling.

Acknolwedgements: I am grateful to the organizers of the XXIII Solvay Conference for giving me the opportunity to present this brief talk in a pleasant and stimulating environment. I also thank N. Seiberg for useful comments on this manuscript. This material is based upon work supported by the National Science Foundation under Grant No. 0243680.

Bibliography

[1] J. M. Maldacena, Adv. Theor. Math. Phys. **2**, 231 (1998) [arXiv:hep-th/9711200].
[2] S. S. Gubser, I. R. Klebanov and A. M. Polyakov, Phys. Lett. B **428**, 105 (1998) [arXiv:hep-th/9802109].
[3] E. Witten, Adv. Theor. Math. Phys. **2**, 253 (1998) [arXiv:hep-th/9802150].
[4] G. 't Hooft, Nucl. Phys. B **72**, 461 (1974).
[5] J. M. Maldacena, Phys. Rev. Lett. **80**, 4859 (1998) [arXiv:hep-th/9803002]; S. J. Rey and J. T. Yee, Eur. Phys. J. C **22**, 379 (2001) [arXiv:hep-th/9803001].
[6] A. M. Polyakov, Nucl. Phys. Proc. Suppl. **68**, 1 (1998) [arXiv:hep-th/9711002].
[7] E. Witten, Adv. Theor. Math. Phys. **2**, 505 (1998) [arXiv:hep-th/9803131].
[8] C. P. Herzog, I. R. Klebanov and P. Ouyang, arXiv:hep-th/0205100.
[9] M. J. Strassler, arXiv:hep-th/0505153.
[10] I. R. Klebanov and M. J. Strassler, JHEP **0008**, 052 (2000) [arXiv:hep-th/0007191].
[11] I. R. Klebanov and E. Witten, Nucl. Phys. B **536**, 199 (1998) [arXiv:hep-th/9807080].
[12] N. Seiberg, Nucl. Phys. B **435**, 129 (1995) [arXiv:hep-th/9411149].
[13] S. S. Gubser, C. P. Herzog and I. R. Klebanov, JHEP **0409**, 036 (2004) [arXiv:hep-th/0405282].
[14] N. Seiberg, Phys. Rev. D **49**, 6857 (1994) [arXiv:hep-th/9402044].
[15] A. Butti, M. Grana, R. Minasian, M. Petrini and A. Zaffaroni, JHEP **0503**, 069 (2005) [arXiv:hep-th/0412187].
[16] A. Dymarsky, I. R. Klebanov and N. Seiberg, arXiv:hep-th/0511254.
[17] S. Kachru, R. Kallosh, A. Linde, J. Maldacena, L. McAllister and S. P. Trivedi, JCAP **0310**, 013 (2003) [arXiv:hep-th/0308055].

5.3.3 *Juan Maldacena: Comments on emergent space-time*

Einstein looked at his equation

$$G_{\mu\nu} = T_{\mu\nu} \tag{1}$$

and he noticed that the left hand side is very beautiful and geometrical. On the other hand, the right hand side is related to the precise dynamics of matter and it depends on all the details of particle physics. Why isn't the right hand side as nice, beautiful and geometric as the left hand side?.

String theory partially solves this problem since in string theory there is no sharp distinction between matter and geometry. All excitations are described by different modes of a string. However, giving a stringy spacetime involves more than fixing the metric, it involves setting the values of all massive string modes. The classical string equations are given by the β functions of the two dimensional conformal field theory [1]

$$\beta_{g_i} = 0 \tag{2}$$

These equations unify gravity and matter dynamics. However, these are just the classical equations and one would like to find the full quantum equations that describe spacetime.

In order to understand the full structure of spacetime we need to go beyond perturbation theory. There are several ways of doing this depending on the asymptotic boundary conditions. The earliest and simplest examples are the "old matrix models" which describe strings in two or less dimensions [2]. We also have the BFSS matrix model which describes 11 dimensional flat space [3]. Another example is the gauge/gravity duality (AdS/CFT)[4, 5]. In all these examples we have a relation which says that an ordinary quantum mechanical system with no gravity is dual to a theory with gravity. Some of the dimensions of space are an emergent phenomenon, they are not present in the original theory but they appear in the semiclassical analysis of the dynamics.

In the gauge theory/gravity duality we have a relation of the form [5]

$$\Psi[ag] \sim e^{(a^D + \cdots)} Z_{Field\ theory}[g] \tag{3}$$

which relates the large a limit of the wavefunction of the universe to the field theory partition function, where a is the scale factor for the metric on a slice of the geometry.

Note that in this relation, the full stringy geometry near the boundary determines the field theory. It determines the lagrangian of the field theory. The full partition function is then equivalent to performing the full sum over interior stringy geometries. In the ordinary ADM parametrization we can think of the dynamical variables of 3+1 dimensional general relativity as given by 3-geometries. The analogous role in string theory is then played by the space of couplings in the field theory, since these are the quantities that the wavefunction depends on. By deforming the examples we know, it seems that it might be possible, in principle, to obtain any field theory we could imagine. In this way we see that the configuration space for a quantum spacetime seems to be related to the space of all possible field theories. This is a space which seems dauntingly large and hard to manage. So, in some sense, the wavefunction of the universe is the answer to all questions. At least all questions we can map to a field theory problem.

After many years of work on the subject there are some things that are not completely well understood. For example, it is not completely clear how locality emerges in the bulk. An important question is the following. What are the field theories that give rise to a macroscopic spacetime?. In other words, we want theories where there is a big separation of scales between the size of the geometry and the scale where the geometric description breaks down. Let us consider an AdS_4 space whose radius of curvature is much larger than the planck scale. Then the corresponding $2 + 1$ dimensional field theory has to have a number of degrees of freedom which goes as

$$c \sim \frac{R_{AdS}^2}{l_{Planck}^2} \tag{4}$$

In addition we need to require that all single particle or "single trace" operators with large spin should have a relatively large anomalous dimension. In other words, if we denote by Δ_{lowest} the lowest scaling dimension of operators with spin larger than two. Then we expect that the gravity description should fail at a distance scale given by

$$L \sim \frac{R_{AdS}}{\Delta_{lowest}} \tag{5}$$

It is natural to think that the converse might also be true. Namely, if we have a theory where all single trace higher spin operators have a large scaling dimension, then the gravity description would be good.

By the way, this implies that the dual of bosonic Yang Mills would have a radius of curvature comparable to the string scale since, experimentally, the gap between the mesons of spin one and spin larger than one is not very large.

One of the most interesting questions is how to describe the interior of black holes. The results in this area are suggesting that the interior geometry arises from an analytic continuation from the outside. Of course, we know that this is how we obtained the classical geometry in the first place. But the idea is that, even in a more precise description, perhaps the interior exists only as an analytic continuation [6]. A simple analogy that one could make here is the following. One can consider a simple gaussian matrix integral over $N \times N$ matrices [2]. By diagonalizing the matrix we can think in terms of eigenvalues. We can consider observables which are defined in the complex plane, the plane where the eigenvalues live. It turns out that in the large N limit the eigenvalues produce a cut on the plane and now these observables can be analytically continued to a second sheet. In the exact description the observables are defined on the plane, but in the large N approximation they can be defined on both sheets.

Faced with this situation the first reaction would be to say that the interior does not make sense. On the other hand we could ask the question: What is wrong with existence only as an approximate analytic continuation?. This might be good enough for the observers living in the interior, since they cannot make exact measurements anyway.

It seems that in order to make progress on this problem we might need to give up the requirement of a precise description and we might be forced to think about a framework, where even in principle, quantities are approximate.

One of the main puzzles in the emergence of space-time is the emergence of time. By a simple analogy with AdS/CFT people, have proposed a dS/CFT [7]. The idea is to replace the formula (3) by a similar looking formula except that the left hand side is the wavefunction of the universe in a lorentzian region, in a regime where it is peaked on a de-Sitter universe. Note that a given field theory is useful to compute a specific amplitude, but in order to compute probabilities we need to consider different field theories at once. For example, we should be able to vary the parameters defining the field theory. In AdS/CFT the way we fill the interior depends on the values of the parameters of the field theory. In this case this dependence translates into a dependence on the question we ask. So, for example, let us suppose that the de-Sitter ground state corresponds to a conformal field theory. If we are interested in filling this de-Sitter space with some density of particles, then we will need to add some operators in the field theory and these operators might modify the field theory in the IR. So they modify the most likely geometry in the past. So it is clear that in this framework, our existence will be part of the input. On the other hand, it is hard to see how constraining this is. In particular, empty de-Sitter space is favored by an exponentially large factor $e^{1/\Lambda}$. On the other hand, it is unclear that requiring our existence alone would beat this factor and produce the much less enthropic early universe that seems to have existed in our past.

Of course, dS/CFT suffers from the problem that we do not know a single example of the duality. Moreover, de-Sitter constructions based on string theory produce it only as a metastable state. In any case, some of the above remarks would also apply if we were to end up with a $\Lambda = 0$ supersymmetric universe in the far future. In that case, we might be able to have a dual description of the physics in such a cosmological $\Lambda = 0$ universe. It seems reasonable to think that these hypothetical dual descriptions would give us the amplitudes to end in particular configurations. In order to compute probabilities about the present we would have to sum over many different future outcomes.

In summary, precise dual descriptions are expected to exist only when the space-time has well defined stable asymptotics. In all other situations, we expect that the description of physics might be fundamentally imprecise. Let us hope that we will soon have a clear example of a description of a cosmological singularity.

Bibliography

[1] J. Polchinski, *String theory*, Cambridge University Press.
[2] For a review see: P. H. Ginsparg and G. W. Moore, arXiv:hep-th/9304011.
[3] T. Banks, W. Fischler, S. H. Shenker and L. Susskind, Phys. Rev. D **55**, 5112 (1997) [arXiv:hep-th/9610043].
[4] J. M. Maldacena, Adv. Theor. Math. Phys. **2**, 231 (1998) [Int. J. Theor. Phys. **38**, 1113 (1999)] [arXiv:hep-th/9711200].
[5] S. S. Gubser, I. R. Klebanov and A. M. Polyakov, Phys. Lett. B **428**, 105 (1998) [arXiv:hep-th/9802109]. E. Witten, Adv. Theor. Math. Phys. **2**, 253 (1998) [arXiv:hep-th/9802150].
[6] L. Fidkowski, V. Hubeny, M. Kleban and S. Shenker, JHEP **0402**, 014 (2004) [arXiv:hep-th/0306170].
[7] E. Witten, arXiv:hep-th/0106109. A. Strominger, JHEP **0110**, 034 (2001) [arXiv:hep-th/0106113].

5.3.4   *Alexander Polyakov: Beyond space-time*

In what follows I shall briefly describe various mechanisms operating in and around string theory. This theory provides a novel view of space-time. I would compare it with the view of heat provided by statistical mechanics. At the first stage the word "heat" describes our feelings. At the second we try to quantify it by using equations of thermodynamics. And finally comes an astonishing hypothesis that heat is a reflection of molecular disorder. This is encoded in one of the most fascinating relations ever, the Boltzmann relation between entropy and probability.

Similar stages can be discerned in string theory. The first is of course the perception of space-time. The second is its description using the Einstein equations. The third is perhaps a possibility to describe quantum space-time by the boundary gauge theory. Let us discuss in more details our limited but important knowledge of the gauge/string correspondence.

5.3.4.1   *Gauge /String correspondence*

It consists of several steps. First we try to describe the dynamics of a non-abelian flux line by some string theory. That means, among other things that the Wilson loop $W(C)$ must be represented as a sum over 2d random surfaces immersed in the flat 4d space-time and bounded by the contour C. Surprisingly, strings in 4d behave as if they are living in the 5d space, the fifth (Liouville) dimension being a result of quantum fluctuations[1] . More detailed analyses shows that while the 4d space is flat, the 5d must be warped with the metric

$$ds^2 = d\varphi^2 + a^2(\varphi)d\overrightarrow{x}^2 \tag{1}$$

where the scale factor $a(\varphi)$ must be determined from the condition of conformal symmetry on the world sheet [2]. This is the right habitat for the gauge theory strings. If the gauge theory is conformally invariant (having a zero beta function) the isometries of the metric must form a conformal group. This happens for the space of constant negative curvature, $a(\varphi) \sim \exp c\varphi$ [3]. The precise meaning of the gauge/strings correspondence [4],[5] is that there is an isomorphism between the single trace operators of a gauge theory, e.g. $Tr(\nabla^k F_{\mu\nu}\nabla^l F_{\lambda\rho}...)$ and the on-shell vertex operators of the string, propagating in the above background. In other words, the $S-$ matrix of a string in the 5d warped space is equal to a correlator of a gauge theory in the flat 4d space. The Yang -Mills equations of motion imply that the single trace operators containing $\nabla_\mu F_{\mu\nu}$ are equal to zero. On the string theory side it corresponds to the null vectors of the Virasoro algebra, leading to the linear relations between the vertex operators. If we pass to the generating functional of the various Yang- Mills operators, we can encode the above relation in the formula

$$\Psi_{WOE}[h_{\mu\nu}(x), ...] = \langle \exp \int dx h_{\mu\nu}(x)T_{\mu\nu}(x) + ...\rangle_{Y.-M} \tag{2}$$

Here at the left hand side we have the "wave function of everything" (WOE). It is obtained as a functional integral over 5d geometries with the metric $g_{mn}(x, y)$, where $y = \exp -c\varphi$, satisfying asymptotic condition at infinity $(y \to 0)$ $g_{\mu\nu} \to \frac{1}{y^2}(\delta_{\mu\nu} + h_{\mu\nu}(x))$. It differs from the "wave function of the universe" by Hartle and Hawking only by the y^{-2} factor. On the right side we have an expression defined in terms of the Yang- Mills only, $T_{\mu\nu}$ being its energy- momentum tensor. The dots stand for the various string fields which are not shown explicitly. An interesting unsolved problem is to find the wave equation satisfied by Ψ. It is not the Wheeler -de Witt equation. The experience with the loop equations of QCD tells us that the general structure of the wave equation must be as following

$$\mathcal{H}\Psi = \Psi * \Psi \tag{3}$$

where $\mathcal{H}$ is some analogue of the loop Laplacian and the star product is yet to be defined. This conjectured non-linearity may lead to the existence of soliton-like WOE-s.

The formula (2), like the Boltzmann formula, is relating objects of very different nature. This formula has been confirmed in various limiting cases in which either LHS or RHS or both can be calculated. I suspect that, like with the Boltzmann formula, its true meaning will still be discussed a hundred years from now.

5.3.4.2   *de Sitter Space and Dyson's instability*

Above we discussed the gauge/ strings duality for the geometries which asymptotically have negative curvature. What happens in the de Sitter case ? It is not very clear. There have been a number of attempts to understand it [6]. We will try here a different approach. It doesn't solve the problem, but perhaps gives a sense of the right direction.

Let us begin with the 2d model, the Liouville theory. Its partition function is given by

$$Z(\mu) = \int D\varphi \exp\{-\frac{c}{48\pi} \int d^2x (\frac{1}{2}(\partial\varphi)^2 + \mu e^\varphi) \tag{4}$$

For large c (the Liouville central charge) one can use the classical approximation. The classical solution with positive μ describes the AdS space with the scalar curvature $-\mu$. By the use of various methods [7] one can find an exact answer for the partition function, $Z \sim \mu^\alpha$ where $\alpha = \frac{1}{12}[c - 1 + \sqrt{(c-1)(c-25)}]$. In order to go to the de Sitter space we have to change $\mu \Rightarrow -\mu$. Then the partition function acquires an imaginary part, $\mathrm{Im} Z \sim \sin \pi\alpha |\mu|^\alpha$. It seems natural to assume that the imaginary part of the Euclidean partition function means that the de Sitter space is intrinsically unstable. This instability perhaps means that due to the Gibbons -Hawking temperature of this space it "evaporates" like a simple black hole. In the latter its mass decreases with time, in the de-Sitter space it is the cosmological constant. If we define the Gibbons- Hawking entropy S in the usual way, $S = (1 - \beta\frac{\partial}{\partial\beta}) \log Z$, we find another tantalizing relation, $\mathrm{Im} Z \sim e^S$, which holds in the classical limit, $c \to \infty$. Its natural interpretation is that the decay rate of the dS space is proportional to the number of states, but it is still a speculation, since the precise meaning of the entropy is not clear. For further progress the euclidean field theory, used above, is inadequate and must be replaced with the Schwinger-Keldysh methods.

In higher dimensions we can try once again the method of analytic continuation from the AdS space. The AdS geometry is dual to a conformally invariant gauge field theory. In the strong coupling limit (which we consider for simplicity only) the scalar curvature of the AdS, $R \propto \frac{1}{\sqrt{\lambda}}$ ($\lambda = g_{YM}^2 N$). So, the analytic continuation we should be looking for is $\sqrt{\lambda} \Rightarrow -\sqrt{\lambda}$. In order to understand what it means in the gauge theory, let us notice that in the same limit the Coulomb interaction of two charges is proportional to $\sqrt{\lambda}$ [8]. Hence under the analytic continuation we get a theory in which the same charges attract each other. Fifty years ago Dyson has shown that the vacuum in such a system is unstable due to creation of the clouds of particles with the same charge. It is natural to conjecture that Dyson's instability of the gauge theory translates into the intrinsic instability of the de Sitter space. Once again the cosmological constant evaporates.

5.3.4.3 *Descent to four dimensions*

Critical dimension in string theory is ten . How it becomes four ? If we consider type two superstrings, the 10d vacuum is stable, at least perturbatively., and stays 10d. Let us take a look at the type zero strings, which correspond to a non-chiral GSO projections. These strings contain a tachyon, described by a relevant operator of the string sigma model. Relevant operators drive a system from one fixed point to another. According to Zamolodchikov's theorem, the central charge must decrease in the process. That means that the string becomes non-critical and the Liouville field must appear. The Liouville dimension provides us with the emergent "time" in which the system evolves and changes its effective dimensionality (the central charge). As the "time" goes by, the effective dimensionality of the system goes down. If nothing stops it, we should end up with the $c = 0$ system which has only the Liouville field. It is possible,however, that non-perturbative effects would stop this slide to nothingness [10]. In four dimensions we have the B -field instantons, described by the formula (at large distances) $(dB)_{\mu\nu\lambda} = q\epsilon_{\mu\nu\lambda\rho}\frac{x_\rho}{x^4}$. In the modern language they correspond to the NS5 branes. These instantons form a Coulomb plasma with the action $S \sim \sum \frac{q_i q_j}{(x_i - x_j)^2}$. As was explained in [10] , the Debye screening in this plasma causes "string confinement", turning the string into a membrane. Formally this is described by the relation

$$\langle \exp i \int B_{\mu\nu} d\sigma_{\mu\nu} \rangle \sim e^{-aV} \tag{5}$$

where we integrate over the string world sheet and V is the volume enclosed by it. There is an obvious analogy with Wilson's confinement criterion. While the gravitons remain unaffected, the sigma model description stops being applicable and hopefully the sliding stops at 4d.

5.3.4.4 *Screening of the cosmological constant*

Classical limits in quantum field theories are often not straightforward. For example, classical solutions of the Yang - Mills theory describing interaction of two charges have little to do with the actual interaction. The reason is that because of the strong infrared effects the effective action of the theory has no resemblance to the classical action. In the Einstein gravity without a cosmological constant the IR effects are absent and the classical equations make sense. This is because the interaction of gravitons contain derivatives and is irrelevant in the infrared.

The situation with the cosmological term is quite different, since it doesn't contain derivatives. Here we can expect strong infrared effects [11] , see also [9] for the recent discussion.

Let us begin with the 2d model (3) . The value of μ in this lagrangian is subject to renormalization. Perturbation theory generates logarithmic corrections to this quantity. It is easy to sum up all these logs and get the result $\mu_{ph} = \mu(\frac{\Lambda}{\mu_{ph}})^{\beta}$, with $\beta = \frac{1}{12}[c - 13 - \sqrt{(c-1)(c-25)}]$.Here Λ is an UV cut-off while the physical (negative) cosmological constant μ_{ph} provides a self-consistent IR cut-off. We see that in this case the negative cosmological constant is anti-screened.

In four dimensions the problem is unsolved. For a crude model one can look at the IR effect of the conformally flat metrics. If the metric $g_{\mu\nu} = \varphi^2 \delta_{\mu\nu}$ is substituted in the Einstein action S with the cosmological constant Λ,the result is $S = \int d^4x[-\frac{1}{2}(\partial\varphi)^2 + \Lambda\varphi^4]$. There is the well known non-positivity of this action. This is an interesting topic by itself, but here we will not discuss it and simply follow the prescription of Gibbons and Hawking and change $\varphi \Rightarrow i\varphi$. After that we obtain a well defined φ^4 theory with the coupling constant equal to Λ. This theory has an infrared fixed point at zero coupling, meaning that the cosmological constant screens to zero.

There exists a well known argument against the importance of the infrared effects. It states that in the limit of very large wave length the perturbations can be viewed as a change of the coordinate system and thus are simply gauge artefacts.This argument is perfectly reasonable when we discuss small fluctuations at the fixed background (see [12] for a different point of view). However in the case above the effect is non-perturbative- it is caused by the fluctuation of the metric near zero, not near some background. In this circumstances the argument fails. Indeed, if we look at the scalar curvature, it has the form $R \sim \varphi^{-3}\partial^2\varphi$. We see that while for the perturbative fluctuations it is always small because of the second derivatives, when φ is allowed to be near zero this smallness can be compensated. In the above primitive model the physical cosmological constant is determined from the equation $\Lambda_{ph} = \frac{const}{\log \frac{1}{\Lambda_{ph}}}$ which always has a zero solution. One would expect that in the time-dependent formalism we would get a slow evaporation instead of this zero. The main challenge for these ideas is to go beyond the conformally flat fluctuations. Perhaps gauge/ strings correspondence will help.

I am deeply grateful to Thibault Damour for many years of discussions on the topics of this paper. I would also like to thank David Gross and Igor Klebanov for useful comments.

This work was partially supported by the NSF grant 0243680. Any opinions, findings and conclusions or recommendations expressed in this material are those of the authors and do not necessarily reflect the views of the National Science Foundation.

Bibliography

[1] A. M. Polyakov, "Quantum Geometry Of Bosonic Strings," Phys. Lett. B **103** (1981) 207.

[2] A. M. Polyakov, "String theory and quark confinement," Nucl. Phys. Proc. Suppl. **68** (1998) 1 [arXiv:hep-th/9711002].

[3] J. M. Maldacena, "The large N limit of superconformal field theories and supergravity," Adv. Theor. Math. Phys. **2** (1998) 231 [Int. J. Theor. Phys. **38** (1999) 1113] [arXiv:hep-th/9711200].

[4] S. S. Gubser, I. R. Klebanov and A. M. Polyakov, "Gauge theory correlators from non-critical string theory," Phys. Lett. B **428** (1998) 105 [arXiv:hep-th/9802109].

[5] E. Witten, "Anti-de Sitter space and holography," Adv. Theor. Math. Phys. **2** (1998) 253 [arXiv:hep-th/9802150].

[6] A. Strominger, "The dS/CFT correspondence," JHEP **0110** (2001) 034 [arXiv:hep-th/0106113].

[7] V. G. Knizhnik, A. M. Polyakov and A. B. Zamolodchikov, "Fractal Structure Of 2d-Quantum Gravity," Mod. Phys. Lett. A **3** (1988) 819.

[8] J. M. Maldacena, "Wilson loops in large N field theories," Phys. Rev. Lett. **80** (1998) 4859 [arXiv:hep-th/9803002].

[9] R. Jackiw, C. Nunez and S. Y. Pi, "Quantum relaxation of the cosmological constant," Phys. Lett. A **347** (2005) 47 [arXiv:hep-th/0502215].

[10] A. M. Polyakov, "Directions In String Theory," Phys. Scripta **T15** (1987) 191.

[11] A. m. Polyakov, 'Phase Transitions And The Universe," Sov. Phys. Usp. **25** (1982) 187 [Usp. Fiz. Nauk **136** (1982) 538].

[12] N. C. Tsamis and R. P. Woodard, "Quantum Gravity Slows Inflation," Nucl. Phys. B **474** (1996) 235 [arXiv:hep-ph/9602315].

5.4 Discussion

G. 't Hooft I would like to make sort of a claim or statement and then a question. Actually it bears on Nati's talk, but also others have mentioned emergent space and emergent time, and I claim that any theory you have allows a rigorous definition of time, not a fuzzy one, and even a rigorous definition of space, and not a fuzzy definition of space. And the argument goes as follows. Assume you have some theory that is supposed to explain some phenomenon. A priori there was no space, no time in the question you've been asking. You just have a theory. Then the theory will contain variables and equations, and a lot of prescriptions how to solve these equations, if it is a good theory. And I claim that, as soon as you have indicated the order by which you have to solve the equations, that order defines causality in your theory, and that defines a notion of time. So time is basically the order by which you have to solve the equations. If you think a little bit, that's exactly for instance how a theory of the planetary system works. The time, the notion of time among the planets is the order by which you solve the equations. If you solve the equations in the wrong order, you might have forgotten that two planets might collide, and then you get impossible answers. So you have to know, exactly that time is the order by which you solve the equations. And, so that is a rigorous definition, there's no way to fool around with that, because if you solve them in the wrong order, you might get the wrong answer. Similarly however, you can also make a rigorous definition of space. And that is because, well, I must assume some form of reduction of a theory into simple equations. If you write down infinitely complicated equations, you don't really know what you're doing, you have to reduce them to simple equations. And then you can ask, two sets of variables, how many equations are they away from each other? And that defines a distance between variables, and that eventually defines space. If you think a little bit, that's the way our present space-time seems to work, that two systems are far away if you have to solve differential equations very very many times before you reach from one point to the other point. So I think that any theory should contain some notion of space as well as time, and in a discussion during the break, the question was asked: does this defines a continuous space-time? And I would say no, most time and space would be discrete in this sense, but two variables like that, connected with by an equation, that defines them to be nearest neighbours, that defines a distance one. And then, so any theory in some sense looks like a lattice.

J. Harvey I cannot resist comparing, that if this defines not only the order of time, but the rate at which time proceeds, then I am sure time proceeds much faster in your part of the world than in my part of the world.

G. 't Hooft But anyway, the question comes then, that if you would drop the notion of reduction, then what are you doing, and is that not a direct contradiction in terms, that we do want theories to be based on simple equations, therefore reduction, in that sense, seems to be absolutely necessary to me, and if you do not have that, should you then not do something else, go into music instead of theoretical physics? That is the question.

J. Polchinski So I would like to respond to Abhay's comments from earlier and ask a question. In canonical general relativity, you write the wave function in terms of geometries at one time, solve the hamiltonian constraint, and time emerges as correlations in that wave functions. But when you talk about emergent gauge symmetry and AdS/CFT, there is something more that happens. Because there's a set of variables which are actually completely neutral under the gauge symmetry, you don't have a hamiltonian constraint, you have variables that satisfy it trivially, and they're related in a complicated way to the other ones. But you can solve the theory, and write the observables entirely in terms of the gauge invariant observables, the ones on which the hamiltonian constraint is already solved. And the question then is: is there some analogue of this which is known, say, in other approaches to quantum gravity?

J. Harvey Does anyone want to answer that question or respond to it? Nati?

N. Seiberg I think the answer is exactly the comment I made to Steve, that in the matrix model description of two dimensional gravity this is exactly what happens.

E. Rabinovici In regard to emerging space-times I have a comment. I think we should attempt to use methods which we learned from statistical mechanics, appropriately modified for gravity, to study possible phases of gravity. And one phase which I think is necessary, we should study more, is that in which α' is infinity, or in other words where the string scale vanishes. And I think once we understand that, it could help also understand more the emergence of space-time.

G. Dvali I just wanted to comment that all these important questions about emerging nature of gravity and space-time at short distances, in the UV, probably should be also asked about large distances, in the infrared, because after all, we only understand, experimentally at least, we only understand the nature of space-time and gravity at intermediate distances, and we have no idea what is happening beyond the 10^{28} centimeters, and we know that something is going on there, the universe is accelerating. Normally we are attributing it to a cosmological constant, but it may very well be that string theory encodes new far infrared scale, and so the nature of space-time and gravity gets dramatically modified there. So space-time may emerge in the UV, and also it happens something in the IR. This question also should be studied.

J. Harvey I think that is sort of along the lines of what Nati called the old approach to the cosmological constant problem, that there's some confusion between UV and IR, which I think is perhaps old, but not completely forgotten.

J. Maldacena Yes, I was going just to mention that this question of very long distances outside our horizon—so even if we had a precise description, sort of reductionist description in the far future where the universe is infinitely large and so on, it might involve regions outside our horizon, and then when we ask a question about our universe, we need to sum over everything that's going on in the outside, so we have to sum over theories and so on. That would be probably an important part.

T. Damour I am confused about in what sense really dynamical gravity emerges in AdS/CFT, and the apparent contradiction with the theorem of Weinberg and Witten. So, in what we saw from Sasha and Juan, the $h_{\mu\nu}$ on the boundary does not satisfy any constraint, so it's not really dynamical; how will the Wheeler-de Witt equation come out? Or let me ask a more practical question, are there correlators, multi-correlators in super Yang-Mills, where I see the massless spin two pole of a dynamical graviton emerging, in super Yang-Mills?

J. Maldacena The answer is no, because the graviton is not a massless particle in four dimensions. So it is a massless particle in five dimensions. So we don't have a local stress tensor in five dimensions, we have a local stress tensor in four. And so the gravity that emerges is the five dimensional gravity, and this is an important point.

G. Gibbons I have a question really for Tom Banks, which is that the formalism he outlined seems to take causal structures primary and given once and for all. So do you envisage that the causal structure varies and fluctuates, or that we have always a fixed causal structure? And if the latter, one normally says that nine tenths of the metric, in dimension four, are given by the causal structure, so you also seem not to be allowing the gravitational field to fluctuate, if you took the view that the causal structure was fixed.

T. Banks In this formalism, the causal structure is put in, but the variables that are going to describe the geometry are quantum fluctuating, so it's a quantum mechanical causal structure. You get a geometry that comes out of it in the large area limit which is unique, and indeed that causal structure will be fixed, but it will only be fixed in the large area limit where things are approximately semiclassical.

G. Gibbons So for you locality is really not a problem, it is fixed. We always have a local theory if I understand you ... and it makes sense to say things commute at space-like separations...

T. Banks There are certainly no operators here that you can define at local points that commute in space-like separations in this formalism. Perhaps we should, I should talk to you privately about this.

J. Harvey David, did you have something you wanted to say?

D. Gross Yes. I think Gerard correctly pointed out some of the things one would like space and time to satisfy in an emergent scenario. The problem is that it might not be obvious how to do that when one looks for the principle that will lead to an ordering and one time. Perhaps some kind of reductionism, although given the ultraviolet-infrared connection we are not even sure of this. Since we don't know what that principle is then how do we know that, if space-time is emergent, one time will emerge? One question about space-time that has not come up here, which has always intrigued me, is why can we easily imagine alternate topologies, alternate dimensions of space, different than what we see around us, but it seems impossible to imagine more than one time. Is a single time anthropically selected? (This is a modern strategy for eliminating things you don't understand.) Or is it simply impossible to imagine physics with more than one time? So when you give up the foundational setting of physics, as people are struggling to do, you really would like to know what principles lead to a unique causal ordering.

The other comment I wanted to make was about locality in string theory and its connection with causality or the analyticity of the S-matrix. This is something we might understand. After all, we have the S-matrix in string theory, and we can ask why is it analytic? Why is string theory causal even though strings are non-local and do not interact locally, at least if you define space-time in terms of the center of mass of the string (i.e. choose a gauge in which the center of mass of the string is identified with space-time). In string field theory, for example, the fields depend not on points in space-time, but on loops, and in terms of the center of mass of the string, as Nati pointed out, the interations of such fields are then non-local. But this is not the way to define space-time in string field theory. In fact, locality becomes manifest and one can derive the analytic properties of the S-matrix if you either work in light-cone gauge, where strings interact at the same light cone time at a single point, or covariantly, if you define space-time to as the midpoint of the string, where the interaction takes place (at least in open string field theory). In both of these cases one can understand, in string perturbation theory, how locality and causality emerge from the interaction of non-local objects such as strings. On the other hand we do not know how to address what happens non-perturbatively. If we use the AdS/CFT duality to non-perturbatively define ten dimensional string theory we are hard pressed to recover locality in the bulk. In addition, in cases we now begin to understand, we see non-local structures emerge from string theory, such as non-commutative field theories, as a description of dual theory on the boundary. There is a whole program, which I would urge people to follow, to explore, in the context of field theory, what kinds of non-localities are allowable and controllable and sensible, expanding what we already know about the non-local field theories that are healthy boundary theories dual to string theories in the bulk.

J. Harvey One of the topics that Nati raised which I must say confuses me as well, is in situations where—string theory on a circle for example—where in different limits you would use one description or its T-dual. It's very tempting to think that you should have a formulation where you have both x-left and x-right, but it seems difficult to have locality in both and so which is one supposed to choose, or what principle determines that. I . . . Do you feel like that's . . . that there is a clear understanding of that, because I find that also very confusing?

D. Gross It is clear, because of this ambiguity, that locality or even what we mean by space-time is something like gauge invariance. It is a description that is inherently ambiguous and there are different descriptions which are useful for different purposes.

J. Harvey Right, but perhaps we are missing the additional gauge degrees of freedom that allow us to project in a clear way onto the different local descriptions right now. You could imagine that there's a formulation where—I guess people have attempted this—where both the variable and the T-dual are there at the same time but there is some additional redundancy.

M. Gell-Mann When I left this field and stopped following what was going on in detail, people had proposed a version of string theory in which there's another variable running along the string, and when you have that, then you can say that in a string vertex, the old string and the two new ones, or the two old ones and the one new one, are laid along one another, so that there is exact locality for every point on the string. It's not a question of the center of mass at all. And I assume that in the intervening years that hasn't disappeared. It's a much more satisfactory way of treating locality.

J. Harvey Alright. As a moderator I don't feel obliged to answer any questions, so if anybody else would like to answer that I'll. . . the author of a textbook, or. . .

G. Veneziano As far as I know, Murray is right, I mean, I thought that in — at least in some versions of string field theory — that is exactly what you do. You put a local coupling in terms of strings, namely when three strings overlap completely, then they interact. This is an invariant local concept, I think.

D. Gross That is absolutely wrong. It is nice to be able to make absolute statements. It is not true that strings interact when they overlap— were it the case that string theory interaction consisted of a vertex where strings totally overlapped, it would be infinitely more nonrenormalizable than ordinary quantum field theory, require an infinite number of constants, have no relation to two-dimensional geometry, and be totally different than the string theory that we know and love. Instead strings meet at one point (in light cone gauge) or overlap on half their length (in the covariant open string field theory approach). The problem is that, unless care is taken to define time carefully, this interaction need not be local in time.

G. Veneziano That is what I thought was the Witten open string field theory action, that it had really overlapping strings.

Note by the editors The discussion between Gross and Veneziano continued over lunch during which the misunderstandings were elucidated. Summarizing: while a local (say ϕ^3) interaction in field theory means that three fields interact when their coordinates all coincide ($x_1 = x_2 = x_3$), three strings interact when every point on one string also belongs to another string. Forcing all three strings to overlap completely by imposing $x_1(\sigma) = x_2(\sigma) = x_3(\sigma)$ would indeed lead to incurable UV divergences. Instead, the condition for having a string interaction is more like momentum conservation, but in coordinate space, i.e. $x_1 + x_2 + x_3 = 0$ (where string bits of opposite orientation are counted with a relative minus sign). This is how string theory provides its compromise between locality and non-locality.

G. Veneziano May I take advantage to... I am just wondering, in view also of what Sasha Polyakov has emphasized, namely that the AdS/CFT correspondence is between an on-shell theory in the bulk and an off-shell theory on the boundary, whether we should really see this as a correspondence, or, you know, just as a tool, OK? After all, suppose on the boundary we manage to have QCD, just QCD, no weak interactions, just QCD, then on the boundary itself, we just may be interested in the only observable, which is the S-matrix on the boundary. And that S-matrix on the boundary would not be in itself sufficient to determine what goes on in the bulk. In other words, it looks to me that this on- versus off-shell duality may mean that actually the boundary field theory is a tool to—the off-shell boundary field theory is just a tool rather than an equivalent thing.

A. Polyakov Well, you know, before you turned on gravity, you have a choice, then off-shell quantities are more or less well-defined. You don't have to consider necessarily the S-matrix. I think this – what we were taught in the days that field theory was despised, that the only thing which makes sense is the S-matrix, which is actually true in the theory of gravity, almost true. But in normal field theory, it is not true, so I think it's quite appropriate that since we make the contact between theory of gravity and the theory without gravity, on the gravity side we must have only on-shell amplitude, while on the theory without gravity side we may have all possible correlation functions, not necessarily on-shell.

J. Harvey It has always seemed to me that AdS/CFT should, you know, be perhaps a precise statement, but should allow for inexact statements. I mean, after all if we had discovered that we live in anti-de Sitter space with a very small cosmological constant rather than de Sitter space, and we went to the experimentalists in Fermilab and told them that what they observe and measure is not real, because it's not defined at infinity, I think they would regard us as rather useless. So it is clear that there has to be room for a description that is, you know, an exceedingly good approximation to observables defined at surfaces at infinity that are defined in a local way in the bulk. But how to actually do this within the machinery we currently have seems to me rather problematic.

D. Gross This indeed is a very deep problem, because we can imagine compact spaces in general relativity, and then we have no local gauge-invariant observables and no place for a holographic description in terms of something which doesn't include gravity, which is well-defined. I find the discussion of compact universes extremely puzzling. Nati said that he has no problem thinking about closed universes, but how do you think about it, Nati, or how does anyone think about what are physical observables in such situations?

N. Seiberg Maybe I should clarify what I have in mind. I have a lot of problems thinking about it, but I don't see any obstacle to their existence.

D. Gross Agreed, but given that they might exist, how do you think about them?

N. Seiberg Well, we have one example of this in two dimensions, which is the quantization of the world-sheet of the string. This is a very concrete example.

J. Maldacena Yes, it seems that in order to have a description of these closed universes, you have to allow yourself the possibility of not having an exact description, that the fundamental description will be fundamentally not precise, I think.

E. Silverstein In fact this can be borne out by holographically dualizing closed universes in the same way that we do for Randall-Sundrum, where indeed the dual description also has lower-dimensional gravity, and you can continue that down to two dimensional gravity plus a large amount of matter and obtain at least a simplification of the problem, but one which illustrates the limitations that you all are talking about.

A. Ashtekar David, what is wrong with gauge-fixing? So if you had closed universes, right, there is gauge-fixed description, and then of course there has to be consistency checks that different choices of gauges will give you the same...

D. Gross As Nati emphasized, gauge symmetries are redundancies in our description of nature, and presumably there are, as the AdS/CFT duality beautifully illustrates, formulations of generally covariant theories where there is no general covariance needed since there are only physical degrees of freedom used to describe the system, so...

A. Ashtekar Maybe I am saying the same.

D. Gross AdS/CFT is the one description of quantum gravity that is best defined in our toolbox, and using the gauge theory description of quantum gravity in 10 dimensions there is no such thing a gauge fixing, because there is no gauge to fix.

B. Greene So just back to the general question of emerging space-time, I wonder, and I am not sure about this, I wonder if it is worth trying to sharpen what one means by emerging space-time, and perhaps Gary's question highlights one instance. You can imagine a situation as you were describing, Chern-Simons theory with a zero metric, where you do have some background coordinate grid and then you can imagine the metric emerging as opposed to the coordinate system emerging, so one can ask: do we talk about emerging topology, emerging differential structure, do we talk about emerging complex structure, and then do we talk about emerging geometrical structure on top of those structures? I mean, I have always wondered: do exotic differential structures have any real role in anything that we're talking about? Could we see that if we spoke not just about emerging x and t, but emerging topological/differential structure, we'd see that maybe the exotic structures are there in some meaningful way and they need to be taken account of. And then one can talk about emerging geometrical structure on top of those. I do not know if that is a worthwhile framework to think about it, but it would help sharpen, I think, what we mean by emerging spacetime.

J. Harvey Well, it is certainly an interesting question whether things can emerge that we don't think describe reality and why they do not emerge.

T. Banks I want to emphasize a point that was made both by Juan and Nati about the question of trying to understand what we need to do to have a macroscopic spacetime, a spacetime with low curvature that's well-described by gravity. And I would claim that in the well-understood examples beyond $1 + 1$ dimensions, we always need to have supersymmetry that's either exact or restored asymptotically in the low-curvature region. The landscape proposal gives examples which claim that this is not a...

J. Harvey In order to have moduli spaces, you mean?

T. Banks No, well, in the BFSS matrix model, in order to have moduli spaces so that you can talk about large distance scattering, you need to have exact supersymmetry. In AdS/CFT, in all the examples that we really understand, in order to have a low-curvature AdS space, in the sense that Juan discussed, we have to have exact supersymmetry. There are claims in the literature about examples where that's not true, I think it's extremely important for us to try to find the conformal field theories that supposedly describe completely non-supersymmetric, very low curvature AdS space.

J. Harvey That is a good point.

S. Weinberg I would like to offer a remark that is so reactionary that I might be ejected from the room.

J. Harvey It is almost time for lunch, so do not worry.

S. Weinberg Listening to the discussion this morning, which I found very stimulating, I was nevertheless reminded of the kind of discussion that went on in the late 1930's and early 1940's about the problems of fundamental physics. There were internal problems of not knowing how to do calculations, and external problems of anomalies in cosmic rays, because although they didn't know it, they were confusing pions and muons, and most people at that time thought that fundamental new ideas were needed, that we had to go beyond quantum field theory as it had been constructed by Heisenberg and Pauli and others, and have something entirely new, something perhaps non-local, or a fundamental length. It turned out that the solution was to stick to quantum field theory, and that it worked. It occasionally occurs to me, well, maybe that is the solution now, that there is not an emergent space-time, that we just have three space and one time dimension, and that the solution is quantum field theory. Now why do you think that is not true? Well, obviously the reason, the most obvious reason is that gravity has problems that you get into, problems of ultraviolet divergences. That is really the wrong way to look at the problem, because if you allow all possible terms in the Lagrangian, with arbitrary powers of the curvature, you can cancel the divergences the same way you do in quantum electrodynamics. But then you say: oh, but the problem is that you have an infinite number of free constants, and the theory loses all predictive power when you go to sufficiently high energy. Well, that is not necessarily true, although it might be true. It might be that there is a fixed point in the theory, that is that there is a point in the infinite-dimensional space of all these coupling constants where the beta function vanishes. And furthermore, when that happens, you actually expect that the surface of trajectories which are attracted into that fixed point as you go to high energy is finite-dimensional. So the theory would in fact have precisely as much predictive power as an ordinary renormalizable field theory, although much more difficult to calculate since the fixed point would not be near the origin. Even so, you would then say: oh, but even so, this theory has a lot of free parameters, maybe a finite number, but where do they come from? What we were hoping for was that string theory would tell us how to calculate everything. And even there, that might not be true, it might be that the surface of trajectories that are attracted to the fixed point is one-dimensional. And we know that there are examples of extremely complicated field theories in which in fact there is a non-Gaussian fixed point with a one-dimensional attractive surface — just a line of trajectories that are ultravioletly attracted to the fixed point and therefore avoid problems when you go to large energy. That is shown by the existence of second order phase transitions. Basically, the condition is that the matrix of partial derivatives of all the various beta functions with respect to all the various coupling parameters should have a finite number of negative eigenvalues, and in the case of a second order phase transitions, in fact, there is just one negative eigenvalue: that is the condition. So it is possible that that is the answer. I suspect it is not. I suspect that these really revolutionary ideas are going to turn out to be necessary. But I think we should not altogether forget the possibility that there is no revolution that's needed, and that good old quantum field theory, although with a non-Gaussian fixed point, is the answer.

I. Klebanov I just had a comment on Brian's question. Of course in AdS/CFT, not all of spacetime is emergent: $3 + 1$, say, dimensions are put in from the beginning, but the other six emerge. And for example in the story I discussed, you can see the emergent Calabi-Yau with the complex structure epsilon emerging from the infrared effects of the field theory. And then I have a brief unrelated comment about locality. I think there are some speculations that you can formulate string theory just on a lattice, and string theory completely erases this lattice, as long as the lattice spacing is small enough. Namely, it's not just an approximation, it's the same theory. And one simple example is for example the $c = 1$ matrix model, where you discretize this one dimension, you can just show that for a lattice spacing small enough, you get exactly equivalent theory to a continuous dimension. So I think that's a relevant picture for locality.

J. Harvey I think we're probably a few minutes over and should wrap this up. So I think...

M. Douglas I had a very short postscript to Steve Weinberg—it's short. So, we used to be very confident that there would not be non-trivial quantum field theories in greater than four dimensions and now we believe there are non-trivial fixed point theories with lots of supersymmetry in six. So similarly, maybe the idea that gravity stops at two in quantum field theory will go up to four.

S. Weinberg Thank you, Mike.

J. Harvey Alright, on that—on that point we'll wrap things up. I guess lunch is in the usual place and the picture is at 1:15, is that correct?

Session 6

Cosmology

Chair: *Stephen Shenker*, Stanford, USA
Rapporteur: *Joseph Polchinski*, UCSB, USA
Scientific secretaries: *Christiane Schomblond* (Université Libre de Bruxelles) and *Peter Tinyakov* (Université Libre de Bruxelles)

S. Shenker As I was writing up the program of the cosmology session of today, I had one of these epiphanies that are such a nice part of our subject. I realized in describing the structure of our little session that phenomena which are familiar from biology seem to be taking place: there is this remarkable phenomenon called "ontology recapitulates phylogeny" where the structure of the embryo when it is growing seems to reproduce the entire history of the species that is developing. And here in our little session we see the whole history of the Universe being recreated. We are likely to start with a Big Bang. Then there will be a rapid period of high temperature in our discussion, probably optically opaque. I do not know what this coffee break is, maybe you can help me with that. Then we return to a prolonged period of inflation which will exit into a period of reheating, again probably optically opaque, and then we will basically return to the period of structure formation. Now, any good analysis like this, you test by trying to apply it outside the domain of its validity. And although it is not written on this schedule, after this discussion what takes place is that David Gross will give some closing remarks. So I tried to place Davis Gross' presentation in this framework and it could be that this presentation will be a Big Crunch. But thinking more carefully, it seems more likely that it will be one of those phases in the evolution of the Universe that seems to go on almost forever.

Laughter.

Alright, I have had my fun. I turn things over to Polchinski.

6.1 Rapporteur talk: The cosmological constant and the string landscape, by Joseph Polchinski

6.1.1 *The cosmological constant*

I would like to start by drawing a parallel to an earlier meeting — not a Solvay Conference, but the 1947 Shelter Island conference. In both cases a constant of nature was at the center of discussions. In each case theory gave an unreasonably large or infinite value for the constant, which had therefore been assumed to vanish for reasons not yet understood, but in each case experiment or observation had recently found a nonzero value. At Shelter Island that constant was the Lamb shift, and here it is the cosmological constant. But there the parallel ends: at Shelter Island, the famous reaction was "the Lamb shift is nonzero, therefore we can calculate it," while today we hear "the cosmological constant is nonzero, therefore we can calculate *nothing*." Of course this is an overstatement, but it is clear that the observation of an apparent cosmological constant has catalyzed a crisis, a new discussion of the extent to which fundamental physics is predictable. This is the main subject of this report.

In the first half of my talk I will review why the cosmological constant problem is so hard. Of course this is something that we have all thought about, and there are major reviews.[1] However, given the central importance of the question, and the flow of new ideas largely stimulated by the observation of a nonzero value, we should revisit this. One of my main points is that, while the number of proposed solutions is large, there is a rather small number of principles and litmus tests that rule out the great majority of them.

In recent years the cosmological constant has become three problems:

(1) Why the cosmological constant is not large.
(2) Why it is not zero.
(3) Why it is comparable to the matter energy density *now* (cosmic coincidence).

I will focus primarily on the first question — this is hard enough! — and so the question of whether the dark energy might be something other than a cosmological constant will not be central.

In trying to understand why the vacuum does not gravitate, it is useful to distinguish two kinds of theory:

(1) Those in which the energy density of the vacuum is more-or-less uniquely determined by the underlying theory.
(2) Those in which it is not uniquely determined but is adjustable in some way.

---

[1] For a classic review see [1]. For more recent reviews that include the observational situation and some theoretical ideas see [2, 3]. A recent review of theoretical ideas is [4]. My report is not intended as a comprehensive review of either the cosmological problem or of the landscape, either of which would be a large undertaking, but a discussion of a few key issues in each case.

I will discuss these in turn.

6.1.1.1 *Fixed-Λ theories*

The basic problem here is that we know that our vacuum is a rather nontrivial state, and we can identify several contributions to its energy density that are of the order of particle physics scales. It is sufficient to focus on one of them; let us choose the electron zero point energy, since we know a lot about electrons. In particular, they are weakly coupled and pointlike up to an energy scale M of at least 100 GeV. Thus we can calculate the electron zero point energy up to this scale from the graphs of Fig. 1 [5],

$$\rho_V = O(M^4) + O(M^2 m_e^2) + O(m_e^4 \ln M/m_e) , \qquad (1)$$

which is at least 55 orders of magnitude too large.

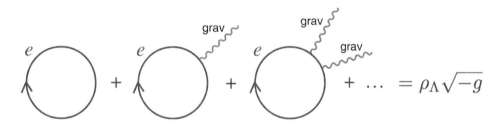

Fig. 6.1 An electron vacuum loop and its coupling to external gravitons generate an effective cosmological constant.

So we must understand why this contribution actually vanishes, or is cancelled. To sharpen the issue, we know that electron vacuum energy does gravitate in some situations. Fig. 2a shows the famous Lamb shift, now coupled to an external gravi-

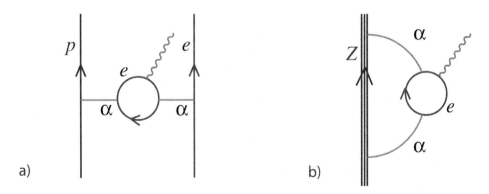

Fig. 6.2 a) The Lamb shift, coupled to an external graviton. b) A loop correction to the electrostatic energy of a nucleus, coupled to an external graviton.

ton. Since this is known to give a nonzero contribution to the energy of the atom, the equivalence principle requires that it couples to gravity. The Lamb shift is very small so one might entertain the possibility of a violation of the equivalence principle, but this is a red herring, as there are many larger effects of the same type.

One of these is shown in Fig. 2b, a loop correction to the electrostatic energy of the nucleus. Aluminum and platinum have the same ratio of gravitational to inertial mass to one part in 10^{12} [6, 7]. The nuclear electrostatic energy is roughly 10^{-3} of the rest energy in aluminum and 3×10^{-3} in platinum. Thus we can say that this energy satisfies the equivalence principle to one part in 10^9. The loop graph shifts the electrostatic energy by an amount of relative order $\alpha \ln(m_e R_{\mathrm{nuc}})/4\pi \sim 10^{-3}$ due to the running of the electromagnetic coupling. Thus we know to a precision of one part in 10^6 that the effect shown in Fig. 2b actually exists. In fact, the effect becomes much larger if we consider quark loops rather than electrons, and we do not need precision experiments to show that virtual quarks gravitate, but we stick with electrons because they are cleaner [8].

We can think of Fig. 2 to good approximation as representing the shift of the electron zero point energy in the environment of the atom or the nucleus. Thus we must understand why the zero point energy gravitates in these environments and not in vacuum, again given that our vacuum is a rather complicated state in terms of the underlying fields. Further, if one thinks one has an answer to this, there is another challenge: why does this cancellation occur in our particular vacuum state, and not, say, in the more symmetric $SU(2) \times U(1)$ invariant state of the weak interaction? It cannot vanish in both because the electron mass is zero in the symmetric state and not in ours, and the subleading terms in the vacuum energy (1) — which are still much larger than the observed ρ_V — depend on this mass. Indeed, this dependence is a major contribution to the Higgs potential (though it is the top quark loop rather than the electron that dominates), and they play an important role in Higgs phenomenology.

I am not going to prove that there is no mechanism that can pass these tests. Indeed, it would be counterproductive to do so, because the most precise no-go theorems often have the most interesting and unexpected failure modes. Rather, I am going to illustrate their application to one interesting class of ideas.

Attempts to resolve the Higgs naturalness problem have centered on two mechanisms, supersymmetry and compositeness (technicolor). In the case of the cosmological constant much attention has been given to the effects of supersymmetry, but what about compositeness, technigravity? If the graviton were composite at a scale right around the limit of Cavendish experiments, roughly 100 microns, would this not cut off the zero point energy and leave a remainder of order $(100\,\mu)^{-4}$, just the observed value [9, 10]? Further this makes a strong prediction, that deviations from the inverse square law will soon be seen.

In fact, it can't be that simple. When we measure the gravitational force in Cavendish experiments, the graviton wavelength is around $100\,\mu$. When we measure the cosmological constant, the graviton wavelength is around the Hubble scale, so there is no direct connection between the two. Moreover, we already know, from the discussion of Fig. 2, that the coupling of gravity to off-shell electrons is unsuppressed over a range of scales in between $100\,\mu$ and the Hubble scale, so whatever is affecting the short-distance behavior of gravity is not affecting longer scales. We can also think about this as follows: even if the graviton were composite one would not expect the graphs of Fig. 1 to be affected, because all external fields are much softer than $100\,\mu$. In order to be sensitive to the internal structure of a particle we need a hard scattering process, in which there is a large momentum transfer to the particle [11]. Further, the large compact dimension models provide an example where gravity is modified at short distance, but the electron zero point loop is not cut off. Thus there is no reason, aside from numerology, to expect a connection between the observed vacuum energy and modifications of the gravitational force law.

Ref. [12] tries to push the idea further, defining an effective theory of 'fat gravity' that would pass the necessary tests. This is a worthwhile exercise, but it shows just how hard it is. In order that the vacuum does not gravitate but the Lamb shift and nuclear loops do, fat gravity imposes special rule for vacuum graphs. The matter path integral, at fixed metric, is doubly nonlocal: there is a UV cutoff around $100\,\mu$, and in order to know how to treat a given momentum integral we have to look at the topology of the whole graph in which it is contained. Since the cosmological constant problem really arises only because we know that some aspects of physics are indeed local to a much shorter scale, it is necessary to derive the rules of fat gravity from a more local starting point, which seems like a tall order. To put this another way, let us apply our first litmus test: what in fat gravity distinguishes the environment of the nucleus from the environment of our vacuum? The distinction is by fiat. But locality tells us that that the laws of physics are simple when written in terms of local Standard Model fields. Our vacuum has a very complicated expression in terms of such fields, so the rules of fat gravity do not satisfy the local simplicity principle.

The nonlocality becomes sharper when we look at the second question, that is, for which vacuum is the cosmological constant small? The rule given is that it is the one of lowest energy. This sounds simple enough, but consider a potential with two widely separated local minima. In order to know how strongly to couple to vacuum A, the graviton must also calculate the energy of vacuum B (and of every other point in field space), and if it is smaller take the difference. Field theory, even in some quasilocal form, can't do this — there are not enough degrees of freedom to do the calculation. If the system is in state A, the dynamics at some distant point in field space is irrelevant. Effectively we would need a computer sitting at every spacetime point, simulating all possible vacua of the theory. Later we will mention a context in which this actually happens, but it is explicitly nonlocal in a strong way.

The failure of short-distant modifications of gravity suggests another strategy: modify gravity at very long distances, comparable to the current Hubble scale, so that it does not couple to vacuum energy at that scale. There is of course a large literature on long-distance modifications of gravity; here I will just point out one problem. If we have zero point energy up to a cutoff of $M \sim 100$ GeV, the radius of curvature of spacetime will be of order M_P/M^2, roughly a meter. So modifications of gravity at much longer distances do not solve the problem, the universe curls up long before it knows about the modification. It is possible that the spacetime curvature decays away on a timescale set by the long-distance modifications, but this would imply a large and uncanceled cosmological constant until quite recently.[2] These problems have already been discussed in Ref. [13], which argues that long distance modifications of gravity can account for the cosmological constant only in combination with acausality.

In another direction, it is tempting to look for some sort of feedback mechanism, where the energies from different scales add up in a way that causes the sum to evolve toward zero. The problem is that only gravity can measure the cosmological constant — this term in the action depends only on the metric — so that the contribution from a scale M is only observed at a much lower scale M^2/M_P, and we cannot cancel $O(M^4)$ against $O(M^8/M_P^4)$. In another language, the cosmological constant has scaling dimension zero and we want to increase it to dimension greater than four; but gravity is clearly classical over a wide range of scales so there is no possibility of this.

Again, there is no proof that some fixed-Λ solution does not exist; perhaps our discussion will spur some reader into looking at the problem in a new way. In fact there is at least one idea that is consistent with our tests: a symmetry *energy* $\rightarrow -energy$. This requires a doubling of degrees of freedom, so the electron loop is cancelled by a mirror loop of negative energy. This idea is discussed as an exact symmetry in ref. [14] and as an approximate symmetry not applying to gravity in ref. [15]; the two cases are rather different because the coordinate invariance is doubled in the first. It might be that either can be made to work at a technical level, and the reader is invited to explore them further, but I will take this as a cue to move on to the next set of ideas.

6.1.1.2   *Adjustable-Λ theories*

Many different mechanisms have been put forward that would avert the problems of the previous section by allowing the cosmological constant to adjust in some way; that is, the vacuum energy seen in the low energy theory is not uniquely determined by the underlying dynamics. A partial list of ideas includes

- Unimodular gravity (see Ref. [1] for a discussion of the history of this idea,

---

[2]One might consider models where this decay occurs in an epoch before the normal Big Bang, but this runs into the empty universe problem to be discussed in Sec. 1.2.

which in one form goes back to Einstein).

- Nonpropagating four-form field strengths [16, 17].
- Scalar potentials with many minima [19, 18, 20].
- A rolling scalar with a nearly flat potential [21]; the potential must be very flat in order that the vacuum energy be constant on shorter than cosmological times, and it must have a very long range to span the necessary range of energies.
- Spacetime wormholes [22–25].
- The metastable vacua of string theory [26–32].
- Self-tuning (an undetermined boundary condition at a singularity in the compact dimensions) [33, 34].
- Explicit tuning (i.e. an underlying theory with at least one free parameter not determined by any principle).

The possible values of ρ_V must either be continuous, or form a sufficiently dense discretuum that at least one value is as small as observed. It is important to note that *zero* cannot be a minimum, or otherwise special, in the range of allowed values. The point is that the electron zero point energy, among other things, gives an additive shift to the vacuum energy; if the minimum value for ρ_V were zero we would have to revert to the previous section and ask what it is cancels the energy in this true vacuum.

In this adjustable scenario, the question is, what is the mechanism by which the actual small value seen in nature is selected? In fact, one can identify a number of superficially promising ideas:

- The Hartle-Hawking wavefunction [35]

$$|\Psi_{HH}|^2 = e^{3/8G^2\rho_V} \qquad (2)$$

strongly favors the smallest positive value of the cosmological constant [36, 37].
- The de Sitter entropy [38]

$$e^S = e^{3/8G^2\rho_V} = |\Psi_{HH}|^2 \qquad (3)$$

would have the same effect, and suggests that the Hartle-Hawking wavefunction has some statistical interpretation in terms of the sytem exploring all possible states.
- The Coleman-de Lucchia amplitude [39] for tunneling from positive to negative cosmological constant vanishes for some parameter range, so the universe would be stuck in the state of smallest positive energy density [18, 40].

These ideas are all tantalizing — they are tantalizing in the same way that supersymmetry is tantalizing as a solution to the cosmological constant problem. That is, they are elegant explanations for why the cosmological constant might be small or zero under some conditions, but not in our particular rather messy universe. Supersymmetry would explain a vanishing cosmological constant in a sufficiently supersymmetric universe, and these mechanisms would explain why it vanishes in an *empty* universe.

To see the problem, note first that the above mechanisms all involve gravitational dynamics in some way, the response of the metric to the vacuum energy. This is as it must be, because again only gravity can measure the cosmological constant. The problem is that in our universe the cosmological constant became dynamically important only recently. At a redshift of a few the cosmological constant was much smaller than the matter density, and so unmeasurable by gravity; at the time of nucleosynthesis (which is probably the latest that a tunneling could have taken place) today's cosmological constant would have been totally swamped by the matter and radiation densities, and there is no way that these gravitational mechanisms could have selected for it.[3] This is the basic problem with dynamical selection mechanisms: only gravity can measure ρ_V, and it became possible for it to do so only in very recent cosmological times. These mechanisms can act on the cosmological constant only if matter is essentially absent.

Another selection principle sometimes put forward is 'existence of a static solution;' this comes up especially in the context of the self-tuning solutions. As a toy illustration, one might imagine that some symmetry acting on a scalar ϕ forced ρ_V to appear only in the form $\rho_V \phi^4$.[4] If we require the existence of a static solution for ϕ then we must have $\rho_V = 0$. Of course this seems like cheating; indeed, if we can require a static solution then why not just require a flat solution, and get $\rho_V = 0$ in one step? In fact these are cheating because they suffer from the same kind of flaw as the dynamical ideas. In order to know that our solution is static on a scale of say 10^{10} years, we must watch the universe for this period of time! The dynamics in the very early universe, at which time the selection was presumably made, have no way to select for such a solution: the early universe was in a highly nonstatic state full of matter and energy.

Of course these arguments are not conclusive, and indeed Steinhardt's talk presents a nonstandard cyclic cosmological history that evades the above no-go argument. If one accepts its various dynamical assumptions, this may be a technically natural solution to the cosmological constant problem. Essentially one needs a mechanism to fill the empty vacuum with energy after its cosmological constant has relaxed to near zero; it is not clear that this is in fact possible.

In the course of trying to find selection mechanisms, one is struck by the fact that, while it is difficult to select for a single vacuum of small cosmological constant, it is extremely easy to identify mechanisms that will populate *all* possible vacua — either sequentially in time, as branches of the wavefunction of the universe, or as

---

[3]This might appear to leave open the possibility that the vacuum energy is at all times of the same order as the matter/radiation density. Leaving aside the question of how this would appear phenomenologically as a cosmological constant, the simplest way to see that this does not really address the problem is to note that as the matter energy goes to zero at late times then so will the vacuum energy: this violates the principal that zero is not a special value. By contrast, the dynamical mechanisms above all operate for a ρ_V-spectrum that extends to negative values.

[4]For example, such a form arises at string tree level, though it is not protected against loop corrections. An exact but spontaneously broken scale invariance might appear to give this form, but in that case a Weyl transform removes ϕ from both the gravitational action and the potential.

different patches in an enormous spatial volume. Indeed, this last mechanism is difficult to evade, if the many vacua are metastable: inflation and tunneling, two robust physical processes, will inevitably populate them all [41–43, 28].

But this is all that is needed! Any observer in such a theory will see a cosmological constant that is unnaturally small; that is, it must be much smaller than the matter and energy densities over an extended period of the history of the universe. The existence of any complex structures requires that there be many 'cycles' and many 'bits': the lifetime of the universe must be large in units of the fundamental time scale, and there must be many degrees of freedom in interaction. A large negative cosmological constant forces the universe to collapse to too soon; a large positive cosmological constant causes all matter to disperse. This is of course the argument made precise by Weinberg [44], here in a rather minimal and prior-free form.[5]

Thus we meet the anthropic principle. Of course, the anthropic principle is in some sense a tautology: we must live where we can live.[6] There is no avoiding the fact that anthropic selection must operate. The real question is, is there any scientific reason to expect that some additional selection mechanism is operating?

Staying for now with the cosmological constant (other parameters will be discussed later), the obvious puzzle is the fact that the cosmological constant is an order of magnitude smaller than the most likely anthropic value. This is an important issue, but to overly dwell on it reminds me of Galileo's reaction to criticism of his ideas because a heavier ball landed slightly before a lighter one (whereas Aristotle's theory predicted a much larger discrepancy):

> Behind those two inches you want to hide Aristotle's ninety-nine *braccia* [arm lengths] and, speaking only of my tiny error, remain silent about his enormous mistake.

The order of magnitude here is the two inches of wind resistance, the ninety-nine *braccia* are the 60 or 120 orders of magnitude by which most or all other proposals miss. This order of magnitude may simply be a 1.5-sigma fluctuation, or it may reflect our current ignorance of the measure on the space of vacua.

If there is a selection mechanism, it must be rather special. It must evade the general difficulties outlined above, and it must select a value that is *almost exactly the same* as that selected by the anthropic principle, differing by one order of magnitude out of 120. Occam's razor would suggest that two such mechanisms be replaced by one — the unavoidable, tautological, one. Thus, we should seriously consider the possibility that there is no other selection mechanism significantly constraining the cosmological constant. Equally, we should not stop searching for such a further principle, but I think one must admit that the strongest reason for

---

[5]For further reviews see Refs. [1, 45, 46].

[6]Natural selection is a tautology in much the same sense: survivors survive. But in combination with a mechanism of populating a spectrum of universes or genotypes, these 'tautologies' acquire great power.

expecting to find it is not a scientific argument but a psychological one:[7] we wish fundamental theory to be as predictive as we have long assumed it would be.

The anthropic argument is not without predictive power. We can identify a list of post- or pre-dictions, circa 1987:

(1) The cosmological constant is not large.
(2) The cosmological constant is not zero.
(3) The cosmological constant is similar in order of magnitude to the matter density.
(4) As the theory of quantum gravity is better understood, it will provide a micro-physics in which the cosmological constant is not fixed but environmental; if this takes discrete values these must be extremely dense in Planck units.
(5) Other constants of nature may show evidence of anthropic constraints.

Items 2 and 3 are the second and third parts of the cosmological constant problem; we did not set out to solve them, but in fact they were solved before they were known to be problems — they are predictions. Item 4 will be discussed in the second half of the talk, in the context of string theory. Item 5 is difficult to evaluate, but serious arguments to this effect have long been made, and they should not be dismissed out of hand.

Let us close this half of the talk with one other perspective. The cosmological constant problem appears to require some form of UV/IR feedback, because the cosmological constant can only be measured at long distances or late times, yet this must act back on the Lagrangian determined at short distance or early times. We can list a few candidates for such a mechanism:

- String theory contains many examples of UV/IR mixing, such as the world-sheet duality relating IR poles in one channel of an amplitude to the sum over massive states in another channel, and the radius-energy relation of AdS/CFT = duality. Thus far however, this is yet one more tantalizing idea but with no known implications for the vacuum energy.
- Bilocal interactions. The exact *energy* → −*energy* symmetry [14] and the worm-hole solution [24, 25] put every point of our universe in contact with every point of another. This ties in with our earlier remarks about the computational power of quantum field theory: here the calculation of the true vacuum energy is done in the entire volume of the second spacetime.
- The anthropic principle. Life, an IR phenomenon, constrains the coupling constants, which are UV quantities.
- A final state condition. At several points — in the long distance modification of gravity, and in the dynamical mechanisms — things would have gone better if we supposed that there were boundary conditions imposed in the future and not just initially. Later we will encounter one context in which this might occur.

---

[7] Again, the Darwinian analogy is notable.

To conclude, we have identified one robust framework for understanding the vacuum energy: (1) Stuff gravitates, and the vacuum is full of stuff. (2) Therefore the vacuum energy must have some way to adjust. (3) It is difficult for the adjustment to select a definite small value for the vacuum energy, but it is easy to access all values, and this, within an order of magnitude, accounts for what we see in nature. We have also identified a number of other possible hints and openings, which may lead the reader in other directions.

6.1.2 *The string landscape*

6.1.2.1 *Constructions*

Now let us ask where string theory fits into the previous discussion. In ten dimensions the theory has no free parameters, but once we compactify, each nonsupersymmetric vacuum will have a different ρ_V. It seems clear that the cosmological constant cannot vary continuously. Proposed mechanisms for such variation have included nondynamical form fields and a boundary condition at a singularity, but the former are constrained by a Dirac quantization condition, and the latter will undoubtedly become discrete once the internal dynamics of the 'singularity' are taken into account. (A rolling scalar with a rather flat potential might provide some effective continuous variation, but the range of such a scalar is very limited in string theory).

Given a discrete spectrum, is there a dense enough set of states to account for the cosmological constant that we see, at least 10^{60} with TeV scale supersymmetry breaking or 10^{120} with Planck scale breaking?[8] The current understanding, in particular the work of KKLT [31], suggests the existence of a large number of metastable states giving rise to a dense discretuum near $\rho_V = 0$. A very large degree of metastability is not surprising in complicated dynamical systems — consider the enormous number of metastable compounds found in nature. As a related example, given 500 protons, 500 neutrons, and 500 electrons, how many very long-lived bound states are there? A rough estimate would be the number of partitions of 500, separating the protons into groups and then assigning the same number of neutrons and electrons to each group; there is some overcounting and some undercounting here, but the estimate should be roughly correct,

$$P(n) \sim \frac{1}{4n\sqrt{3}} e^{\pi\sqrt{2n/3}} , \quad P(500) \sim 10^{22} . \tag{1}$$

The number of metastable states grows rapidly with the number of degrees of freedom.

In string theory, replace *protons*, *neutrons*, and *electrons* with *handles*, *fluxes*, and *branes*. There are processes by which each of these elements can form or decay, so it seems likely that most or all of the nonsupersymmetric vacua are unstable,

---

[8]These numbers would have to be larger if the probability distribution has significant fluctuations as recently argued in Ref. [47].

and the space of vacua is largely or completely connected. Thus all states will be populated by eternal inflation, if any of the de Sitter states is. The states of positive ρ_V would also be populated by any sort of tunneling from nothing (if this is really a distinct process), since one can take the product of an S^4 Euclidean instanton with any compact space.

The number 500 has become a sort of a code for the landscape, because this is the number of handles on a large Calabi-Yau manifold, but for now it is an arbitary guess. It is still not certain whether the number of vacua in string theory is dense enough to account for the smallness of the cosmological constant, or even whether it is finite (it probably becomes finite with some bound on the size of the compact dimensions: compact systems in general have discrete spectra[9]).

The nuclear example has a hidden cheat, in that a small parameter has been put in by hand: the action for tunnelling of a nucleus through the Coulomb barrier is of order $Z_1 Z_2 (m_{\mathrm p}/m_{\mathrm e})^{1/2}$, and this stabilizes all the decays. String theory has no such small parameter. One of the key results of KKLT is that in some regions of moduli space there are a few small parameters that stabilize all decays (see also Ref. [48]). Incidentally, the stability of our vacuum is one reason to believe that we live near some boundary of moduli space, rather than right in the middle where it is particularly hard to calculate: most likely, states right in the middle of moduli space decay at a rate of order one in Planck units.

How trustworthy are the approximations in KKLT? A skeptic could argue that there are no examples where they are fully under control. Indeed, this is likely to inevitable in the construction of our vacuum in string theory. Unlike supersymmetric vacua, ours has no continuous moduli that we can vary to make higher-order corrections parametrically small, and the underlying string theory has no free parameters. It could be that our vacuum is one of an infinite discrete series, indexed by an integer which can be made arbitrarily large, and in this way the approximations made parametrically accurate, but in the KKLT construction this appears not to be the case: the flux integers and Euler number are bounded. For future reference we therefore distinguish *series* and *sporadic* vacua, by analogy to finite groups and Lie algebras; perhaps other constructions give series of metastable nonsupersymmetric vacua.

The KKLT construction has something close to a control parameter, the supersymmetry breaking parameter w_0. In an effective field theory description we are free to vary this continuously and then the approximations do become parametrically precise; in this sense one is quite close to a controlled approximation. In specific models the value of w_0 is fixed by fluxes, and it is a hard problem (in a sense made precise in Ref. [49]) to find vacua in which it is small. Thus, for now the fourth prediction from the previous section, that string theory has enough vacua to solve the cosmological constant problem, is undecided and still might falsify the whole idea.

---

[9]See the talk by Douglas for further discussion of this and related issues.

Underlying the above discussion is the fact that we still have no nonperturbative construction of string theory in any de Sitter vacua, as emphasized in particular in Refs. [50, 51]. As an intermediate step one can study first supersymmetric AdS vacua, where we do understand the framework for a nonperturbative construction, via a dual CFT. The KKLT vacua are built on such AdS vacua by exciting the system to a nonsupersymmetric state. The KKLT AdS vacua are sporadic, but there are also series examples with all moduli fixed, the most notable being simply $AdS_5 \times S^5$, indexed by the five-form flux. Thus far we have explicit duals for many of the series vacua, via quiver gauge theories, but we do not yet have the tools to describe the duals of the sporadic vacua [52]. The KKLT construction makes the prediction that there are $10^{O(100)}$ such sporadic CFTs — a surprising number in comparison to the number of sporadic finite groups and Lie algebras, but indeed 2+1 dimensional CFTs appear to be much less constrained. It may be possible to count these CFTs, even before an explicit construction, through some index; see Ref. [53] for a review of various aspects of the counting of vacua.

Beyond the above technical issues, there are questions of principle: are the tools that KKLT use, in particular the effective Lagrangian, valid? In many instances these objections seem puzzling: the KKLT construction is little more than gluino condensation, where effective Lagrangian methods have long been used, combined with supersymmetry breaking, which can also be studied in a controlled way. It is true that the KKLT construction, in combination with eternal inflation, is time-dependent. However, over much of the landscape the scale of the time-dependence is well below the Planck scale, because the vacuum energy arises from a red-shifted throat, and so the landscape is populated in the regime where effective field theory is valid.

A more principled criticism of the use of effective Lagrangians appears in Refs. [50, 51]; I will try to paraphrase this here. It is not precisely true that the nonsupersymmetric KKLT states (or any eternally inflating states) are excitations of AdS vacua. That is, it is true locally, but the global boundary conditions are completely different. Normally one's intuition is that the effective Lagrangian is a local object and does not depend on the boundary conditions imposed on the system, but arguments are given that this situation is different. In particular one cannot tunnel among inflating states, flat spacetimes, and AdS states in any direction (for example, tunneling from eternal inflation to negative cosmological constant leads to a crunch); thus these are in a sense different theories. This is also true from a holographic point of view: the dual Hamiltonians that describe inflating, flat, and AdS spaces will inevitably be completely different (as one can see by studying the high energy spectrum). Is there then any reason to expect that constructions of an effective action, obtained from a flat spacetime S-matrix, have any relevance to an eternally inflating system?

I believe that there is. The entire point of holography and AdS/CFT duality is that the bulk physics is emergent: we obtain the same bulk physics from many different Hamiltonians. We can already see this in the AdS/CFT context, where many different quiver gauge theories, even in different dimensions, give the same IIB bulk string theory, and local experiments in a large AdS spacetime are expected to give the same results as the same experiments in flat spacetime. Thus there is no argument in principle that these do not extend to the inflating case. Also, while holography does imply some breakdown of local field theory, it does so in a rather subtle way, as in phase correlations in Hawking radiation. By contrast, the expectation value of the energy-momentum tensor in the neighborhood of a black hole (i.e. the total flux of Hawking radiation) appears to be robust, and the quantities that enter into to construction of string vacua are similar to this.

However, for completeness we mention the possible alternate point of view [54]: that the landscape of metastable dS vacua has no nonperturbative completion, or it does have one but is experimentally ruled out by considerations such as those we will discuss. Instead there is a completely separate sector, consisting of theories with finite numbers of states, and if these lead to emergent gravity it must be in a stable dS spacetime.

6.1.2.2 *Phenomenological issues*

Thus far we have dwelt on the cosmological constant, but the string landscape implies that other constants of nature will be environmental to greater or lesser extents as well. In this section we discuss a few such parameters, especially those which appear to be problematic for one reason or another.

$\theta_{\mathbf{QCD}}$ Why is $\theta_{\rm QCD}$ of order 10^{-9} or less? This strong CP problem has been around for a long time in gauge theory, and several explanations have been proposed — an axion, a massless up quark, and models based on spontaneous CP violation. However, it has been argued that none of these are common in the string landscape; for example, the first two require continuous symmetries with very tiny explicit breakings, and this appears to require fine tuning. Further, it is very hard to see any anthropic argument for small $\theta_{\rm QCD}$; a larger value would make very little difference in most of physics. Thus we would conclude that the multiverse is full of bubbles containing observers who see gauge theories with large CP-violating angles, and ours is a one-in-a-billion coincidence [50].

Of course, this is a problem that is to some extent independent of string theory: the axion, for example, has always been fishy, in that one needed a global symmetry that is exact *except for* QCD instantons. The string landscape is just making sharper an issue that was always there.

String theory does come with a large number of potential axions. In order that one of these solve the strong CP problem it is necessary that the potential energy from QCD instantons be the dominant contribution to the axion potential; any non-QCD contribution to the axion mass must be of order $10^{-18}\,{\rm eV} \times (10^{16}\,{\rm GeV}/f_{\rm a})$ or less. This is far below the expected scale of the moduli masses, so appears to imply a substantial fine tuning (even greater than the direct tuning of θ_{QCD}) and so rarity in the landscape.

However, the landscape picture also suggests a particular solution to this problem. In order to obtain a dense enough set of vacua, the compact dimensions must be topologically complex, again with something around 500 cycles. Each cycle gives rise to a potential axion, whose mass comes from instantons wrapping the cycle (we must exclude would-be axions which also get mass from other sources, such as their classical coupling to fluxes). Generically one would expect some of these cycles to be somewhat large in string units; for example, one might expect the whole compact space to have a volume that grows as some power of the number of handles. The axions, whose masses go as minus the exponential of the volume, would be correspondingly light. Thus, compactifications of large topological complexity may be the one setting in which the QCD axion *is* natural, the smallness of θ_{QCD} being an indirect side effect of the need for a small cosmological constant. More generally, it will be interesting to look for characteristic properties of such topologically complex compactifications.

This example shows that even with anthropic selection playing a role, mechanism will surely also be important.

The baryon lifetime This is a similar story to θ_{QCD} [50]: as far as we understand at present, the baryon lifetime is longer than either anthropic argument or mechanism can account for, so that bubbles with such long-lived baryons would be rare in the multiverse. This problem is lessened if supersymmetry is broken at high energy. This is an significant challenge to the landscape picture: it is good to have such challenges, eventually to sharpen, or to falsify, our current understanding.

The dark energy parameter w A naive interpretation of the anthropic principle would treat the dark energy equation of state parameter w as arbitrary, and look for anthropic constraints. However, in the string landscape a simple cosmological constant, $w = -1$, is certainly favored. With supersymmetry broken, the scalar potential generically has isolated minima, with all scalars massive. In order to obtain a nontrivial equation of state for the dark energy we would need a scalar with a mass of order the current Hubble scale. Our discussion of axions indicates a mechanism for producing such small masses, but it would be rather contrived, for no evident reason, that the mass would be of just the right scale as to produce a nontrivial variation in the current epoch.

Three generations Three generation models appear to be difficult to find in string theory. A recent paper quantifies this [55]: in one construction they are one in a billion, even after taking into account the anthropic constraint that there be an asymptotically free group so that the long distance physics is nontrivial. It is then a puzzle to understand how we happen to live in such a vacuum. One conjecture is that all constructions thus far are too special, and in the full landscape three generations is not rare. Again, explaining three generations is equally a problem for any hypothetical alternate selection mechanism — another challenge to sharpen our understanding.

Q I am not going to try to discuss this parameter in detail; I am only going to use it to make one rhetorical point. The anthropic bound on Q, which is the normalization of the primordial temperature fluctuations, has been quoted as [56]

$$10^{-6} < Q < 10^{-4} \, , \tag{2}$$

and it is interesting that the observed value is in the middle, not at either end. What would we expect from the landscape?

A string theorist would note that the anthropic bound is on $\rho_V Q^{-3}$ [44], and so by making Q a factor of 10 larger we can multiply ρ_V by 1000, and there will be many more vacua with this larger value of Q. A cosmologist would note that a smaller Q would imply a flatter potential and so more inflation, and therefore much more volume and many more galaxies. Thus the cosmologist and the string theorist agree that we should be on the end of the anthropic range, but they disagree on which end.

This is a caricature, of course — there are other considerations, and model-dependencies [50, 57–60]. I use it to make two points: first, it is a puzzle that we are in the middle of the anthropic range, yet another thing to understand. Second, the string theorist and the cosmologist each look at part of the measure, but it is clear that we are far short of the whole picture. (For reviews of the counting and the volume factors see Refs. [53] and [61] respectively.)

ρ_V Can we understand understand the number 283, as in

$$\rho_V = e^{-283.2} M_{\mathrm{P}}^4 \ ? \tag{3}$$

I quote it in this way, as a natural log, to emphasize that we are to think about it completely free of all priors (such as the fact that we have ten fingers). Thus, there may be an anthropic relation between $\rho_V / M_{\mathrm{P}}^4$ and $M_{\mathrm{weak}}/M_{\mathrm{P}}$, for example, but we should not make any assumption about the latter. It should be possible to calculate the number 283, at least to some accuracy. We know that it has to be big, to get enough bits and cycles, but why is 100 not big enough, and why is 1000 not better?

One possibility, the best from the point of view of string theory, is that $\rho_V / M_{\mathrm{P}}^4$ has its original purely in microphysics; that it, that it is close to the smallest attainable value, set by the density of the discretuum. The other extreme is that it is almost purely anthropic — that 283, plus or minus some uncertainty, really gives the best of all attainable worlds, and any attempt to vary parameters to give a larger or smaller value inevitably makes things worse. Certainly, knowing where we sit between these two extremes is something that we must eventually understand in a convincing way.

Other questions An obvious question is whether we can understand the supersymmetry-breaking scale (see [63] and references therein). Is low energy supersymmetry, or some alternative [64, 65], favored? Will we figure this out before the LHC tells us?

Another potentially telling question [66]: are there more coincidences like the cosmic coincidence of ρ_V, such as the existence of two different kinds of dark matter with significant densities?

6.1.2.3 *What is string theory?*

Of course, this is still the big question. We have learned in recent years that the nonperturbative construction of a holographic theory is very sensitive to the global structure of spacetime. Thus, the current point of view, the chaotically inflating multiverse, casts this question in a new light. It is also another example of how the landscape represents productive science: if we ignore this lesson, ignore chaotic inflation, we may be trying to answer the wrong question.

Before addressing the title question directly, let us discuss one way in which it bears upon the previous discussion. We touched briefly on the issue of the measure. This has always been a difficult question in inflationary cosmology. Intuitively one would think that the volume must be included in the weighting, since this will be one factor determining the total number of galaxies of a given type. However, this leads to gauge dependence [67] and the youngness paradox [68]. Further, this would imply that the vacuum of highest density plays a dominant role, whereas the de Sitter entropy would suggest almost the opposite, that when the system is in a state of high vacuum energy it has simply wandered into a subsector of relatively few states. Further, the idea of counting separately regions that are out of causal contact is contrary to the spirit of the holographic principle.

There have been attempts to modify the volume weighting to deal with some of the paradoxes (for a recent review see Ref. [61]), but as far as I know none as yet take full advantage of the holographic point of view, and none is widely regarded as convincing. Providing a compelling understanding of the measure is certainly a goal for string theory. It is possible that this can be done by some form of holographic reasoning, even without a complete nonperturbative construction. It is perhaps useful to recall Susskind's suggestion, that the many worlds of chaotic inflation are the same as the many worlds of quantum mechanics. This can be read in two directions: first, that chaotic inflation is the origin of quantum mechanics — this seems very ambitious; second, that the many causal volumes in the chaotic universe should just be seen as different states within the wavefunction of a single patch — this is very much in keeping with holography. It is also interesting to note that the stochastic picture presented in Ref. [67] has a volume-weighted probability that seems to have a youngness paradox, and an unweighted one that seems to connect with the Hartle-Hawking and tunneling wavefunctions, and possibly with a thermodynamic picture.

Now, what is the nonperturbative construction of these eternally inflating states? The lesson from AdS/CFT is that the dual variables that give this construction live at the boundary of spacetime. In the context of eternal inflation, the only natural boundaries lie to the future, in open FRW universes (and possibly also in time-reversed universes to the past) [32, 69, 70].

This is much like AdS/CFT with timelike infinity replacing spatial infinity, and so it suggests that time will be emergent. Let us interpose here one remark about emergent time (see also the presentations by Seiberg and by Maldacena at this meeting). Of course in canonical general relativity there is no time variable at the start, it emerges in the form of correlations once the Hamiltonian constraint is imposed. This sounds like emergent time, but on the other hand it is just a rewriting of the covariant theory, and one would expect emergent time to be something deeper.

To see the distinction between emergent time in these two senses let us first review emergent gauge symmetry. In some condensed matter systems in which the starting point has only electrons with short-ranged interactions, there are phases where the electron separates into a new fermion and boson [71, 72],

$$e(x) = b(x)f^\dagger(x) \ . \tag{4}$$

However, the new fields are redundant: there is a gauge transformation

$$b(x,t) \to e^{i\lambda(x,t)}b(x,t) \ , \quad f(x,t) \to e^{i\lambda(x,t)}f(x,t) \ , \tag{5}$$

which leaves the physical electron field invariant. This new gauge invariance is clearly emergent: it is completely invisible in terms of the electron field appearing in the original description of the theory (this "statistical" gauge invariance is not to be confused with the ordinary electromagnetic gauge invariance, which does act on the electron.) Similarly, the gauge theory variables of AdS/CFT are trivially invariant under the bulk diffeomorphisms, which are entirely invisible in the gauge theory (the gauge theory fields do transform under the asymptotic symmetries of $AdS_5 \times S^5$, but these are ADM symmetries, not gauge redundancies).

Thus, in the case of emergent time we look for a description of the theory in which time reparameterization invariance is invisible, in which the initial variables are trivially invariant. It is not a matter of solving the Hamiltonian constraint but of finding a description in which the Hamiltonian constraint is empty. Of course we can always in general relativity introduce a set of gauge-invariant observables by setting up effectively a system of rods and clocks, so to this extent the notion of emergence is imprecise, but it carries the connotation that the dynamics can be expressed in a simple way in terms of the invariant variables. The AdS/CFT duality solves this problem by locating the variables at spatial infinity, and in the present context the natural solution would be to locate them at future infinity. That is, there some dual system within which one calculates directly the outgoing state in the FRW patches, some version of the Hartle-Hawking wavefunction perhaps. To access our physics in a nonsupersymmetric and accelerating bubble would then require some holographic reconstruction as in the bulk of AdS/CFT. Certainly such a picture would cast a very different light on many of the questions that we have discussed; it does suggest a possible mechanism for 'post-selection' of the cosmological constant.

It would be useful to have a toy model of emergent time. The problem with the string landscape is that all states mix, and one has to deal with the full problem; is there any isolated sector to explore?

6.1.3 *Conclusions*

A few closing remarks:

- The extent to which first principles uniquely determine what we see in nature is itself a question that science has to answer. Einstein asked how much choice God had, he did not presume to know the answer.
- That the universe is vastly larger than what we see, with different laws of physics in different patches, is without doubt a logical possibility. One might argue that even true this is forever outside the domain of science, but I do not think it is up to us to put a priori bounds on this domain. Indeed, we now have five separate lines of argument (the predictions near the end of Sec. 1) that point in this direction. Our current understanding is not frozen in time, and I expect that if this idea is true (or if it is not) we will one day know.
- A claim that science is less predictive should be subjected to a correspondingly higher level of theoretical skepticism. Our current picture should certainly be treated as tentative, certainly until we have a nonperturbative formulation of string theory.
- The landscape opens up a difficult but rich spectrum of new questions, e.g. [73].
- There are undoubtedly many surprises in the future.

Let me close with a quotation from Dirac:

> One must be prepared to follow up the consequences of theory, and feel that one just has to accept the consequences no matter where they lead.

and a paraphrase:

> One should take seriously all solutions of one's equations.

Of course, his issue was a factor of two, and ours is a factor of 10^{500}.

Acknowledgments: I would like to thank Nima Arkani-Hamed, Lars Bildsten, Raphael Bousso, Michael Dine, Michael Douglas, Eliezer Rabinovici, and Eva Silverstein for helpful discussions. This work was supported in part by NSF grants PHY99-07949 and PHY04-56556.

Bibliography

[1] S. Weinberg, *The Cosmological Constant Problem*, Rev. Mod. Phys. **61** (1989) 1.
[2] S. M. Carroll, *The cosmological constant*, Living Rev. Rel. **4** (2001) 1. URL http://www.livingreviews.org/lrr-2001-1 . [arXiv:astro-ph/0004075].

[3] T. Padmanabhan, *Cosmological constant: The weight of the vacuum*, Phys. Rept. **380** (2003) 235 [arXiv:hep-th/0212290].

[4] S. Nobbenhuis, *Categorizing different approaches to the cosmological constant problem*, arXiv:gr-qc/0411093.

[5] S. R. Coleman and E. Weinberg, *Radiative Corrections As The Origin Of Spontaneous Symmetry Breaking*, Phys. Rev. D **7** (1973) 1888.

[6] P. Roll, R. Krotkov, and R. Dicke, Ann. Phys. (N. Y.) **26** (1964) 442.

[7] V. Braginsky and V. Panov, Zh. Eksp. Teor. Fiz. **61** (1971) 873.

[8] J. Bagger *et al.* [American Linear Collider Working Group], *The case for a 500-GeV e+ e- linear collider* arXiv:hep-ex/0007022.

[9] S. R. Beane, *On the importance of testing gravity at distances less than 1-cm*, Gen. Rel. Grav. **29** (1997) 945 [arXiv:hep-ph/9702419].

[10] R. Sundrum, *Towards an effective particle-string resolution of the cosmological constant problem*, JHEP **9907** (1999) 001 [arXiv:hep-ph/9708329].

[11] E. Rutherford, *The scattering of the alpha and beta rays and the structure of the atom*, Proc. Manchester Lit. and Phil. Soc. IV, **55** (1911) 18.

[12] R. Sundrum, *Fat Euclidean gravity with small cosmological constant*, Nucl. Phys. B **690** (2004) 302 [arXiv:hep-th/0310251].

[13] N. Arkani-Hamed, S. Dimopoulos, G. Dvali and G. Gabadadze, *Non-local modification of gravity and the cosmological constant problem*, arXiv:hep-th/0209227.

[14] A. D. Linde, *The Universe Multiplication And The Cosmological Constant Problem*, Phys. Lett. B **200**, 272 (1988).

[15] D. E. Kaplan and R. Sundrum, *A symmetry for the cosmological constant*, arXiv:hep-th/0505265.

[16] A. Aurilia, H. Nicolai and P. K. Townsend, *Hidden Constants: The Theta Parameter Of QCD And The Cosmological Constant Of N=8 Supergravity*, Nucl. Phys. B **176**, 509 (1980).

[17] M. J. Duff and P. van Nieuwenhuizen, *Quantum Inequivalence Of Different Field Representations*, Phys. Lett. B **94**, 179 (1980).

[18] L. F. Abbott, *A Mechanism For Reducing The Value Of The Cosmological Constant*, Phys. Lett. B **150**, 427 (1985).

[19] W. A. Bardeen, S. Elitzur, Y. Frishman and E. Rabinovici, *Fractional Charges: Global And Local Aspects*, Nucl. Phys. B **218**, 445 (1983).

[20] T. Banks, M. Dine and N. Seiberg, *Irrational axions as a solution of the strong CP problem in an eternal universe*, Phys. Lett. B **273**, 105 (1991) [arXiv:hep-th/9109040].

[21] T. Banks, *T C P, Quantum Gravity, The Cosmological Constant And All That...*, Nucl. Phys. B **249**, 332 (1985).

[22] S. R. Coleman, *Black Holes As Red Herrings: Topological Fluctuations And The Loss Of Quantum Coherence*, Nucl. Phys. B **307**, 867 (1988).

[23] S. B. Giddings and A. Strominger, *Loss Of Incoherence And Determination Of Coupling Constants In Quantum Gravity*, Nucl. Phys. B **307**, 854 (1988).

[24] T. Banks, *Prolegomena To A Theory Of Bifurcating Universes: A Nonlocal Solution To The Cosmological Constant Problem Or Little Lambda Goes Back To The Future*, Nucl. Phys. B **309**, 493 (1988).

[25] S. R. Coleman, *Why There Is Nothing Rather Than Something: A Theory Of The Cosmological Constant*, Nucl. Phys. B **310**, 643 (1988).

[26] A. D. Sakharov, *Cosmological Transitions With A Change In Metric Signature*, Sov. Phys. JETP **60**, 214 (1984) [Zh. Eksp. Teor. Fiz. **87**, 375 (1984 SOPUA,34,409-413.1991)].

[27] L. Smolin, *The Fate of black hole singularities and the parameters of the standard*

models of particle physics and cosmology, arXiv:gr-qc/9404011.

[28] R. Bousso and J. Polchinski, *Quantization of four-form fluxes and dynamical neutralization of the cosmological constant,* JHEP **0006**, 006 (2000) [arXiv:hep-th/0004134].

[29] J. L. Feng, J. March-Russell, S. Sethi and F. Wilczek, *Saltatory relaxation of the cosmological constant,* Nucl. Phys. B **602**, 307 (2001) [arXiv:hep-th/0005276].

[30] E. Silverstein, *(A)dS backgrounds from asymmetric orientifolds,* arXiv:hep-th/0106209.

[31] S. Kachru, R. Kallosh, A. Linde and S. P. Trivedi, *De Sitter vacua in string theory,* Phys. Rev. D **68**, 046005 (2003) [arXiv:hep-th/0301240].

[32] L. Susskind, *The anthropic landscape of string theory,* arXiv:hep-th/0302219.

[33] N. Arkani-Hamed, S. Dimopoulos, N. Kaloper and R. Sundrum, *A small cosmological constant from a large extra dimension,* Phys. Lett. B **480**, 193 (2000) [arXiv:hep-th/0001197].

[34] S. Kachru, M. B. Schulz and E. Silverstein, *Self-tuning flat domain walls in 5d gravity and string theory,* Phys. Rev. D **62**, 045021 (2000) [arXiv:hep-th/0001206].

[35] J. B. Hartle and S. W. Hawking, *Wave Function Of The Universe,* Phys. Rev. D **28**, 2960 (1983).

[36] E. Baum, *Zero Cosmological Constant From Minimum Action,* Phys. Lett. B **133**, 185 (1983).

[37] S. W. Hawking, *The Cosmological Constant Is Probably Zero,* Phys. Lett. B **134**, 403 (1984).

[38] G. W. Gibbons and S. W. Hawking, *Cosmological Event Horizons, Thermodynamics, And Particle Creation,* Phys. Rev. D **15**, 2738 (1977).

[39] S. R. Coleman and F. De Luccia, *Gravitational Effects On And Of Vacuum Decay,* Phys. Rev. D **21**, 3305 (1980).

[40] J. D. Brown and C. Teitelboim, *Neutralization Of The Cosmological Constant By Membrane Creation,* Nucl. Phys. B **297**, 787 (1988).

[41] A. D. Linde, *Chaotic Inflation,* Phys. Lett. B **129**, 177 (1983).

[42] A. Vilenkin, *The Birth Of Inflationary Universes,* Phys. Rev. D **27**, 2848 (1983).

[43] A. D. Linde, *Eternally Existing Selfreproducing Chaotic Inflationary Universe,* Phys. Lett. B **175**, 395 (1986).

[44] S. Weinberg, *Anthropic Bound On The Cosmological Constant,* Phys. Rev. Lett. **59**, 2607 (1987).

[45] A. Vilenkin, *Anthropic predictions: The case of the cosmological constant,* arXiv:astro-ph/0407586.

[46] S. Weinberg, *Living in the multiverse,* arXiv:hep-th/0511037.

[47] D. Schwartz-Perlov and A. Vilenkin, *Probabilities in the Bousso-Polchinski multiverse,* arXiv:hep-th/0601162;
P. Steinhardt, private communication.

[48] A. R. Frey, M. Lippert and B. Williams, *The fall of stringy de Sitter,* Phys. Rev. D **68**, 046008 (2003) [arXiv:hep-th/0305018].

[49] F. Denef and M. R. Douglas, *Computational complexity of the landscape. I,* arXiv:hep-th/0602072.

[50] T. Banks, M. Dine and E. Gorbatov, *Is there a string theory landscape?* JHEP **0408**, 058 (2004) [arXiv:hep-th/0309170].

[51] T. Banks, *Landskepticism or why effective potentials don't count string models,* arXiv:hep-th/0412129.

[52] E. Silverstein, *AdS and dS entropy from string junctions or the function of junction conjunctions,* arXiv:hep-th/0308175.

[53] M. R. Douglas, *Basic results in vacuum statistics,* Comptes Rendus Physique **5**, 965

(2004) [arXiv:hep-th/0409207].

[54] T. Banks, *Some thoughts on the quantum theory of stable de Sitter space*, arXiv:hep-th/0503066.

[55] F. Gmeiner, R. Blumenhagen, G. Honecker, D. Lust and T. Weigand, *One in a billion: MSSM-like D-brane statistics*, JHEP **0601**, 004 (2006) [arXiv:hep-th/0510170].

[56] M. Tegmark and M. J. Rees, *Why is the CMB fluctuation level* 10^{-5}? Astrophys. J. **499**, 526 (1998) [arXiv:astro-ph/9709058].

[57] M. L. Graesser, S. D. H. Hsu, A. Jenkins and M. B. Wise, *Anthropic distribution for cosmological constant and primordial density perturbations*, Phys. Lett. B **600**, 15 (2004) [arXiv:hep-th/0407174].

[58] B. Feldstein, L. J. Hall and T. Watari, *Density perturbations and the cosmological constant from inflationary landscapes*, Phys. Rev. D **72**, 123506 (2005) [arXiv:hep-th/0506235].

[59] J. Garriga and A. Vilenkin, *Anthropic prediction for Lambda and the Q catastrophe*, arXiv:hep-th/0508005.

[60] M. Tegmark, A. Aguirre, M. Rees and F. Wilczek, *Dimensionless constants, cosmology and other dark matters*, Phys. Rev. D **73**, 023505 (2006) [arXiv:astro-ph/0511774].

[61] A. Vilenkin, *Probabilities in the landscape*, arXiv:hep-th/0602264.

[62] N. Arkani-Hamed, S. Dimopoulos and S. Kachru, *Predictive landscapes and new physics at a TeV*, arXiv:hep-th/0501082.

[63] M. Dine, D. O'Neil and Z. Sun, *Branches of the landscape*, JHEP **0507**, 014 (2005) [arXiv:hep-th/0501214].

[64] N. Arkani-Hamed and S. Dimopoulos, *Supersymmetric unification without low energy supersymmetry and signatures for fine-tuning at the LHC*, JHEP **0506**, 073 (2005) [arXiv:hep-th/0405159].

[65] P. J. Fox *et al.*, *Supersplit supersymmetry*, arXiv:hep-th/0503249.

[66] A. Aguirre and M. Tegmark, *Multiple universes, cosmic coincidences, and other dark matters*, JCAP **0501**, 003 (2005) [arXiv:hep-th/0409072].

[67] A. D. Linde, D. A. Linde and A. Mezhlumian, *From the Big Bang theory to the theory of a stationary universe*, Phys. Rev. D **49**, 1783 (1994) [arXiv:gr-qc/9306035].

[68] A. H. Guth, *Inflation and eternal inflation*, Phys. Rept. **333**, 555 (2000) [arXiv:astro-ph/0002156].

[69] B. Freivogel and L. Susskind, *A framework for the landscape*, Phys. Rev. D **70**, 126007 (2004) [arXiv:hep-th/0408133].

[70] R. Bousso and B. Freivogel, *Asymptotic states of the bounce geometry*, arXiv:hep-th/0511084.

[71] A. D'Adda, M. Luscher and P. Di Vecchia, *A 1/N Expandable Series Of Nonlinear Sigma Models With Instantons*, Nucl. Phys. B **146**, 63 (1978).

[72] G. Baskaran and P. W. Anderson, *Gauge Theory Of High Temperature Superconductors And Strongly Correlated Fermi Systems*, Phys. Rev. B **37**, 580 (1988).

[73] K. Intriligator, N. Seiberg and D. Shih, arXiv:hep-th/0602239.

6.2 Discussion

E. Silverstein You have said that you think that dimensionality is conserved and that the supercritical string theories do not mix with the rest. I do not understand that comment because we know of such transitions.

J. Polchinski I think they might mix in one direction, so if you go back in time the dimension rises and you do not mix with the low-dimension stuff. That was what I was sort of hoping. I was hoping that there is some kind of scaling in the large D limit.

J. Maldacena The supercritical strings have a tachyon, so this is not a well-defined boundary condition in the future.

E. Silverstein It is not true that every supercritical strings have a tachyon. Some models do, and some models do not, just like with other string theories. It is possible to project out tachyons in supercritical strings.

A. Strominger A question to Polchinski: Why do you think it is possible to decide how predictive science is? Are you asserting that, if you have an anthropic explanation for something, it is possible to rule out that there is another explanation that we have missed? Of course, if you were able to predict everything, you would know that science was totally predictive. But if you fail to do that, how can you even imagine ruling out some clever bleb field method of understanding things that we just had not thought about?

J. Polchinski I think it is fair to say that we do not need to be skeptical until we answer the question "what is string theory?", a question which has a very long future.

A. Strominger What is string theory and whether string theory is related to our world? Fair enough.

J. Polchinski There is structure and correlations that you could not anticipate. The cosmological constant is a smoking gun, there may be others. There is still room for skepticism. It's a Goedel thing. There are undecidable questions. We should not assume a priori how predictive science is, it is not something we can know.

A. Strominger Right, but we can try as hard as we can to predict everything that we are able to.

J. Polchinski We can try relaxing different prejudices and see where each of them leads.

B. Greene Is it not important to draw a distinction between correlations and explanations? I agree that one can find interesting correlations between various features of the theory, but is that an explanation? No, it is just an interesting correlation.

J. Polchinski It is true, but it may be the way nature is. You cannot exclude that this is the way nature is.

D. Gross I just wanted to ask about the many worlds equals many bubbles. You

went over that very rapidly, and since I spoke for Hartle, and one of the things
on one of the slides was "many worlds not equal to many bubbles", so I am
asking his question: why is that?

J. Polchinski I put that in there partly because Hartle did, but also because
Susskind said it at "Strings" and other people said it in other language. You
can read that in two directions. The more ambitious direction is to say that
eternal inflation explains quantum mechanics. I do not think anybody believes
that. The less ambitious way to read it is simply that holography tells you that
you do not talk about things that are out of causal contact, they are folded back
into your own wave function, and so in that sense the many worlds of eternal
inflation really are the many worlds of quantum mechanics.

M. Gell-Mann In Everett's work of the late 1950's when he was a graduate student of Wheeler, he formulated the ideas that were ancestrial to a lot of things
that we do today. Bryce De Witt called them — or somebody else, not Everett
— called them "many worlds". What they are, and what they continue to be
today, is a set of ideas about multiple alternative histories of a Universe. If
you want to believe that for every such history there is a universe somewhere,
then if the universes do not communicate with one another, it does not change
anything. But the many-worlds idea is not in any way important for this idea of
many histories of the Universe. The other idea having to do with the budding
off of new universes in a multi-verse is completely different. That might have
some actual consequences for the individual universe we study. If it used to be
part of a bigger system which broke up, the state vector would become a density
matrix of a certain kind, for example. There might be some consequences. But
referring to the other theory, the many-histories idea, as many worlds is just
misleading, I think.

N. Seiberg I really enjoyed your nuclear physics analogy. I would like to pursue
it a little bit further. If at the time people faced nuclear physics with all its
resonances, they had adopted the anthropic principle, you could have repeated
your talk with minor changes at that time and it would have had a very negative
effect on the development of science. The standard model would not have been
discovered. Returning to our time, there could be all sorts of rich physics that
we should understand, and adapting the anthropic principle will prevent us
from finding it. I think it is premature to declare defeat.

M. Douglas May I answer this? This is sort of a standard answer. There are
different metaphors that I like, such as the one that says that we live in a big
crystal and we have to figure out which one it is. The situation here is sort
of historically reverse. Back then you could have said that the real interesting
thing is to identify the protons and quarks and so forth, while here we are
actually working backwards. We really believe in the fundamental nature of
the supersymmetric pieces, and now we are trying to assemble this into the
complicated mishmash that looks more like our world. From that point of

view we have already answered the prettiest part of the problem, and now we have this more difficult, chemical, condensed matter problem of reproducing the nature of the world we see. I am not saying this is necessarily the case, but it would be an equally valid analogy to the one that you are suggesting.

E. Rabinovici I want to point out that in field theories that are scale-invariant and finite, the vacuum energy is the same if scale invariance is spontaneously broken or not, and in particular it can be zero. I appreciate that you view this spontaneous breaking of the scale invariance as fine-tuning, therefore it is not on your list, but I beg to differ.

J. Polchinski I think that your selection principle is what I call the static solution principle where you insist that the Lagrangian be such as to give you a static solution. I think that is not a selection principle, and in particular, if you think about it, it is like the dynamical solutions where to know that you would have a static solution you would have to look at your system for all of time. It is not something that can constrain what state the system is at the beginning of time. So it really is the same problem as in other dynamical ideas. That also applies to self-tuning which uses the same strategy.

V. Rubakov I wanted to make a comment on your argument against the dynamical relaxation mechanisms. Actually the fact that the cosmological constant is so small might tell us that the cosmological evolution is quite different from what we think. This was called by Graham Ross "déjà-vu Universe", meaning that there could be a state of the Universe some time in the past which was very similar to the Universe we are now living in, and at that time one or another relaxation mechanism worked. Or, even more, maybe there is strong non-locality and some relaxation mechanism works at very large length scales, while beyond our small part of the Universe, the Universe is just empty. Then all this criticism will not work.

A. Polyakov A couple of comments, nothing to do with philosophy. There are two mechanisms which are probably worth having in mind. They may work eventually. First, as far as the strong CP problem is concerned, there is at least one model in which the solution of the problem comes from the infrared corrections and is completely analogous to the behavior of the theta-angle in the quantum Hall effect. Namely, the behavior is that you take an arbitrary bare θ of the order of 1 and, as you go to the infrared, it tends to zero. Which means that it actually predicts that if you go back to higher energy, the renormalization group flow enhances the CP violation. I certainly do not have any realistic model of field theory with this feature, but some highly non-trivial Yang-Mills theories have it, so it is worth keeping in mind. Although it actually plays for the anthropic principle, which I do not want.

The second thing is that there is also another model, I think, in which we can at least see a tendency of the cosmological constant to be screened by the renormalization group flow also. And it is very natural, because the cosmological term is the only term in the Lagrangian which does not have derivatives, so it is important in the infrared. There could be a phenomenon similar to the Landau zero charge picture when it is screened out as you go to large distances. So that is another possibility, but I admit it does not solve the coincidence problem. But it is still interesting to keep in mind.

N. Arkani-Hamed I just want to make some general comments about the angst of predictivity in this picture. I think that part of the problem is that we continue to confuse prediction of parameters with the more traditional accomplishment of physics which is to predict and understand new dynamics. I do not think you can even start talking about environmental or anthropic or whatever constraints on parameters until you understand the dynamics well enough even to be able to figure out what the relevant parameters are and how they vary. No one would have been tempted to try to explain the phenomena of nuclear physics anthropically. There is clearly major dynamics that was not understood, and there was not even a question of talking about parameters to be tuned or not tuned. Similarly, if we could go up to the string scale or the Planck scale we would experimentally see what is going on, we would see weakly coupled strings if we happened to be in the weakly coupled sector of the theory, and all of that would be absolutely wonderful. It is only because of the practical difficulty of not being able to do that, that we have psychologically replaced it with being able to predict all of the parameters, which of course was never promised. While in the history of physics, understanding dynamics always has pushed us forward, focusing on parameters has some times pushed us forward, and sometimes has not been the right question. The classic example of Kepler trying to predict the distance of the planets from the Sun, is an example of something that was just the wrong question.

I feel that what is different about our situation today is that, at least in our understanding of long distance theories in terms of effective Lagrangians, we are finally at a point where with a finite small number of parameters, we can imagine in a controlled way what happens to physics as those parameters are changed. Our questions are questions about those parameters. That dynamics, at least at large distances, is basically governed by special relativity and quantum mechanics, so that dynamics is under control in the infrared. Now, in that context some parameters can be environmental, but not all of them. So the kind of response that anything that you do not understand, there is an anthropic explanation for it, is simply not true. There is no anthropic explanation for V_{cb}, there is no anthropic explanation for bottom quark mass, there is no anthropic explanation for θ_{QCD} or why there are three generations. And no one would even attempt to come up with them, there are some things that are clearly irrelevant to infrared physics. There are few other parameters, the relevant operators and perhaps some other parameters, which are of great importance for infrared physics and may have an environmental explanation, or not. I want to remind, as Galison very nicely talked about in the first talk in this conference, and Polchinski was mentioning as well, that the issue of how much we get to predict is really not up to us. The angst over not having one world or maybe 10^{500} worlds pales in comparison to the angst that the Laplacian determinists must have felt when they were told that they could not predict the position and velocity of every particle deep into the future from measuring the positions and velocities of every particle now. We are talking about a far less drastic reduction in the degree of predictivity than was already suffered in this transition from classical to quantum mechanics.

Finally I just want to make a concrete point when we talk about the cosmological constant as the "central engine" that is motivating us to think about all of these issues. It is really true that throughout particle physics, and other parts of cosmology, there are, at a much smaller and less dramatic level, but certainly present, many little tunings like this on which the existence of interesting atoms, more complex structures, stars etc., really critically depend. And if you really believe that there is a unique theory, and a unique vacuum, then you really believe that there is a formula involving pure numbers that sets each and every one of those constants. And in such a situation it would really be shocking that so many of them ended up having just the right values that they had to have in order to allow us to exist.

I can just make one last comment. The situation is a lot like in biology. Not everything is selected for. Some things are the way they are because they are selected for, some things are the way they are because they cannot be any other way. It is the analog of environmental selection versus symmetries and dynamics, and one of the characteristics of biological creatures is that somethings are exquisitely designed, and other things are just sort of random, and the standard model looks a lot like that. There are some things that are exquisitely adjusted like the vacuum energy, there are all these irrelevant things like the third generation, V_{cb}, and the bottom quark mass which may just be incidental things that came along for the ride, correlated with other things that happened to be selected for. I think we would all agree that if this was not the picture of the world, we would have an easier job. But it does not strike me as a particularly awkward thing, and certainly not worse than the classical to quantum transition, at least as the issue of predictivity is concerned.

6.3   Prepared Comments

6.3.1   *Steven Weinberg*

Well, I was asked to talk for ten minutes and I don't have any positive new ideas to offer; so I am going to make some remarks of the opposite sign.

First of all, this is a small addendum to Joe's talk. I have a worry about the anthropic prediction or argument about the vacuum energy: that anthropic considerations may not really explain quite why it is as small as it is. If you fix the fluctuations at early times and suppose they don't scan and then calculate what is the average vacuum energy density that would be seen by an astronomer, in any part of the multiverse, weighting, and here Alan's point on how to weight things comes in, but if you do something for want of anything better, you weight the different subuniverses according to the fraction of baryons that find themselves in galaxies that are large enough to hold on to heavy elements after the first generation of stars, then you find that the average density that will be seen by all these astronomers throughout the multiverse, the vacuum energy density, is about 13 times the energy density of matter in our universe at the present time, not that that's a fundamental unit, but it just happens to be a convenient unit. In fact, experimentally, the number is not 13, it is 2.3 and you can ask what is the probability of getting a vacuum energy that small. The answer is: it is about 13%, 13% of all astronomers weighted the way I described will see a vacuum energy as small as we see it. Well, that's not so bad: I mean, 13% I could live with, those are the breaks. But this hinges on an assumption that, in order to hold on to heavy elements, the size of a fluctuation in the co-moving radius projected to the present has to be 2 Mpc's or greater and the answer is quite sensitive to that: if you reduce it to 1 Mpc, then the probability goes from 13% down to 7%. This is a difficult astrophysical question which is beyond my pay grade but, it really is important for astrophysicists to settle the question of how large fluctuations have to be to hold on to their heavy elements. And I just wanted to give you that to worry about a little bit.

Now, Alan has talked about the wonderful agreement of theory and observation for the microwave anisotropy, I could not agree more, it's wonderful, while we have been ringing our hands, the real cosmologists have been in hog heaven and, as Alan pointed out, everything, all the agreement that we see, not only for the microwave background but also for large scale structure, which continues the curve up to larger and larger values of L, large than can be reached by studying the microwave background, all this agreement flows from the assumption that the perturbations before they reenter the horizon are adiabatic, Gaussian and scale invariant. With that and just adjusting the overall scale, you fit these curves. So the wonderful shape of the curves does not really tell you very much about the early universe, it tells you the perturbations, when they are outside the horizon, are adiabatic, Gaussian and scale invariant. Now, that's usually interpreted in terms of a single scalar field rolling down a potential and the first caution I would like to offer is that this outcome is actually much more robust than that and much more generic and so that there isn't that much reason to believe in this very simple picture.

First of all, that the perturbations are adiabatic. By the way, in practice, as far as the microwave background is concerned, that means that $\delta\rho/(\rho+p)$ (ρ being the energy density and p being the pressure) is the same for the cold dark matter and the photon-baryon plasma, and that is verified to a fair degree of accuracy, although is is not very accurate right now. Well, that is extremely easy to achieve: it's automatic if you have a single scalar field rolling slowly down a potential; it's not automatic if you have many scalar fields, as you might expect, but if after inflation all these scalar fields dump their energy into a heat bath and if at that time, because baryon-number has not yet been generated, there were no non zero conserved quantum numbers, then of course automatically the perturbations must be adiabatic; that's almost trivial. What is a little bit less trivial is that later on, when the cold dark matter and then the neutrinos decouple from the photon-baryon plasma, the perturbations remain adiabatic. So that, it is by no means true that if you have many scalar fields you expect non adiabatic perturbations to be observed at the present time. Now, it is possible that you can get non adiabatic perturbations, there are the so called curvaton models, where you carefully arrange that some of the scalar fields that were present during inflation do not dump their energy into the heat bath but survive for some reason and these provide a model for non adiabatic perturbations. I think that the generic case is that you get adiabatic perturbations. That is true even if you have things much weirder than scalar fields: as long as after inflation you have a heat bath with no non zero conserved quantum numbers, then, even later when you no longer have local thermal equilibrium, you still have purely adiabatic perturbations.

That they are Gaussian, well, that follows from the fact that the perturbations are small, we know that experimentally, and that there was a time (this is true of a lot of theories although not all theories) in the very very early universe, when the physical wave number was large compared to the expansion rate, that the fields behaved like free fields. It's easy to arrange theories of many kinds including multiple scalar field theories in which that is true and if it is true, then you get Gaussian perturbations.

Scale invariance? Well, there are lots of theories that give you scale invariance. I made that remark at a meeting in Santa Barbara and Andrei Linde challenged me to thing of others. Of course, one example is multiple scalar fields all rolling slowly down a potential, but I could not really come up with any alternative but, Neal Turok here, just the other day, pointed out that in the oscillating or bouncing cosmology that he was suggesting you do get scale invariant perturbations. And scale invariance after all, scale invariant perturbations are pretty ubiquitous in nature, communications engineers call them $1/f$ noise and they are used to $1/f$ noise, even though it has nothing to do with inflation.

So, I would say that what you really need in order to settle these questions and to really get a handle on what was happening during inflation, which I don't think we have now, is to observe the effect of the tensor modes, the gravitational waves. Many people have said this, it's hardly an original observation. Fortunately, the Europeans are going ahead with the Planck satellite which may be able to detect the effect of tensor modes on the polarization of the microwave background because they are not wasting their money on manned space flight the way America and Russia are.

Now, I have talked about what are the necessary conditions for what we see: adiabatic, scale invariant, Gaussian perturbations. What about sufficient? The question here comes from the quantum corrections. We normally say that the quantum corrections are small. Why do we say they are small? Because the measure of smallness, the factor that you get every time you add a loop to a diagram, is something like $G H^2$, where G is Newton's constant and H is the Hubble constant, at the time the perturbation left the horizon. Experimentally, we know $G H^2 \simeq 10^{-12}$ so that is why quantum corrections are small and you don't have to worry about them. But, is it really true that the quantum corrections only depend on what was happening at the time the perturbations left the horizon? The calculations that have generally been done have been purely classical, for instance Maldacena calculated corrections, non Gaussian terms, that were corrections to the usual results, but that corresponded to a tree graph in which you have 3 lines coming into a vertex: that was not really a specifically quantum effect. When you include quantum effects, you begin to worry because the Lagrangian, after all, contains terms with positive powers of the Robertson-Walker scale factor a. For example, for a scalar field with a potential, you get an a^3 just from the square root of the determinant and even without a potential, just from the $(\bar{\nabla}\phi)^2$-term you get a factor of a. Now, there are lots of complicated cancellations which deal with this and, in fact, you can show that there are lots of theories in which the same result applies: the quantum corrections depend only on what was happening at the time of horizon exit and therefore they are small and therefore we don't worry about them.

It is clear though that there are other theories where that is not true. In particular... A kind of theory where it is true is a minimally coupled massless scalar field which has zero vacuum expectation value (not the inflaton but an additional scalar field with zero vev). If it does not have any potential, then the quantum effects caused by loops of that particle, to any order, do not produce any effects that grow with a as you go to late times in inflation. But, if you add a potential for the scalar, $V(\phi)$, then you get terms that do.

Last week I thought I was going to come here and show you a theory in which you get positive powers of a so that all bets are off and that the corrections become very large at late time. And just last week I was able to prove a theorem that, in fact, in every theory that I am able to think of, the corrections, when they are there, grow only like $\log a$. So, I am afraid, I don't have anything exciting to announce. Although there are quantum effects which do not depend only on what is happening at the time of horizon exit, they do not grow any faster than $\log a$. Which is a pity but that seems to be the way it is.

This material is based upon work supported by the National Science Foundation under Grant Nos. PHY-0071512 and PHY-0455649 and with support from The Robert A. Welch Foundation, Grant No. F-0014, and also grant support from the US Navy, Office of Naval Research, Grant Nos. N00014-03-1-0639 and N00014-04-1-0336, Quantum Optics Initiative.

6.3.2 *Renata Kallosh: Inflationary models as a test of string theory*

Our Universe is an Ultimate Test of the Fundamental Physics. High-energy accelerators will probe the scale of energies way below GUT scales. Cosmology and astrophysics are the only known sources of data in the gravitational sector of the fundamental physics (above GUT, near Planck scale). After the supernovae obser-

Early Universe Inflation	Late-time Acceleration
Near de Sitter space	Near de Sitter space
13.7 billion years ago	Now
During 10^{-35} sec	During few billion years
$V \sim H^2 M_P^2$	$V \sim H^2 M_P^2$
$H_{\text{infl}} \leq 10^{-5} M_P$	$H_{\text{accel}} \leq 10^{-60} M_P$

$$\frac{\dot{a}}{a} = H \approx \text{const}$$

$$\frac{\ddot{a}}{a} > 0, \qquad a(t) \sim e^{Ht}$$

vations and particularly after the release of the 1st year WMAP data on CMB in 2003 it become clear that any fundamental theory which includes gravity as well as particle physics has to address these data. The theory is expected to explain the origin of the near de Sitter space both during inflation as well as during the current acceleration. The most recent new observations from the Boomerang [1] are in agreement with the so-called *standard cosmological model* supported by the first set of WMAP data. This is ΛCDM model, in which the universe is spatially flat, it has a mysterious combination of matter we know it ($\sim 5\%$), cold dark matter ($\sim 25\%$) and dark energy ($\sim 70\%$). The model is using just few parameters to explain the large amount of cosmological observations. These parameters are suggested by the inflationary cosmology [2] which plays a significant role in ΛCDM standard model. The model also incorporates the current acceleration of the universe, see Table I.

String theory is the best candidate for the unified theory of all fundamental interactions. It has been realized over the last few years that the long standing difficulties in explaining cosmological observations may be resolved due to the current progress in string theory.

Quite a few models of inflation were derived since 2003 within compactified string theory with stabilized moduli. The inflaton field, whose evolution drives inflation is the only field which is not stabilized before the exit from inflation. Each of these models relies on particular assumptions. Some of these models have clear predictions for observables and therefore are FALSIFIABLE by the future observations. We will shortly comment here on few recently constructed models of inflation where, under clearly specified assumptions, one can predict three important observables.

(1) Tilt of the primordial spectrum of fluctuations, n_s
(2) The tensor to scalar ratio, $r = \frac{T}{S}$
(3) Light cosmic strings produced by the end of inflation

One can approximate the spectrum of the scalar and tensor perturbations of the metric by a power-law, writing

$$\Delta_{\mathcal{R}}^2(k) = \Delta_{\mathcal{R}}^2(k_*) \left[\frac{k}{k_*}\right]^{n_s-1} \quad , \quad \Delta_h^2(k) = \Delta_h^2(k_*) \left[\frac{k}{k_*}\right]^{n_t} \quad , \quad r = \frac{\Delta_h^2(k_*)}{\Delta_{\mathcal{R}}^2(k_*)}$$

where n_s, n_t are known as the scalar spectral index and the gravitational spectral index, respectively, and k_* is a normalization point, r is the tensor/scalar ration , the relative amplitude of the tensor to scalars modes. The observations require n_s close to one, which corresponds to the perturbations in the curvature being independent of scale. The deviation of the spectral index from one, $n_s - 1$, is a measure of the violation of the scale invariance of the spectrum of primordial fluctuations.

The only known at present viable mechanism for generating the observed perturbations is the inflationary cosmology, which posits a period of accelerated expansion in the Universe's early stages. In the simplest class of inflationary model the dynamics are equivalent to that of a single scalar field ϕ slowly rolling on an effective potential $V(\phi)$. Inflation generates perturbations through the amplification of quantum fluctuations, which are stretched to astrophysical scales by the rapid expansion. The simplest models generate two types of density perturbations which come from fluctuations in the scalar field and its corresponding scalar metric perturbation, and gravitational waves which are tensor metric fluctuations. Defining slow-roll parameters, with primes indicating derivatives with respect to the scalar field, as

$$\epsilon = \frac{m_{\mathrm{Pl}}^2}{16\pi} \left(\frac{V'}{V}\right)^2 \quad ; \quad \eta = \frac{m_{\mathrm{Pl}}^2}{8\pi} \frac{V''}{V} \, ,$$

the spectra can be computed using the slow-roll approximation ($\epsilon, |\eta| \ll 1$). In each case, the expressions on the right-hand side are to be evaluated when the scale k is equal to the Hubble radius during inflation. The spectral indices and tensor to scalar ratio follow

$$n_s \simeq 1 - 6\epsilon + 2\eta \quad ; \quad n_t \simeq -2\epsilon \, , \quad r \simeq 16\epsilon \simeq -8n_t \, ,$$

The last relation is known as the consistency equation for the single field inflation models, it becomes an inequality for multi-field inflationary models.

It has been recognized recently that cosmic strings give a potentially large window into the string theory [3]. CMB observations put a stringent constraint on cosmic strings produced at the end of inflation: only very light cosmic strings with the tension $(G\mu)_{obs} < 5 \cdot 10^{-7}$ are consistent with the data. The KKLMMT model [4] of stringy inflation is based on the throat geometry with the highly warped region and therefore can easily explain the existence of light cosmic strings. At present, however, no evidence is available for such cosmic strings, see [5] for the recent Hubble Space Telescope observation which proved that the object CSL-1 is not a lensing of a galaxy by a cosmic strings, contrary to previous expectations.

Inflationary models in string theory have few clear predictions for the primordial spectral index and for the tensor to scalar ratio. A significant amount of gravitational waves is expected to take place in chaotic models of inflation [6]. However, most models of modular and brane inflation in string theory known at present do not predict any significant amount of gravitational waves. Still there are models, like [7] where in the context of string theory there is a possibility to explain the primordial gravitational waves in case they will be actually detected.

With regard to the spectral index n_s the situation is developing in a rather interesting way. The 1st year WMAP data alone suggest

$$n_s = 0.99 \pm 0.04 \ ,$$

the combination of the data from WMAP+CBI+ACBAR+2dFGRS gives

$$n_s = 0.97 \pm 0.03 \ .$$

Moreover, there is an indication from the most recent release of the Boomerang data [1] that the central value of n_s may be moving downwards towards

$$n_s \approx 0.96$$

and it will most interesting to know what emerges from the new WMAP data[10].

The inflationary models in string theory in some cases have a clear computable prediction for n_s. For example, a racetrack model of modular inflation [8] predicts $n_s = 0.95$. This is a model with one Kähler modulus where the system has a saddle point and inflates due to axion-inflaton into a stabilized KKLT string flux vacua with de Sitter minimum [9] to account of the current acceleration of the universe. Another model of hybrid inflation, the so-called D3/D7 brane inflation [10], predicts $n_s = 0.98$ under a condition of the softly broken shift symmetry protecting the near-flat inflationary potential in this model.

Planck satellite (2008?) is expected to provide the precision data on spectral index n_s at the level of 0.5% ! This will help to focus on those models of inflation

---

[10]On March 16 2006, the WMAP three year data release took place at

http://lambda.gsfc.nasa.gov/.

The new data strongly support the standard cosmological model and favor an inflationary model $\lambda\phi^2$ of [6]. The new value of the spectral index is $n_s = 0.95 \pm 0.02$ in agreement with [1]. From all known at present models of stringy inflation with a clear prediction for the spectral index, the racetrack model [8] seems to give the best fit to the data.

in string theory which will support the data from Planck and other sets of future observations.

- If the inflationary models are derived in string theory by reliable methods and assumptions stated clearly
- If the models have unambiguous prediction for observables
- When the precision data will come in we will be able to test the string theory assumptions underlying the derivation of the corresponding "best fit data" inflationary models.

New cosmological data will be coming during the next 10-20 years "fast and furious"!

Bibliography

[1] C. J. MacTavish *et al.*, *Cosmological parameters from the 2003 flight of BOOMERANG*, astro-ph/0507503.

[2] A. H. Guth, *"The Inflationary Universe: A Possible Solution To The Horizon And Flatness Problems"*, Phys. Rev. D **23** (1981) 347; A. D. Linde, "A New Inflationary Universe Scenario: A Possible Solution Of The Horizon," *Flatness, Homogeneity, Isotropy And Primordial Monopole Problems*, Phys. Lett. B **108**, 389 (1982); A. Albrecht and P. J. Steinhardt, *Cosmology For Grand Unified Theories With Radiatively Induced Symmetry Breaking*, Phys. Rev. Lett. **48** (1982) 1220.

[3] J. Polchinski, *Introduction to cosmic F- and D-strings*, hep-th/0412244.

[4] S. Kachru, R. Kallosh, A. Linde, J. Maldacena, L. McAllister and S. P. Trivedi, *Towards inflation in string theory*, JCAP **0310** (2003) 013 hep-th/0308055.

[5] M. V. Sazhin, M. Capaccioli, G. Longo, M. Paolillo and O. S. Khovanskaya, *The true nature of CSL-1*, astro-ph/0601494.

[6] A. D. Linde, *Chaotic Inflation*, Phys. Lett. B **129** (1983) 177.

[7] M. Alishahiha, E. Silverstein and D. Tong, *DBI in the sky*, Phys. Rev. D **70** (2004) 123505 hep-th/0404084; S. Dimopoulos, S. Kachru, J. McGreevy and J. G. Wacker, *N-flation*, hep-th/0507205.

[8] J.J. Blanco-Pillado, C.P. Burgess, J.M. Cline, C. Escoda, M. Gómez-Reino, R. Kallosh, A. Linde and F. Quevedo, *Racetrack inflation*, JHEP **0411** (2004) 063 hep-th/0406230; *Inflating in better racetrack*, hep-th/0603132.

[9] S. Kachru, R. Kallosh, A. Linde and S. P. Trivedi, *De Sitter vacua in string theory*, Phys. Rev. D **68** (2003) 046005 hep-th/0301240.

[10] K. Dasgupta, C. Herdeiro, S. Hirano and R. Kallosh, *D3/D7 inflationary model and M-theory*, Phys. Rev. D **65** (2002) 126002 hep-th/0203019.

6.3.3 *Andrei Linde: Eternal inflation in stringy landscape and the anthropic principle*

In the beginning of the 80's, when the inflationary theory was first proposed, one of its main goals was to explain the amazing uniformity of our universe. Observations told us that the universe looks the same everywhere and that the physical laws in all of its parts are the same as in the vicinity of the solar system. We were looking for a unique and beautiful theory that would unambiguously predict all properties of our universe, including the observed values of all parameters of all elementary particles, not leaving any room for pure chance.

However, most of the parameters of elementary particles look more like a collection of random numbers than a unique manifestation of some hidden harmony of Nature. But there was one important property shared by many of these parameters: Changing them in any substantial way would lead to the universe where we could not exist. This fact is the foundation of the cosmological anthropic principle [1]. This principle is based on a simple fact: We can observe the universe with a given set of properties only if these properties are compatible with our very existence.

Whereas this fact is certainly correct, many scientists are still ashamed of using the anthropic principle. It is often associated with the idea that the universe was created many times until the final success. It was not clear who did it and why was it necessary to make the universe suitable for our existence. There were some attempts to relate the anthropic principle to the many-world interpretation of quantum mechanics, or to quantum cosmology, but these attempts looked esoteric, and they did not explain why all parts of the universe have similar properties. Indeed, it seemed to be much simpler to have conditions required for our existence in a small vicinity of the solar system rather than in the whole universe.

Fortunately, most of the problems associated with the anthropic principle were resolved more than 20 years ago with the invention of inflationary cosmology. First of all, in the context of inflationary cosmology, nice conditions in a small vicinity of the solar system imply similar conditions in the observable part of the universe, thus removing the most difficult objection against the anthropic principle. Also, in the context of chaotic inflation [2] there is no need to assume that initial conditions were the same in all parts of the universe. If initial conditions were different in different parts of the universe (or in different universes), then a generic inflationary universe should consist of many exponentially large regions containing matter in all of its possible states, with scalar fields rolled down to all possible minima of their energy density, and with space with all of its possible types of compactification. This observation provided the first scientific justification of the anthropic principle [3, 4].

The situation becomes even more interesting when one takes into account quantum fluctuations produced during inflation. It was shown in [5] that even if the universe started with the same initial conditions everywhere, e.g. in the $SU(5)$-symmetric minimum of the $SU(5)$ SUSY, inflationary fluctuations lead to jumps of the scalar fields from one minimum of its potential to another, which divides the universe into exponentially large domains with matter in all possible states corresponding to all minima of the $SU(5)$ SUSY, including our $SU(3) \times SU(2) \times U(1)$ minimum.

These observations merged into one coherent picture after the discovery of eternal inflation in the context of chaotic inflation scenario [6].[11] According to this scenario, some parts of the universe continue eternally jumping at density which may be as high as the Planck density. Inflationary fluctuations produced in this regime are powerful enough to jump over any barrier, and divide the universe into exponentially large domains in which not only the scalar fields but even the type of compactification and the effective dimension of our space-time may change [10].

A similar regime may exist in the theories with many different local de Sitter minima even if inflation near these minima is not of the slow-roll type [11]: The field may tunnel from an upper minimum to the lower minimum and back. A combination of this effect and the effect discovered in [8, 3, 9, 6] provided a necessary background for the string landscape scenario [12].

This scenario is based on the recent discovery of the mechanism of moduli stabilization in string theory [13], which allowed to describe inflation and the present stage of acceleration of the universe. Once this mechanism was found, it was realized that the total number of possible metastable de Sitter vacua in string theory is enormously large, perhaps 10^{100} or 10^{1000} [14]. During inflation our universe becomes divided into exponentially large domains of 10^{1000} different types, which is a perfect setup for the anthropic principle.

The large set of stringy vacua introduces an incredibly large set of *discrete* parameters. However, some of the parameters of our universe are determined not by the final values of the fields in the minima of their potential related to the string theory landscape, but by the dynamical, time-dependent values which they were taking at different stages of the evolution of the universe. This introduces a large set of *continuous* parameters which may take different values in different parts of the universe.

One example of a continuous parameter is the ratio n_γ/n_B. Its observed values is about 10^{-10}. In some cases, the reason why this number is so small is pretty obvious, but in the original version of the Affleck-Dine scenario a typical value of this parameter was $O(1)$ [15]. The ratio n_γ/n_B in this scenario is determined by the angle between two scalar fields soon after inflation. This angle is a free parameter which may take different values in different parts of the universe due to inflationary fluctuations of these fields [16]. It was argued in [16] that the process of galaxy formation strongly depends on the ratio n_γ/n_B. Therefore even if though the total volume of the parts of the universe with $n_\gamma/n_B = O(1)$ is 10^{10} times greater than

---

[11]The regime of eternal inflation was known to exist in old inflation [7] and in new inflation [8, 3, 9], but none of these papers except [3] mentioned the relation of this regime to the anthropic principle.

the total volume of the parts with $n_\gamma/n_B = 10^{-10}$, we can live only in the parts with $n_\gamma/n_B \sim 10^{-10}$ [16].

The second such example, which is in fact very similar, is the ratio of the baryonic matter density to the cold dark matter density, $\zeta = \rho_{CDM}/\rho_B \sim 5$. In the theory with light axions, $m_a \ll 10^{-5}$ eV, the natural value of this ratio would be much smaller than 0.2, which was considered as a strong evidence that the axions do in fact have mass $m_a \sim 10^{-4} - 10^{-5}$ eV [17]. The resolution of the problem was very similar to the one mentioned above: In inflationary cosmology with $m_a \ll 10^{-5}$ eV the universe consists of many different exponentially large regions with different values of ζ. The prior probability of formation of the region with different $\sqrt{\zeta}$ after a period of inflation does not depend on ζ. However, the existence of galaxies and stars of our type would be much less probable for $\sqrt{\zeta}$ one or two orders of magnitude greater than its present value. This provides an anthropic explanation of the presently observed value of ρ_{CDM}/ρ_B [18, 19].

One of the most spectacular applications of the anthropic principle is the cosmological constant problem. Naively, one could expect vacuum energy to be equal to the Planck density, $\rho_\Lambda \sim 10^{94} g/cm^3$, whereas the recent observational data show that $\rho_\Lambda \sim 10^{-29} g/cm^3$, which is about 0.7 of the total energy density of the universe ρ_0. Why is it so small but nonzero? Why ρ_Λ nearly coincides with ρ_0?

The first anthropic solution to the cosmological constant problem in the context of inflationary cosmology was proposed in 1984 in [20]. The vacuum energy density can be a sum of the scalar field potential $V(\phi)$ plus the energy of fluxes $V(F)$. I argued that quantum creation of the universe is not suppressed if it is created at the Planck energy density, $V(\phi) + V(F) = 1$, in Planck units. Eventually the field ϕ rolls to its minimum at some value ϕ_0, and the vacuum energy becomes $\rho_\Lambda = V(\phi_0) + V(F)$. Since initially $V(\phi)$ and $V(F)$ with equal probability could take any values with $V(\phi) + V(F) = 1$, we get a flat probability distribution to find a universe with a given value of the cosmological constant $\rho_\Lambda = V(\phi_0) + V(F)$. Finally, I argued that life would be possible only for $-\rho_0 \lesssim \rho_\Lambda \lesssim \rho_0$. This fact, in combination with inflation, which makes such universes exponentially large, provides a possible solution to the cosmological constant problem.

In the next couple of years after my work, several other anthropic solutions to the cosmological constant problem were proposed [21]. All of them took for granted that life is possible only for $-\rho_0 \lesssim \rho_\Lambda \lesssim \rho_0$. The fact that ρ_Λ could not be much smaller than $-\rho_0$ was indeed quite obvious, since such a universe would rapidly collapse. Meanwhile the constraint $\rho_\Lambda \lesssim \rho_0$ was much less trivial; it was fully justified only few years later, in a series of papers starting from the famous paper by Weinberg [22].

I would be able to continue this discussion, describing the constraints on the amplitude of density perturbations, on the dimensionality of the universe, on the electron and proton masses, on the expectation value of the Higgs field, etc. What we see here is that many properties of our universe become less mysterious if one try to relate them to the fact of our own existence. We still have to learn how to calculate the probabilities in the eternally inflating universe. We still need to find a full string theory description of particle phenomenology and estimate the total number of vacua which can describe our world. But it will be very difficult to turn back and unlearn what we just learned. Now that we have found that there exist simple anthropic solutions to many problems of modern physics, one would need either to find an alternative solution to all of these problems, or to learn how to live in the democratic world where the freedom of choice applies even to our universe.

Bibliography

[1] J.D. Barrow and F.J. Tipler, *The Anthropic Cosmological Principle* (Clarendon Press, Oxford, 1986).

[2] A. D. Linde, *Chaotic Inflation*, Phys. Lett. **129B** (1983) 177.

[3] A.D. Linde, *Nonsingular Regenerating Inflationary Universe*, Print-82-0554, Cambridge University preprint, 1982.

[4] A.D. Linde, *The New Inflationary Universe Scenario*, In: *The Very Early Universe*, ed. G.W. Gibbons, S.W. Hawking and S.Siklos, Cambridge University Press (1983), pp. 205-249.

[5] A. D. Linde, *Inflation Can Break Symmetry In SUSY*, Phys. Lett. B **131** (1983) 330.

[6] A. D. Linde, *Eternally Existing Self-reproducing Chaotic Inflationary Universe*, Phys. Lett. B **175** (1986) 395.

[7] A. H. Guth, *The Inflationary Universe: A Possible Solution To The Horizon And Flatness Problems*, Phys. Rev. D **23** (1981) 347.

[8] P. J. Steinhardt, *Natural Inflation*, In: *The Very Early Universe*, ed. G.W. Gibbons, S.W. Hawking and S.Siklos, Cambridge University Press (1983).

[9] A. Vilenkin, *The Birth Of Inflationary Universes*, Phys. Rev. D **27** (1983) 2848.

[10] A. D. Linde and M. I. Zelnikov, *Inflationary Universe With Fluctuating Dimension*, Phys. Lett. B **215** (1988) 59.

[11] K. M. J. Lee and E. J. Weinberg, *Decay Of The True Vacuum In Curved Space-Time*, Phys. Rev. D **36** (1987) 1088.

[12] L. Susskind, *The anthropic landscape of string theory*, arXiv:hep-th/0302219.

[13] S. Kachru, R. Kallosh, A. Linde and S. P. Trivedi, *De Sitter vacua in string theory*, Phys. Rev. D **68** (2003) 046005 [arXiv:hep-th/0301240].

[14] R. Bousso and J. Polchinski, *Quantization of four-form fluxes and dynamical neutralization of the cosmological constant*, JHEP **0006** (2000) 006 [arXiv:hep-th/0004134]; M. R. Douglas, *The statistics of string / M theory vacua*, JHEP **0305** (2003) 046 [arXiv:hep-th/0303194].

[15] I. Affleck and M. Dine, *A New Mechanism For Baryogenesis*, Nucl. Phys. B **249** (1985) 361.

[16] A. D. Linde, *The New Mechanism Of Baryogenesis And The Inflationary Universe*, Phys. Lett. B **160** (1985) 243.

[17] J. Preskill, M. B. Wise and F. Wilczek, *Cosmology Of The Invisible Axion*, Phys. Lett. B **120** (1983) 127; L. F. Abbott and P. Sikivie, *A Cosmological Bound On The Invisible Axion*, Phys. Lett. B **120** (1983) 133; M. Dine and W. Fischler, *The Not-So-Harmless Axion*, Phys. Lett. B **120** (1983) 137.

[18] A. D. Linde, *Inflation And Axion Cosmology*, Phys. Lett. B **201** (1988) 437.

[19] M. Tegmark, A. Aguirre, M. Rees and F. Wilczek, *Dimensionless constants, cosmology and other dark matters*, Phys. Rev. D **73** (2006) 023505 [arXiv:astro-ph/0511774].

[20] A. D. Linde, *The Inflationary Universe*, Rept. Prog. Phys. **47** (1984) 925.

[21] A. D. Sakharov, *Cosmological Transitions With A Change In Metric Signature*, Sov. Phys. JETP **60** (1984) 214 [Zh. Eksp. Teor. Fiz. **87** (1984) 375]; T. Banks, *TCP, quantum gravity, the cosmological constant, and al l that ...*, Nucl. Phys. B**249** (1985) 332; A.D. Linde, *Inflation And Quantum Cosmology*, Print-86-0888 (June 1986) in 300 Years of Gravitation, ed. by S.W. Hawking and W. Israel, Cambridge University Press, Cambridge (1987).

[22] S. Weinberg, *Anthropic Bound On The Cosmological Constant*, Phys. Rev. Lett. **59** (1987) 2607.

6.4 Discussion

G. 't Hooft I would like to have an answer to the question: Is this notion of the landscape the same as we have heard earlier today? Because here it seems that you can travel on a spacelike orbit. Let us imagine that we can move in this universe on a spacelike orbit, just to watch around us. The question is: would we be able (even in principle) then to enter into another part of the universe where the standard model is a different one and, if so, what would that transition look like? Would we pass through a membrane, or would we pass through the horizon of a black hole? What is it that separates these different kinds of universes? In a picture Linde had these bubbles connected by little throats, and I wonder if you would travel through such a throat what would the transition be like? Is there a moment when you would say "Hey, now we have 4 generations!", or 25 generations, or something that changes?

A. Linde Let me answer the first question, "What I am going to see at the boundary?" That depends on the boundary. Usually, you have boundaries as follows: you have one minimum and another minimum and between the minima you know that you must go up the hill. Typically then, these two minima are divided by domain walls. The sizes of two sides are exponentially large. If these domains are both de Sitter, you are in a hopeless situation, because if you will be traveling here this part will be running away from you because of de Sitter expansion. If you are sitting in Minkowski space, which is just by chance, then you will have a possibility to travel here for some time, and after that you would see the wall and if you are young and stupid still at that time, then you will go through the wall and die because your particles will not exist in that part of the universe.

G. 't Hooft The question was not really whether one could travel there on a time-like geodesic, but suppose — in our imagination — one goes over a space-like geodesic, or a space-like orbit, or even back in time. At some point you should see a transition when you go from one universe to the next. So if you go back in time or on a space-like orbit, then, as you say, you would go through some membrane or something. I want to know what that thing is like.

A. Linde Are you are asking about the process, for example, that happens if I am sitting at this point, and quantum fluctuations just push me to some place?

S. Shenker I think that what 't Hooft is asking is the following. Suppose you pick some time variable, does not matter which. Look at a fixed-time space slice and then go along. What happens when you jump from one bubble to the next?

A. Linde That is the thing that I answered, exactly. The image which I showed you is the result of a computer simulation of one particular time slice, a space-like hypersurface.

G. 't Hooft I just want to have you pin down somehow: What would this membrane look like? A membrane separating two different kinds of universes: would that be a running away horizon? I am not quite happy with it, but that could be a conceivable answer. So, is it a horizon which runs away from us or is there really a membrane? Would it be something like in science fiction, a kind of wall at the end of the universe and you just have to go through it to see the other part of the universe, anything like that?

A. Linde As I said, this is a model-dependent issue, and it also depends on the nature of the state. If you are sitting in de Sitter space, you never even come to this part of the Universe. It will run away from you. If you are sitting in Minkowski space, then you would have two possibilities: the one is that the boundary travels away from you with the speed equal to the speed light and then you will never touch it. There is another possibility that this part travels with the speed of light towards you, and then you will never see it because at the moment you see it, how to say, there will be nobody to respond and nobody to report. NSF will not support your further work.

G. 't Hooft Maybe the question to the other people here is the following: is that the same notion of landscape as seen in other talks?

A. Linde We are talking about the same landscape, but this landscape is an "animal with many faces". All of these things are very much different. The parts of the universe which are de Sitter have some properties, parts which are Minkowski are rare animals, and they have some other properties. Parts which we are supposed to associate with anti-de Sitter, they are actually not anti-de Sitter but collapsing Friedman Universes. All of them can be part of this picture.

S. Shenker One of the things about these issues I find the most interesting (that is probably because I just started thinking about them) is this issue that Guth, Linde and others discussed, about putting a measure on the space of "pocket universes". Now that it seems more and more likely that string theory contains some kind of landscape, and it seems that there is a well-defined quantum gravity. For this mysterious bubbling phenomenon, there should be some kind of question about how likely is every kind of bubbling universe. And it is incredibly hard, as Guth mentioned, to figure out what that question means. Now, we think we have a well posed theory. Either this question is, for some reason, completely nonsensical or we should be able to sharpen it up. This problem is one to think about. It has the psychological advantage that you can think about issues of landscape and bubbles and never have to think about anthropics, which I find psychologically appealing.

M. Douglas I agree with the importance of this. There is another version of this question that I have read in the existing literature. It just seems that if at some point of your calculation you get infinity over infinity, or a limit of quantities which is going to become infinity over infinity, then your definition is inherently ambiguous, and at that point you have already lost. Was there any kind of suggestion or hope in the existing works, of a definition that would make at least a denominator finite? I see that the infinity comes in because of the growth of volume. The definition of inflation is this growth of the volume, and any limit would make that infinite, but somehow one has to avoid that.

A. Linde Well, if we would know the answers, we would probably have written about them a long time ago. But I would like just to give some analogies. The first analogy is that there is a question "What is the time of the greatest productivity of a person?" and another question is "What is the time when a typical physicist produces his best work?" The typical time, or age, when the best works are produced by physicists is between 20 and 30. Which would mean that all of us must just retire now in shame because we all are out of maximal productivity. On the other hand, when Ginzburg discussed this question at our seminars (and he was already from my perspective at that time very old — which is my age right now), he said the following: this is a question of statistics. If you are interested in what is the typical age physicists produce their best works, you measure it weighting it with the total number of scientists (which at that time was exponentially growing). That is why you have this youngness paradox. But then, Ginzburg said, I am interested in my own productivity, and that is a different measure. Now, you ask me which of these questions makes sense and I tell you: both make sense from a statistical point of view. What one should learn from this is not which of these measures is better but which of them has any relation with the anthropic principle. That is one possible answer. Another possible answer is that actually we may take a very humble attitude, and the humble attitude would be like that: all my life, I was wondering why I was born in Moscow; when I was a young pioneer, I was wondering why I am so happy to be born in Moscow where the best children in the world live. When I got older I still asked why I was born on Moscow, anyway. Then, right now, I am saying: these questions may have no meaning because there are much more Chinese. So, if I would just measure the total number of people, then by this measure I would be an exception. But if I know that I am born in Moscow, and I see everybody speaking English around, then I would think that this is something surprising, this is something that I must explain.

So, I must ask conditional probability questions: under given experimental results, e.g. under the given result that I know what is $\Delta\rho/\rho$, do I still find the present value of the cosmological constant surprising? No, I do not. But if you let me consider all possible values of Λ and $\Delta\rho/\rho$ and everything else, I will have huge areas of landscapes with infinite volume and I will have nothing to say. So, in a more humble way, that is exactly what experimentalists do: they make new measurements and after these new measurements they evaluate what is the probability of the next outcome of the next experiment. If you use the anthropic principle in this way, there is a better chance that you will not say anything nonsensical.

F. Englert I would like to make a comment, essentially to convey my uneasiness about the way this anthropic principle is used here. It unavoidably makes one think about the principle of natural selection in biology. Now, I do not want to discuss natural selection in biology but I think the transcription of it from biology to physics is a little bit dangerous. I just want to give a simple example. Suppose that we do not know the theory of gravity and suppose that someone asks why do apples fall on the ground (the apple had a historical importance in gravity, that is why I choose apples). The answer is extremely simple: because, if they do not fall on the ground, they do not give rise to trees and therefore those apples that do not fall on the ground have disappeared. You can very easily generalize that statement: not only apples but, of course, everything falls on the ground because what does not fall on the ground is no more there. The morality of this is that, of course, we would like to say that maybe one should not reason too simplistically on this, but one should better look for a theory of gravitation at that moment. Maybe one should look for a decent theory for explaining that particular element which is the cosmological constant.

B. Greene Just a quick remark relevant to the question that was asked about finding the measure on the landscape. I guess I am not still quite convinced that this is a really interesting question. And the reason I am not convinced is the following. If you would ask the same kind of question in the context of ordinary field theories — look at the landscape of all possible field theories and write some measure on that space — you are never going to find the Standard Model as some generic field theory in this space of field theories. It is a very special field theory, and yet it is the one that is right. Although I understand the motivation for having a measure on the space of the landscape from string theory, that is to have a possible anthropic solution to the cosmological constant, but what if you go beyond that and talk about the rest of phenomenology? The basic question is: since we all know that very special theories are sometimes the right theory, why try to have some sense of genericity as a guide to finding the right model?

S. Shenker Now we are going to have the last couple of talks.

6.4.1 *Paul J. Steinhardt: A modest proposal for solving the cosmological constant problem*

Probably all of the participants at the Solvay Conference have dreamed of solving the cosmological constant (Λ) problem. And probably all have, at one time or another, sought the same solution: a dynamical relaxation mechanism that gradually cancels all contributions to Λ, whether due to physics at the Planck scale, the electroweak scale, the QCD scale, *etc.* In this way, the universe could begin with a natural value for Λ of order the Planck scale but have an exponentially small value today. During the last quarter century, though, a serious roadblocks has been placed in the way of this dream due to a combination of inflationary cosmology and dark energy. Inflationary cosmology requires that the relaxation time be long compared to a Hubble time during the first instants after the big bang so that the universe can undergo the cosmic acceleration necessary to resolve the horizon and flatness problems and generate a nearly-scale invariant spectrum of density fluctuations. After inflation, it is essential that the relaxation time become short compared to a Hubble time in order for primordial nucleosynthesis and galaxy formation to proceed in accordance with observations. The discovery of dark energy, though, means that the universe entered a new period of cosmic acceleration 10 billion years later, so the relaxation time must be long compared to a Hubble time today. The situation seems to call for a relaxation mechanism that transforms magically on cue from slow to fast and back to slow again, a cosmological somersault that appears to be anything but simple.

In this comment, I would like to introduce a suggestion by Neil Turok and myself [1] for reviving the concept of a *simple* dynamical relaxation mechanism. Here, rather than seeking a relaxation time that is sometimes shorter and sometimes longer than a Hubble time, we propose a relaxation time that is *always exponentially long compared to a Hubble time*. (Finding ultra-slow relaxation mechanisms turns out not to be difficult; as illustrated below, some have already been identified in the literature.) In our picture, Λ is decreasing excruciatingly slowly throughout cosmic history at a rate too small to be detected even after 14 billion years. Furthermore, the relaxation process slows downs as Λ approaches zero from above. Hence, most of cosmic history is spent with a small, positive cosmological constant, in accordance with what we observe.

Before describing how the concept works, it is instructive to compare our picture of the cosmological constant with the case of the 'Hubble constant,' H. H is about 10^{-42} GeV, exponentially tiny compared to the QCD, electroweak or Planck scale. If it were truly a constant, physicists would find it hard to understand how its value could emerge from fundamental physics. Yet, this small value is essential if galaxies, stars and planets are ever to form. Some might feel driven to introduce an anthropic principle or multiverse to explain the small value. But, as we already know, this is not necessary. We understand that Einstein's theory of general relativity tells us that the 'Hubble constant' is not a constant after all and that gravity incorporates a dynamical relaxation mechanism that naturally causes H to decrease with time. H was once large – so large that galaxies could not form – but after 5 billion years it reached a value small enough for structure to evolve. Furthermore, the Hubble constant decreases more slowly as its value shrinks, so most of cosmic history is spent with a small positive Hubble constant. So, as far as the Hubble constant is concerned, we live at a typical location in space and time. Its small value today is not considered a deep mystery; it is just a sign that the universe is old compared to a Planck time.

Our proposal for Λ is similar. The key difference is a matter of timescale. The Hubble constant changes by a factor of 10^{100} in 14 billion years. For the cosmological constant, we envisage that it relaxes by 10^{100} in $10^{10^{120}}$ years or more. The essential idea is that Λ is small but nearly constant (compared to a Hubble time) today because it has had an exponentially long period (compared to a Hubble time) to relax.

Of course, to implement this idea, the universe must be much older than 14 billion years. For such an old, expanding universe to have a non-negligible H and matter density, it had better be that H and the matter density can be reset to large values at times during the period that Λ slowly decreases. As one reflects further upon the idea, it becomes apparent that a cyclic model of the type described by Neil Turok [2–4] is ideally suited for this purpose – although it is interesting to note that the model was not designed with this idea in mind.

First, the cyclic model provides more time. Each cycle lasts perhaps a trillion years, but there is no known limit to how many cycles there may have been in the past. So, a universe that is $10^{10^{120}}$ years or longer is quite feasible. Second, the cyclic model provides a mechanism, the periodic bounces between branes, for regularly replenishing the universe with matter and radiation at regular intervals and, consequently, regularly restoring H to a large value. Third, the cyclic model does not include a period of high energy inflation, removing the key roadblock discussed in the introduction. Finally, the cyclic model includes matter fields that live on the branes and couple to the brane metric. According to the model, the branes expand from cycle to cycle; the periodic crunches occur because of a contraction along the extra dimension. Hence, fields on the branes are redshifted from cycle to cycle but are not blue shifted during the periods of contraction (of the extra dimension); this turns out to be useful for maintaining the slow relaxation process for reasons that are explained in Ref. [1].

As for the slow relaxation mechanism, there are various possibilities. For simplicity, I focus here on a concrete example first introduced twenty years ago by L. Abbott [5], but in the wrong context. (Another mechanism with similar properties was introduced by J. Brown and C. Teitelboim a few years later [6].) Abbott proposed relaxing the cosmological constant by adding an axion field ϕ with a tilted 'washboard' potential

$$V(\phi) = M^4 \cos\frac{\phi}{f} + \frac{\epsilon}{2\pi f}\phi, \tag{1}$$

where $M \sim 1$ eV, $f \sim 10^{16}$ GeV, and $\epsilon^{1/4} \sim .1$ meV are sample values that serve the purpose. The gauge interaction provides a natural explanation for the small value of M, analogous to the explanation for the QCD scale, Λ_{QCD}. $\Lambda_{QCD} \sim 100$ MeV is generated dynamically and can be expressed in terms of the Planck mass m_p as $\Lambda_{QCD} \sim m_p \exp(-2\pi/\alpha_{QCD})$, where $\alpha_{QCD} = 0.13$ is not so different from unity. Here we imagine that ϕ lives on the hidden brane and is coupled to hidden gauge fields. A modest difference in the the hidden sector coupling constant, $\alpha_{hidden} \sim 0.09$, suffices to obtain the value of M desired for our model. The tilt come from an interaction that softly breaks the periodic shift-symmetry of the axion. The soft breaking scale, $\epsilon^{1/4} \leq 1$ meV, is comparable to but somewhat smaller than M; it sets the scale of the steps in energy density along the washboard: $V_N - V_{N-1} = \epsilon$. The model is technically naturally in that the coefficients are not subject to large quantum corrections.

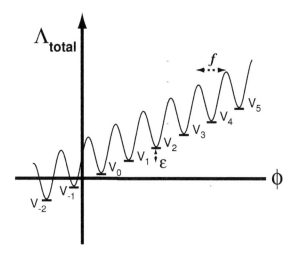

Fig. 6.3   The washboard potential defined in Eq. 2.

The total vacuum energy density is

$$\Lambda_{total} = \lambda_{other} + V(\phi) \tag{2}$$

where λ_{other} incorporates all other contributions from Planck, electroweak, QCD and other non-axionic physics. The washboard potential (Fig. 6.3) has periodically spaced minima at $\phi = 2\pi f N + \phi_0$ where N is an integer and ϕ_0 is the value of ϕ at $V = V_0$. The minima have vacuum density $V_N = V_0 + N\epsilon$ where V_0 is the vacuum density of the minimum with the smallest non-negative Λ_{total}. The potential also has minima $V_{-1}, V_{-2}, \ldots$ with negative vacuum density.

Suppose the universe begins at some minimum with a large positive potential density V_N. The universe begins to work its way down the potential by quantum tunneling through the energy barriers. That is, growing bubbles of vacuum with $V = V_{N-1}$ form in a background with $V = V_N$. The process continues from one minimum to the next until V approaches V_0 and $\Lambda_{total} \leq \epsilon$. Since the average tunneling rate is $\Gamma \sim M^4 e^{-B}$ where $B \sim M^2 f / V_N$, the relaxation rate decreases exponentially as the field tunnels downhill. Hence, the universe spends exponentially more time at stages when the cosmological constant is small and positive. At the last positive minimum V_0, the tunneling time is roughly $10^{10^{120}}$ years. Eventually, bubbles nucleate with $V = V_{-1}$ in their interior, but these are anti-deSitter minima that undergo gravitational collapse in one Hubble time (about 14 billion years). The collapsed regions probably form black holes. But since most of the universe continues to expand at an accelerating rate, these black holes represent an insignificant fraction of the volume. So, for any patch of space, no matter the value of λ_{other}, ϕ relaxes down to and spends most of cosmic time at the minimum with the smallest positive Λ_{total}.

Abbott's notion was to apply this mechanism in a standard big bang universe where it fails utterly. In the long time it takes to tunnel, the universe expands so much that it is completely vacuous by the time ϕ begins to tunnels down the potential. However, the concept dovetails perfectly with the cyclic model.

If the axion and the gauge fields to which it couples live on the hidden brane, their evolution is independent of the cycling motion of the branes along the extra dimension. Also, because the temperature generated at each big crunch/big bang transition is much less than f [4], the axion is not excited by the periodic reheating of the universe. Therefore, the evolution of the axion and the cycling are completely decoupled. The result is a cosmology with two inherent and disparate time scales: the time for a cycle (about a trillion years) and the tunneling time for the axion (about $10^{10^{120}}$ years in the final stages). The universe spends exponentially many cycles at each step down the washboard potential, with increasingly many cycles as V_N approaches zero. Once a region tunnels to a minimum with negative energy density, the cycling becomes unstable [3] and, with ten billion years, the region collapses into a black hole. Meanwhile, most of the universe continues to cycle.

Although different patches of space work their way downhill at different times, the patches are uniform on scales large compared to the Hubble horizon due to the smoothing caused during each cycle. Furthermore, every patch is locally equivalent: if one measures the Hubble constant to be near 10^{42} GeV and the dark energy density to be near $(1 \text{ meV})^4$ *anywhere in the universe*, the physical conditions and astronomical scene should be similar to what we see today. The situation is the opposite of the scene suggested by anthropic/landscape scenario in which most of the universe never looks like what we see within our horizon and is forever inhospitable to the formation of galaxies, stars, planets and life. All other things being equal, a theory that predicts that life can exist almost everywhere is overwhelmingly preferred by Bayesian analysis (or common sense) over a theory that predicts it can exist almost nowhere.

Remarkably, our modest proposal is subject to experimental refutation. Because it incorporates the cyclic picture for generating perturbations, it shares the cyclic predictions of a purely Gaussian spectrum of energy density perturbations and a gravitational wave spectrum with amplitude too tiny to produce a measurable B-mode polarization [4]. There may be other cosmological conundra that can be resolved by having a relaxation time much longer than a Hubble time, and they may lead to further cosmological tests. At this point, these ideas are new and formative, perhaps to be developed and debated at a future Solvay meeting.

Bibliography

[1] P. J. Steinhardt and N. Turok, *Why the cosmological constant is small and positive*, to appear, `hep-th/xxxxxx`.

[2] N. Turok, see contribution to these proceedings.

[3] P.J. Steinhardt and N. Turok, *A cyclic model of the universe*, Science **296**, 1436 (2003).

[4] P.J. Steinhardt and N. Turok, *The cyclic model simplified*, New Astronomy Reviews **49**, 43 (2005).

[5] L. Abbott, *A mechanism for reducing the value of the cosmological constant*, Phys. Lett. B**150**, 427 (1985).

[6] J. D. Brown and C. Teitelboim, *Dynamic neutralization of the cosmological constant*, Phys. Lett. **195 B** 177 (1987).

6.5 Discussion

S. Weinberg This is a trivial remark unless, of course, it is wrong. As I understand it, there is no particular reason in Steinhardt's scenario why the vacuum energy density would be related to the matter density that we see now in the present Universe?

P. Steinhardt No. In the model that I presented, there is nothing that links the total amount of dark matter to the cosmological constant directly. In fact, the ratio decreases from cycle to cycle as the cosmological constant relaxes. However, the universe spends most time when the cosmological constant is small, as we observe it today.

S. Weinberg The anthropic principle (that is the principle that dares not speak its name) does require that the vacuum energy density be of the same order of magnitude, but somewhat larger than the mass energy density that we see now, and in fact even a little larger than we actually are seeing. But yours does not?

P. Steinhardt Right. The amount of dark matter produced at the bounce is presumably the same from cycle to cycle, but the cosmological constant is slowly relaxing away. The ratio of dark matter to dark energy is, therefore, changing with time and we do not have a precise prediction for what the value is for this particular cycle.

A. Linde The equation for the potential which you use for your model contains a cosine term and a linear term. If you remove the cosine term and leave just the linear term, this would be exactly the potential of the model which I used in 1986 to suggest an anthropic solution for the cosmological constant problem, and pretty recently we repeated the discussion of this model in some details with Vilenkin and Garriga, and this model is also extremely similar to the model suggested by Banks at approximately the same time.

G. 't Hooft I wonder whether you are not introducing another kind of unnaturalness in such a system, which is the very small value for the mass term with respect to the kinetic term for such a dilaton-like field. After all you have a variation of the cosmological constant over cosmological scales, so there must be a very small effective mass in that. Is that not an unnatural feature of such a model?

P. Steinhardt It was Abbott's idea that the field be an axion with a mass term generated by the same kinds of instanton effects as the QCD axion. So, the coefficient of the cosine term, for example, would be something of order the Planck scale times $\exp(-2\pi\alpha)$, where α is the strength of the coupling at the Planck scale. For QCD, α is of order 0.1. Here we imagine that, for an axion on the other brane coupled to hidden gauge fields, the coupling might be a bit different. With a value of 0.08, just 20 per cent less, one gets values that work just fine for our scenario. And, because of the axion's special symmetries, the couplings are protected from quantum corrections. So, as Abbott suggested, the potential is technically natural. But, this is a new idea; you could think of lots of different kinds of potentials that may be more appealing; you might consider different kinds of time histories in which you could use the same basic principles of "ultra-slow relaxation" and cyclic evolution. There may be even better ideas out there.

H. Nicolai I have a question to Steinhardt. How do you reconcile the second law of thermodynamics with the cyclic universe?

P. Steinhardt Essentially what happens in this model is that at each stage when you have a collision, you produce a lot of entropy. Then during the subsequent stages of expansion you stretch the branes and you spread the entropy out so that the entropy density becomes exponentially low. And then you have another collision which creates more entropy. So, entropy is actually building up, if you add it up over the entire brane from cycle to cycle. But the entropy density is cycling, and it is the entropy density that is important for cosmology, for an observer like us who can only observe within the horizon. So the entropy is out there, it is just too spread out for us to see it.

F. Wilczek First I would like to say something profound, and then I will illustrate it with something which may or may not be profound. The profound thing I would like to say is that it is sometimes possible to solve some problems without solving all problems. I would like to illustrate that with the case of axion cosmology which was alluded to here. In that case, the assumptions leading to a kind of multi-verse picture, namely what ordinarily might be thought of as universal parameters in fact vary, is much simpler and does not rely on branching universes. All that may or may not be important to determine that measure. It just depends on the fact that the initial misalignment of the axion angle with the QCD angle at the time of the Peccei-Quinn transition carries very little freight in the early Universe, so it is truly random over the multiverse. If inflation intervenes, the measure is absolutely fixed, and there is no question that it does not interfere with any other microphysics. That is a beautiful example when there is no alternative to anthropic reasoning.

T. Banks A question for Steinhardt. In this model, if there is some oscillation of the scalar around its minimum as you approach the crunch, then there is an anti-friction force on this scalar field because the universe is contracting. I would have imagined that if you just run the equations down to the singularity you would find that it wants to jump all over the place around the potential, because the potential becomes irrelevant compared to the anti-friction which is blowing up because $\dot{a}/a$ goes to zero. So, if you could comment on that.

P. Steinhardt The problem you talk about would be a problem if this was a scalar field living in an ordinary bouncing universe like the ideas that people had in the 20s and 30s, because it would have just this problem during the contracting phase: any kinetic energy would be blue-shifted. If you imagine, though, that this field lives on the other brane, in the cyclic model the brane never goes through periods of contraction, it stretches, there is the collision, it stretches more and collides again. So far as that field is concerned, it is essentially seeing an expanding or static background. To say it in a more concrete way, in a 4-dimensional theory what happens is that the field is not just coupled to the scale factor, but also couples to the radion field, and the radion field during the contraction phase exactly compensates for the effect of expansion, so there is no effect, no excitation of the scalar field. Geometrically, this is because these fields are living on the brane which is not contracting.

T. Banks You will have to show me.

P. Steinhardt I wanted to issue a challenge to those pursuing the anthropic principle. We have now measured a number of important parameters about the universe that are important for the existence of life, but we are about to determine more parameters that are important for life: the shape of the power spectrum, for example, which is usually characterized by a spectral index, though it is possible that the spectrum will actually have some bumps and wiggles in it. We are also going to learn something about the reionization epoch. I would like those who are pursuing the anthropic principle to give us a definite prediction before the measurements are made. What are your anthropic expectations? Hopefully, you will converge on a single answer. It does not good if every proponent gives me a different answer. I suspect you do not have an answer at all, but I think it is an appropriate challenge for you to come up with one before the measurements are made so we can see if the anthropic principle has any real utility, and there is a chance for another success story to add to the semi-success of the cosmological constant.

S. Weinberg I think it is an unfair challenge, because we do not know which parameters scan. We do not know which parameters vary from multiverse to multiverse in a smooth way. If you tell us which ones do, we migh tell you what value to expect.

P. Steinhardt Then give me a table of possibilities, depending on your assumptions about which parameters scan.

N. Seiberg I am surprised that I am the only speaker, besides Kallosh, who mentioned the letters "LHC" at this conference. I want to pose the question: how do you think the LHC will change the scene of our field? What fraction of the people will continue studying the kind of physics that we have been discussing here: string theory and cosmology? What fraction will move to more phenomenology-related topics? And also, what kind of impact can we expect from LHC on more fundamental physics, shorter-distance physics?

R. Kallosh My understanding is that if LHC will tell us that there are supersymmetric particles, it would mean that we have this fermionic dimension in space and time, and it will be supergravity instead of gravity. So I would think that we would all tend to consider cosmology in the framework of supergravity and string theory. If they will not see supersymmetry immediately it will not immediately enforce us into it, but we still may do it.

N. Arkani-Hamed I just want to pop onto Seiberg's question. I think it is very interesting that with two years to go until the LHC, the situation is dramatically different from what it was in 1982 with two years to go until the discovery of the W and Z-boson at CERN. I think that 10 years ago, even the phenomenologist among us would have been a lot more confident about what to expect at the LHC than we are now. And as I mentioned in my previous remarks, that is associated with the fact that, quite apart from all the hints from cosmology and the theoretical hints from the string landscape, there is a growing sense of unease with why we have not seen evidence for new physics at the TeV scale. By itself this is not particularly dramatic, but still, something could easily have shown up already. So I think there are three possibilities for what might happen at the LHC and how it might impact the way we think about these questions. One of them is that we see evidence for a really, completely natural theory. And within a natural theory, some nice mechanism built into it would beautifully explain why it is that this expectation we have had for all this time that something would have shown up, was wrong. If that happened I would actually be given some pause and would certainly re-think possible natural solutions to the cosmological constant problem. There is the opposite possibility that the LHC might actually prove that the weak scale is finely tuned. That is a possibility that we have not been contemplating at all, but it is something that may actually happen. And if that happened, there is a variety of models, split supersymmetry being one of the examples but there are others, where you could really be able to prove that the weak scale is very finely tuned. If that happened, while in itself it is not evidence for landscape and anthropic reasoning or anything like that, it would be, I think, another big push in that direction. Now, the most ambiguous thing that can happen, I think, is that we might discover a natural theory and find that it is a little bit tuned. That is possible and basically every attempt to go beyond the Standard Model, when you look at it in detail, the theory is just tuned. Depending on how you talk about it, at the percent level, at five percent, half a percent, it does not matter, but there is something a little bit wrong. It could be that this is just what it is like. In the Standard Model we have several parameters which appear to be a little finely tuned, and that would just be another example. I think that would be the most ambiguous possibility. The natural possibility and the tuned natural possibility would shed no light on the cosmological constant problem, but I think that if we find evidence for tuning of the weak scale at LHC, for me it would be a very powerful evidence that the cosmological constant is also finely tuned. And, whatever the explanation is, we would have to think about it along those lines.

Session 7

Closing remarks, by David Gross

I originally decided not to prepare closing remarks beforehand, but to summarize the mood of this conference and the conclusions which we have reached. Unfortunately, I do not think we have any conclusions and my mood is one of total exhaustion. Nonetheless I will say a few words.

We started this wonderful conference with history and we have been talking a lot about history. We all respect enormously the history of the Solvay Conferences. We look back at them with awe and try to learn from them. But I still wonder why we have talked so much about history. Maybe it is because I invited Peter Galison to start off the conference with the history of the Solvay Conferences; and he did such a marvelous job. But also I think it is characteristic of a period of confusion that we look to the past with the hope of getting guidance or learning lessons. So I will also draw a few historical analogies from which we may learn lessons.

I used to say that the state of string theory is analogous to the state of quantum theory between the Bohr atom (1913) and the development of quantum mechanics(1924), a wonderful period of utter confusion. Physicists had part of the truth and were faced with many paradoxes. If this historical analogy was correct then we are very lucky, because the progress made in that period from the Bohr atom to quantum mechanics relied very little on experiment. I would say that you could have put Arnold Sommerfeld and all of his students on a desert island in 1913, and they would have come up with quantum mechanics. It was inevitable, once they had the semi-classical approximation to quantum mechanics, even though they did not know what quantum mechanics was, that they would figure out the correct theory. I think that is true that experiment played little role after the Bohr atom; so far no historian has contradicted me. Without experiment they might not have known about spin, but they would have come up with quantum mechanics.

But I am beginning to be a little more pessimistic about the state of affairs in string theory. Maybe a better analog is not 1913 but 1911, which happens to be the year of the first Solvay Conference. In 1911, physicists did not have the Bohr atom and they were faced with many sources of confusion. When we look back at the first Solvay Conference there were two big clouds in front of physics. One was "quanta and radiation" (and the first conference was called "Radiation and the Quanta") and the threats to the classical picture — wave-particle duality — that Planck's and Einstein's revolutions entail. However, there was another big problem that hung over their heads. This was the phenomenon of radioactivity that exploded onto the scene of physics at about the same time. This totally unpredictable decay of atoms seemed to threaten determinism. Radioactivity was totally random, unaffected by any physical perturbation of the atomic systems and seemed to violate both determinism and energy conservation. But, if you go back and look at the discussions at the first Solvay Conference they totally focused on quanta and radiation. There was almost no discussion of radioactivity, which was much more of a threat to classical physics.

We too have discussed, during this conference, many areas in which we are confused, and where revolutionary ways of thinking might be called for. One area has to do with space and time and the threats to our classical, quantum mechanical, field theoretical, traditional notions of space and time. We have heard at great length about many new discoveries within quantum gravity and string theory that hint that we are going to have to modify in a profound way our notions of space and time. But it is as unclear today how the final picture will look in 94 years from now, as it was unclear back in 1911 how the dual pictures of waves and particles would be reconciled.

But there is another profound change that is looming over us, one that was discussed in the last session, namely, the scope of science and its predictability: are the basic laws, constants, parameters, and everything else up for grabs? Must we resort to the anthropic principle? At the first Solvay Conference physicists were perplexed by quanta and radiation on the one hand and by radioactivity on the other hand. We have emergent space-time on the one hand and the anthropic principle on the other hand. How should we draw the analogy? I believe that the discussion of emergent space-time is similar to the discussion about quanta and radiation, whereas the anthropic revolution — and I do regard it as a revolutionary change in how we do physics, if true — is more analogous to the way people might have tried to deal with radioactivity back in 1911(which they were smart enough not to mention).

I would like to make it clear that I do not regard the anthopic principle as evil. It can lead to good science. What I found most interesting in Joe (Polchinski)'s talk were the new questions that he was asking in the framework of string theory and cosmology that were stimulated by thinking anthropically. So I don't think anthropic arguments are evil, they can stimulate people to do good physics.

What really bothers me about anthropic arguments was well expressed by Paul (Steinhardt)'s challenge to the people who believe in the anthropic principle to make a prediction; and by Steven (Weinberg)'s answer that we cannot make a prediction until you tell us which parameters are anthropically determined. Let me say why this disturbs me. What disturbs me about anthropic arguments is that they sound too much like "just so" stories. First, you have to know what it is that you are allowed to discuss anthropically. For instance, if you try to argue that the hierarchy scale is something that is determined anthropically, and therefore it is okay to fine tune it — life picks out the scale — and then it turns out that we learn from the LHC that it is determined by supersymmetry, people who believe in anthropic arguments would just say "OK, that is not a parameter that scans". So, my problem with the predictability of anthropic arguments is that it they are incredibly imprecise and easy to squirm out of. It is true, as Steven has emphasized to me, that in science you do not get a choice to decide what it is that science can do or cannot do: Nature decides. True, but we have made incredible progress in science by pushing predictability, and increasingly precise predictability: we do not just estimate that the cosmological constant lies in some range; we calculate the gyromagnetic ratio of the electron with incredible precision. This ever increasingly precise predictability is one of our most special and important scientific tools. I would like not to have to give it up. But it is not up to me, as Steven correctly points out.

The other philosophical aspect of the anthropic principle that bothers me enormously is that, unlike other principles in physics, it thrives on ignorance. Other physical principles in physics get stronger the more we know, yet the anthropic principle get stronger the less we know. Once we know something, especially if we know it precisely, it disappears completely from the realm of anthropic arguments. Anthropic arguments might be necessary, but they fly in the face of the success of physics over the last few centuries. Hopefully they will not be required.

On a less philosophical note I find that the arguments that lead, in string theory, to a multitude (a landscape) of possible universes to be very shaky. The so-called vacua that define the landscape are not vacua at all but metastable states. As such they are really not well defined in a truly precise sense. To define these we would have to know how they were populated, where they come from. To answer this question we are driven to understand eventually the full cosmological history of such states, including the big bang. And although there are very clever and somewhat convincing arguments for many metastable states, there is to date not a single consistent stringy description of cosmology all the way back to the beginning. Could it be that most or all of the landscape is ruled out since it does not fit into a consistent cosmology—we do not know!

Most important, we do not know what string theory really is. When we say "The theory leads to...".., we do not really know what we are talking about. We have many, often totally different, ways of describing approximate solutions to string theory; but what is string theory? We do not know the basic formulation of the "theory", to which all of these different dual descriptions are approximate. I am beginning to wonder whether we might be coming to the conclusion that string theory is inherently incomplete. Originally, many of us believed that string theory represented a very dramatic break with our previous notions of quantum field theory. That was good. We probably needed something that was a serious break with quantum field theory(QFT) to solve the problems of quantum gravity, cosmological singularities, etc. But now we have learned that string theory is not that much of a break with QFT. In fact, our best definitions of string theory are QFTs. Maybe, as Nathan (Seiberg) and Juan (Maldacena) remarked, any QFT is equivalent to a dual description of some kind of string theory. Also, the fact that we have so many different pictures of string theory, perhaps even different theories — we are not sure that they are truly all connected in a dynamical sense — might be a hint that string theory is a "framework" and not a theory, much as quantum field theory is a framework in which other principles (such as symmetry principles or renormalizabiltiy) must be appended to arrive at a theory. Maybe something is truly missing in our understanding of string theory, which we cannot identify? Our present understanding seems to me to be similar to the situation that physicists confronted in 1911. If they had tried in 1911 to deal with the problem of radioactivity, to explain how a nucleus could just sit there and randomly emit particles, they would have had no chance at all. That is probably why they did not discuss it at the first Solvay Conference. They were missing something absolutely fundamental in attacking that problem, namely quantum mechanics. Once quantum mechanics came along, with tunneling, radioactivity fell out. I have the feeling that the cosmological constant and all the bagage that goes along with, such as the the landscape, might be a problem of the same nature as radioactivity was in 1911, and that we are missing perhaps something as profound as they were back then. So, with that note of confusion, I will end.

I would like to end by thanking Marc. I had some role in organizing the scientific content of this conference, but the real hero behind this all is Marc. He has done an incredible job both in reviving the Solvay Intitutes and revitalizing them. Hopefully these wonderful conferences will continue all the way to the next century, so that we can come back in 96 years when all will be clear. I would like to thank Marc and I call for all of us to give him a big hand.

Answer by Marc Henneaux

Thank you very much. When two years ago, we decided to invigorate the Solvay Conferences and to have one in 2005 according to the old format, I think it was at the same time a very risky bet and a dream. The person who really convinced me that the bet was not so risky was David: when I went to see him, he was very supportive and immediately offered a tremendous help. The fact that you have been in Brussels this week shows that indeed the bet was not too risky: we are very grateful that you made the dream come true. And it will go on. Following the success of this conference, there will be other Solvay Conferences in the future — as David just told us, we have at least 96 years to go.

To conclude, I would like to thank all of you, to thank all the chairs, to thank all the rapporteurs who did an excellent job, and especially to thank the chair of this conference, David.

And I would like to also warmly thank Isabelle, Dominique, Stéphanie and Fabienne without whose faithful dedication the organization of the conference would have been a disaster.